Instructor's Guide to Accompany

Spence/Mason
Human Anatomy and Physiology
Third Edition

Robin C. Lenn
Sacramento City College

The Benjamin/Cummings Publishing Company, Inc.
Menlo Park, California • Reading, Massachusetts
Don Mills, Ontario • Wokingham, U.K. • Amsterdam • Sydney
Singapore • Tokyo • Madrid • Bogota • Santiago • San Juan

ISBN 0-8053-7010-2

BCDEFGHIJ-AL- 898

THE BENJAMIN/CUMMINGS PUBLISHING COMPANY, INC.
2727 Sand Hill Road
Menlo Park, California 94025

CONTENTS

Guide to Audiovisual References 1

Film Guide 3

Bibliography 41

Chapter 1 Introduction to Anatomy and Physiology 66

Chapter 2 The Chemical and Physical Basis of Life 72

Chapter 3 The Cell 80

Chapter 4 Tissues 88

Chapter 5 The Integumentary System 96

Chapter 6 The Skeletal System 102

Chapter 7 Articulations 115

Chapter 8 The Muscular System: General Structure and Physiology 122

Chapter 9 The Muscular System: Gross Anatomy 132

Chapter 10 The Nervous System: Its Organization and Components 142

Chapter 11 Neurons, Synapses and Receptors 148

Chapter 12 The Central Nervous System 158

Chapter 13 The Peripheral Nervous System 170

Chapter 14 The Autonomic Nervous System 178

Chapter 15 The Special Senses 184

Chapter 16 The Endocrine System 197

Chapter 17 The Circulatory System: Blood 207

Chapter 18 The Circulatory System: The Heart 215

Chapter 19 The Circulatory System: Blood Vessels 227

CONTENTS

Chapter 20   The Lymphatic System   241

Chapter 21   Defense Mechanisms of the Body   249

Chapter 22   The Respiratory System   259

Chapter 23   The Digestive System: Anatomy
              and Mechanical Processes   273

Chapter 24   The Digestive System: Chemical
              Digestion and Absorption   283

Chapter 25   Metabolism, Nutrition, and
              Temperature Regulation   293

Chapter 26   The Urinary System   305

Chapter 27   Fluid and Electrolyte Balance
              and Acid-Base Regulation   319

Chapter 28   The Reproductive System   333

Chapter 29   Pregnancy, Embryonic Development,
              and Inheritance   349

Sample Multiple Choice Lab Exam   361

Final Examination   363

AUDIOVISUAL REFERENCES

Boucher, Brian, et al. Handbook and Catalogue for Instructional Media Selection Englewood Cliffs, New Jersey: Educational Technology Pub., 1973.

Crowley, Leonard V. Visual Aids in Anatomy and Physiology. Chicago: Yearbook Medical Pub., 1976.

Educational Media Council: Educational Media Index. New York: McGraw-Hill, 1964.

Educators Guide to Free Films. 37th rev. ed. Randolph, Massachusetts: Educators Progress Service, 1977.

Film Review Index. Monterey Park, California: Audio-Visual Associates, 1977.

Hart, T. L., et al. Multi-Media Indexes, Lists and Review Sources: A Bibliographic Guide. New York: Dekker, 1975.

Index to Instructional Media Catalogues. Ann Arbor: Bowker, 1974.

Multi-Media Reviews Index. Ann Arbor: Preirian Press, 1977.

National Information Center for Educational Media (NICEM). Media Indexes. University of Southern California, University Park, California 90007.

Rufsvold, Margaret. Guides to Educational Media. 4th ed. Chicago: American Library Assoc., 1977.

Sive, Mary. Educators' Guide to Media Lists. Littleton, Massachusetts: Libraries Unlimited, 1975.

Taggart, Dorothy T. A Guide to Sources in Educational Media and Technology. Metuchen, New Jersey: Scarecrow, 1975.

Wittrich, Walter A. and Berget, James L. Educators' Guide to Free Audio and Video Materials. 24th rev. ed. Randolph, Massachusetts: Educators Progress Service , 1977.

SOURCES FOR SLIDES, TRANSPARENCIES, FILM STRIPS, FILM LOOPS, AND VIDEOTAPES

Slides

Carman Educational Associates, Box 205, Youngstown, New York 14174.

Science Software Systems, Inc., 11899 West Pico Boulevard, West Los Angeles. California.

Film Strips and Film Loops

BFA Educational Media, 211 Michigan Avenue, Box 1975, Santa Monica, California, 90406

Carolina Biological Supply Co., Burlington, North Carolina 27215.

Scientific American, Inc., 415 Madison Avenue, New York, New York 10017.

Transparencies

Carolina Biological Supply Co., Burlington, North Carolina 27215.

Denoyer-Geppert Audio-Visuals, 5235 Ravenswood Avenue, Chicago, Illinois 60640.

Hammond, Inc., Educational Division, 515 Valley Street, Maplewood, New Jersey 07040.

International Visual Aids Center, 619 Chausee De Mons, Bruxelles 7, Belgium.

Lansford Publishing Co., P.O. Box 8711, 1088 Lincoln Avenue, San Jose, California 95155.

Photo and Sound Co., 870 Monterey Pass Road, Monterey Park, California 91754.

Standard Publishing Co., 8121 Hamilton Avenue, Cincinnati, Ohio 45231.

Video Cassettes

Poly Morph Films, 331 Newbury Street, Boston, Massachusetts 02115.

Public Television Library, 475 L'Enfant Plaza, S.W., Washington, D.C. 20024.

Time-Life Multimedia, Time-Life Building, New York, New York 10020.

FILM GUIDE

The number of films which follows is far greater than could be shown in any one course. However, individual films vary considerably, for example in depth of coverage, interest, accuracy, year of production, and relevancy to the text and individual instructor's teaching style. Ease of availability and cost also enter into the decision as to whether a particular film will be used. The author has obviously not been able to view every one of the following films and the selection of many of the titles is based solely upon the title and/or description in various directories. It is recommended that if a title is of particular interest, check your own college's catalog of films. If the film is not available there, request it from the distributor listed. Each film description includes the following information in the sequence indicated.

1.   Title (in capital letters)

2.   Distributor*, length of film, color (C)/monochrome (B/W), year of release

3.   Description of film content

Omission of any data means that the information was not available. "Out of print" means the film is no longer generally available; however, private film libraries may still have a copy. The producer of these out of print films is occasionally listed to assist you in tracing a copy.

*The distributor code key follows the chapter listings of the film titles.

Suggested 16mm Films

Chapter 1: Introduction to Human Physiology

FUNCTIONS OF THE BODY (UEVA, 15 min., B/W, 1953). Broad coverage of seven organ systems of the body relating their interdependence. Animation and x-ray photography used.

INCREDIBLE VOYAGE (MGHT, 26 min., C, 1969). Probing the human body with an endoscope, microscope, x-rays and time-lapse photography. Techniques of exploring the living human discussed. Part of TV production: The Twenty First Century.

MAN, THE INCREDIBLE MACHINE (NGS, 16 min., C, 1975). A well-made film focusing on human systems: digestion, respiration, speech, motion, skeleton, muscles, skin, hearing, heart and circulation.

SKYLAB MEDICAL EXPERIMENTS (NASA, 31 min., C, 1976). Describes medical experiments performed in Skylab, including lower negative pressure device, ergometer, rotating litter chair, and experiment support system.

Chapter 2: The Chemical and Physical Basis for Life

ENERGETICS OF LIFE (JW, 23 min., C). Describes the sources of metabolic energy, explaining why molecules like ATP have potential energy. Relates sun energy to photosynthesis and glucose formation.

INSIDE THE CELL, PT. 1 - ENZYMES IN INTRACELLULAR CHEMISTRY (USNAC, 40 min., C, 1979). Presents enzyme theory with animated drawings of the better known steps of glucose metabolism as of 1949. Shows techniques used by Nobel Prize winners.

INSIDE THE CELL, PT. 2 - REGULATION OF ENZYMES (USNAC, 43 min., C, 1952). By the use of radioactive acetate ion in rats, explains the factors regulating enzyme action. The action of drugs, antimetabolites and the roles of Krebs' cycle and glycolysis are explored.

MOLECULES AND LIFE (MGH1, 20 min., C, 1971). By the use of selected examples, describes the achievements in molecular biology which have revolutionized our understanding of life. Examines vitamin B12, hemoglobin, myoglobin and proteins in general.

OSMOSIS AND FLUID COMPARTMENTS (WAYNE, 31 min., B/W, 1974). A film designed for professionals which illustrates the principles of diffusion and osmosis. Demonstrates the effects of toxicity on living cells. Identifies the 3 fluid compartments of the body and the major electrolytes and the factors involved in maintaining a constant volume within these spaces.

PROTEIN - STRUCTURE AND FUNCTION (JW, 16 min., C). Describes the structure and function of protein in the body, including the role of molecular shape in enzyme action.

THE ROLE OF WATER IN LIFE (Out of print, 29 min., B/W, 1955). The biological importance of water at the cellular level is described, including its role as a raw material in building compounds, as a component of protoplasm and as an extracellular medium.

THE STRUCTURE OF PROTEIN (BFA, 16 min., C, 1970). Traces the building of amino acids, polypeptides, protein molecules. Shows how DNA information is transferred to RNA which acts as a template for the construction of proteins.

Chapter 3: The Cell

AGING (CRM, 22 min., C, 1973). Examines some popular yet commonly mistaken attitudes toward the aged and their place in society. Discusses two major theories on the aged; the activity theory, the disengagement theory -- and the research being done by Bernice Neugarten, Robert Havinghurst and Sheldon Tobin, of the University of Chicago.

ASSAULT ON LIFE: RECENT ADVANCES IN GENETICS (TL, 50 min., B/W, 1968). An examination of advances in research in genetics and reproduction occurring prior to 1968. Contains cautions on pollution and overpopulation.

BIRTH, LIFE AND DEATH OF CELLS (BFI, 16 min., B/W). Shows the appearance and actions of living cells in tissue culture in time lapse photography.

BLUEPRINT FOR LIFE (MF, 14 min., C, 1970). Examines chromosomes and their relation to certain birth defects. Provides a discussion of heredity and meiosis, fertilization, formation of the zygote, early cleavage and the new individual. Illustrates a number of chromosomal abnormalities including Down's Syndrome, Trisomy 18 and Turner's Syndrome.

CANCER: THE WAYWARD CELL (DA, 24 min., C, 1976). Contains current research on the diagnosis, causes, prevention, treatment and future cures. Includes the thoughts and feelings of a young woman who is dying of cancer.

CELL BIOLOGY - LIFE FUNCTIONS (CIF, 19 min., C, 1965). Explores the structure and function of the cell. Discusses physical and biological processes associated with life.

CELL BIOLOGY: MITOSIS AND DNA (CIF. 16 min., C, 1965). Describes the structure and duplication process of the DNA molecule. Uses microphotography and animation to show the division of a living cell. Illustrates the stages of the mitotic process, and explains the meiotic process.

CELL BIOLOGY, STRUCTURE AND COMPOSITION (CIF, 14 min., C, 1965). Cell structure is described through photomicrography, electron micrography and models. Describes specialization of protoplasm into cell membrane, cytoplasm, nucleus, chromosomes, endoplasmic reticulum and other organelles. Interaction of these structures to life function is explored.

THE CELL: A FUNCTIONING STRUCTURE, PARTS 1 & 2 (CRM, 30 min. each, C, 1972). Organelle structure and function. Also includes microscopy, metabolism, reproduction, gene structure, and gene function. Shows ways cells obtain nourishment, discusses cellular reproduction, genetic code, chloroplast and cellular communication. Features Dr. David Suzuki, Dr. Richard McIntosh and Dr. Donald Fosket.

CENTURY 3: THE GIFT OF LIFE (USNAC, 28 min., C, 1975). Consists of three segments, the first an actual kidney transplant at MGH; the second a review of microscopy developments and gene study at the Rockefeller Institute, with speculation about chromosome manipulation and its possible uses; the third, filmed at the University of Pennsylvania, a discussion of the sociological and moral issues of extending life span.

CHEMICAL MAN (PFP, 8 min., C, 1968). Describes the complexity and beauty of the chemistry of humans at cellular and molecular levels.

CHROMOSOMAL ABNORMALITIES: THE AUTOSOMES (MF, 25 min., C, 1969). Presents patients with autosomal trisomies associated with chromosomes 21, 18 and 13. Illustrates deletions such as structural chromosomal aberrations by the cri-du-chat syndrome and the Philadelphia chromosome associated with chronic myeloid leukemia.

CHROMOSOMAL ABNORMALITIES: THE SEX CHROMOSOMES (MF, 25 min., C, 1969). Illustrates the effects of karyotypes and Barr body status on the histology and clinical findings in XXY Klinefelter's syndrome and its variants. Demonstrates the Lyon hypothesis and its significance in normal and abnormal sex chromosomal constitutions.

CHROMOSOMAL ERRORS (MF, 10 min., C, 1976). Genetic counseling situation used to introduce meiosis, mitosis, fertilization, and fetal development. Human chromosomal defects, such as Down's Trisomy 13 and Klinefelter's, are discussed.

THE CHROMOSOME: GENERAL CONSIDERATION (MF, 14 min., C, March of Dimes, 1969). Demonstrates normal human karyotype preparation, nomenclature and normal variations. Stresses correlation between clinical findings and cytologic observations and emphasizes autosomal aneuploidy.

DEATH OF A CELL (SQ, 15 min., B/W). The anatomical changes that occur as a cell dies as the result of various mechanisms are examined by time-lapse phase-contrast microscopy.

DEVELOPMENT AND DIFFERENTIATION (CRM, McGraw-Hill Films, 20 min., C, 1974). Explains the mechanism by which an embryo develops into a highly organized group of cells. Uses animation to show the roles of DNA and cytoplasm related to cell function.

DNA: BLUEPRINT OF LIFE (JW, 20 min., C, 1968). The DNA molecule, protein structure and protein synthesis within the ribosomes are explained with the use of animated models. The evolutionary importance of these fundamental mechanisms is described.

DNA: MOLECULE OF HEREDITY (EBF, 16 min., C, 1960). George W. Beadle of the California Institute of Technology explains, with the help of photomicrography, animation, models and live subjects, why DNA, a giant molecule in the chromosomes, is the basis of growth, reproduction and mechanism for transporting hereditary information from one generation to the next.

THE DNA STORY (MG, 45 min., C, 1976). From the NOVA series: explores the history of the discovery of DNA discussing the role of both chemistry and physics in determining the structure of the molecule of heredity.

DNA STRUCTURE AND REPLICATION (Out of print, McGraw-Hill, 30 min., B/W, 1960). Lecture given by Dr. G. D. Watson to help undergraduates understand DNA.

THE DOUBLE HELIX. NOBEL LAUREATES SERIES (Out of print, MGHF, 12 min., C, 1971). Interesting explanation of structure and function of DNA. Describes the duplication of this molecule.

ENZYME DEFECTS AND DNA (MF, 15 min., C, 1968). Presents the genetic basis and clinical manifestations of phenylketonuria, galactosemia, pseudo-cholinesterase and hemoglobinopathies to relate the genotype to an observable phenotype. Describes evidence for the helical structure of DNA and how DNA stores, transmits and translates genetic information.

EXTENDING LIFE (BFA, 15 min., C, 1976). First half surveys transplants as a means of prolonging life, demonstrating advances in tissue replacement, and presents the control of human development through genetic manipulation; includes patient testimonials. Second half presents moral and ethical aspects of extending life, interviewing a priest and lay persons. Accompanied by text and discussion guide.

FUNCTION OF DNA AND RNA IN PROTEIN SYNTHESIS (MGHT, 14 min., C). Describes functions of DNA and transfer ribosomal and messenger RNA in protein synthesis. Shows how genetic information is transferred in cell division.

GENE ACTION (EBF, 16 min., C, 1963). Illustrates the biochemical mechanisms of inheritance through simple animation. Shows replication of DNA chromosomes during mitosis and how it serves as a pattern for the messenger RNA. Employs live action shots to show simple organisms.

GENES AND PROTEIN SYNTHESIS (MF, 17 min., C, 1968). Uses sickle cell trait to show one-gene, one-enzyme hypothesis. Describes allelism and nonallelism and biochemical tests.

GENETIC DEFECTS: THE BROKEN CODE (IU, 90 min., C, 1976). Presentation of common genetic detects such as cystic fibrosis, Huntington's chorea, hemophilia, combined immune deficiency, and sickle cell anemia.

GENETICS AND YOU (MF, 11 min., C, 1976). Ethical issues of cloning, controlled mating, breeding, and test-tube babies are presented.

GENETICS: FUNCTIONS OF DNA & RNA (CIF, 13 min., C, 1968). Shows how DNA codes result in specific protein synthesis and how mutation and differentiation occur.

GENETICS: HUMAN HEREDITY (CIF, 16 min., C, 1968). Human genetics studies are limited by the lack of controlled breeding programs. The methodology of human genetics involves a three-pronged attack: biochemical genetics, involving DNA chemistry and the changes caused by genes; nature-nurture studies.

THE GENETICS OF TRANSPLANTATION (MF, 19 min., C, 1975). Demonstrates blood and tissue typing procedures necessary for successful organ transplants. Discusses the major histo-compatibility antigens.

GREGOR MENDEL (UCE, 24 min., C, 1974). Professor Richard M. Eakin impersonates Gregor Mendel in the words, dress and manner of his time. Describes his interest in heredity and his experiments with the pea plant, hypothesizing the existence of "elementen," Mendel's particulate units of heredity.

HISTORY OF GENETICS: THE PHYSICAL BASIS OF INHERITANCE (MF, 19 min., C, 1968). Traces the history of genetic research from Mendel to 1965. Shows human karyotypes, identification of individual mitosis and meiosis.

HUMAN HEREDITY (PE, 23 min., C, 1969). Emphasizes similarities and differences in human beings and functions of DNA and RNA, utilizing hair and skin color and body type. Refers to 48 chromosomes as the human chromosomal complement.

HUMAN TRAITS SHOWING SIMPLE MENDELIAN INHERITANCE (Out of print, 30 min., B/W, 1961).

INTERLOCKING PARTS OF THE CELL MECHANISM (MGHT, 16 min., C, 1967). Describes the organelles or the cell and their functions.

LASER AND LIVING CELLS (TF, 15 min., C). Shows how the laser is being used to explore the minute processes of the living cell.

LEARNING ABOUT CELLS (EBF, 16 min., C, 1976). Basic structures, functions of a cell. Principles of cell growth and specialization described. Role of nucleus is demonstrated.

THE LIFE AND DEATH OF A CELL (UCE, 20 min.). Marvelous film on the life functions of a cell typified by amoeba. Excellent introductory film which points out the properties and functions common to all cells.

THE LIVING CELL (HR, 28 min., C, 1972). Reviews the structure and function of living cells, examining collections of cells via cinemicroscopy and electron micrographs.

THE LIVING CELL: AN INTRODUCTION (EBF, 20 min., C, 1974). Examination of cell structure and biochemical processes. Describes historical background of cell discovery, role of nucleus, and protein synthesis.

THE LIVING CELL: DNA (EBF, 20 min., C, 1976). Through simulated micro-photography, shows how DNA codes and directs the functioning of an organism.

MEIOSIS (MGHT, 14 min., C, 1969). Shows how the number of chromosomes in germ cells is halved. An introductory flow chart shows the fertilization meiosis cycle. Shows formation of polar bodies.

MEIOSIS - IN SPERMATOGENESIS OF THE GRASSHOPPER (PHASE, 19 min., B/W, 1949). Meiosis, in spermatogenesis of the grasshopper, displayed by time-lapse photography using a phase-contrast microscope. One minute of film represents three hours of actual division.

MEIOSIS: SEX CELL FORMATION (EBF, 16 min., C, 1963). Explains meiosis in the formation of sex cells, cells with half the number of chromosomes as were present in the parent cell nucleus. Discusses how meiosis enables organisms to produce offspring with a wide variety of characteristics and how this variability plays a role in natural selection.

MITOSIS (IFB, 9 min., C, 1958). Describes cell division in animation.

MITOSIS (EBF, 24 min., C, 1961). Shows cell division in living cell and demonstrates the effects of chemicals and radiation on dividing cells.

MITOSIS AND MEIOSIS (IU, 17 min., B/W, 1956). Compares mitosis and meiosis. Uses starfish life cycle for the demonstration.

PEDIGREE PATTERNS (MF, 17 min., C, 1968). Describes construction of the human pedigree. Uses pedigrees of several human defects to illustrate dominant and autosomal recessive inheritance.

X-LINKED INHERITANCE (MF, 12 min., C, 1976). Demonstrates inheritance of recessive traits located on the chromosome using color blindness as an example.

Chapter 4: Tissues

TISSUES OF THE HUMAN BODY (CIF, 16 min., C, 1963). By live photography and animation, the tissues of the body are discussed.

Chapter 5: Skin

PIGMENT TRANSFER IN SKIN CELLS (BFA, 5 min., C, 1972). Shows epidermal cells and melanocytes in tissue culture; in diagram and in vitro, we see pigment transferred from the melanocyte to the epidermal cell.

SKIN (Out of print, 20 min., B/W, 1970). Shows role of skin as a protection against microorganisms and heat loss. Reveals research in cell division in the skin.

THE SKIN AS A SENSE ORGAN (IFB, 12 min., C, 1975). Microscopic examination of sensory receptors of the skin. Shows reflex arc.

THE SKIN, from the series How Your Body Works (TL, 20 min., B/W, 1971). Properties of the human skin and the ways it serves to protect the human body.

SPLIT SKIN GRAFTS (IEFL, 10 min., C, 1962). The removal of a split skin graft with an electric dermatome is demonstrated. The graft is used to cover an infected stab wound. The results are followed up one year later.

YOUR SKIN (Out of print, 15 min., C, 1964). Describes structure and function of skin and of sebum melanin and sweat. Emphasizes the role of cleansing the skin to wash away sweat, sebum and microorganisms.

Chapter 6: Skeleton

BODY FRAMEWORK, PART 1 (ETF, 15 min., B/W, 1962). Describes the structure and function of the skeleton including chemical composition, growth and repair of the main varieties of joints.

FRACTURES: AN INTRODUCTION (JJ, 28 min., C, 1953). Live-action and animated anatomical drawings are used to describe how fractures occur and how they are repaired.

HOW THE BODY MOVES: THE SKELETON (PMR, 20 min., B/W, 1971). Shows how skeletal muscle is attached to bone and thereby can bring about movement. Describes function of bone: calcium storage, relation to breathing, support, movement, protection. Traces the development of bones.

HUMAN BODY: SKELETON (CIF, 10 min.). Support, movement and protection aspects of the complex skeletal system is illustrated by cinefluorography.

THE SKELETON (EBF, 13 min., B/W, 1953). Through animation, illustrates the composition, growth, movement and functions of the skeletal system.

THE SPINAL COLUMN (EBF, 11 min., B/W, 1956). Through a variety of techniques, the structure and function of the spinal column are presented.

Chapters 8, 9: Muscles

BODY FRAMEWORK, PART 2 (ETF, 15 min., B/W, 1962). Through photography and animation, presents the structure and use of muscles. Importance of exercise and good posture habits is stressed.

EXERCISE PHYSIOLOGY (SRBO, 17 min., C, 1974). Physiology of exercise. Part of a PE coach film series.

HOW THE BODY MOVES: THE MUSCLES, from the series HOW YOUR BODY WORKS (PMR, 20 min., B/W, 1971). Describes structure and function of muscle. Illustrates concept of antagonistic pairs of muscles. Differentiates voluntary and involuntary muscles.

THE HUMAN BODY - MUSCULAR SYSTEM (CIF, 14 min., C, 1962). Describes the function of the three varieties of muscle in the body.

MECHANICS OF HUMAN MOTION - THROWING (USC, 21 min., B/W). A study of human motion--throwing--involving both high speed photography (5000 frames/sec.) and x-ray motion pictures. Helpful in teaching students how to analyze motion.

MICROELECTRODES IN MUSCLE (JW, 19 min., C, 1970). Through the use of frog muscle, the electrical changes in the transmembrane potential during the passage of an action potential and the effects of changing the quality of the external solution.

MUSCLE (CRM, 25 min., C, 1972). Description of the dynamics of muscle contraction using the sliding filament theory. Role of membrane potential and chemical pumps examined.

MUSCLE: A STUDY OF INTEGRATION (CRM, 25 min., C, 1972). Uses animation to show muscle tissue and processes involved in muscle contraction. Describes the three types of muscle, explains the roles of membrane potential and chemical pumps, presents the sliding filament theory of muscle contraction and explains electronic feedback therapy for alleviating problems caused by muscular tension. Excellent film.

MUSCLE: CHEMISTRY OF CONTRACTION (EBF, 15 min., C, 1969). Describes the internal structure of muscles and explains muscle contraction with the use of a plastic model of thick and thin filaments within a single sarcomere.

MUSCLE DYNAMICS OF CONTRACTION (EBF, 21 min., C, 1969). Reveals dynamics of muscle contraction in the activities of people. Shows electron micrographs of muscle cells describing the action of the whole muscle in terms of the activity of its cellular units.

MUSCLE: ELECTRICAL ACTIVITY OF CONTRACTION (EBF, 9 min., C). Shows that all muscle activity is initiated by nerve impulses and that electrical stimuli can evoke muscle contraction.

MUSCLE CONTRACTION AND OXYGEN DEBT (Out of print, MGHT, 16 min., C, 1967). Describes muscle action and explores relationship between glycogen, lactic and pyruvic acids.

MUSCULAR CONTRACTION UNDER THE MICROSCOPE (CU, 27 min., B/W, 1965). A. F. Huxeley explains the sliding filament theory and why a stretched muscle develops less tension than one of normal length.

MYOGLOBIN (SB, 10 min., C, 1971). The structure of the myoglobin molecule which enhances the pickup of oxygen from the blood is analyzed by computer animation.

WHAT MAKES MUSCLE PULL - THE STRUCTURAL BASIS OF CONTRACTION (JW, 9 min., C, 1972). Relates muscle structure, muscle protein and function and shows how energy transfer occurs, by animation.

Chapter 10: Nervous System Organization

FUNDAMENTALS OF THE NERVOUS SYSTEM (EBF, 17 min., C, 1962). Functions of the nervous system explained through photomicrography, animation, and medical demonstrations. Division of the nervous system and major tasks performed.

THE HUMAN BODY - NERVOUS SYSTEM (CIF, 13 min., C, 1958). The principal organs of the nervous system are shown with explanation of the basic function of the nervous system. Main brain areas are described.

THE NERVOUS SYSTEM IN MAN (IU, 18 min., C, 1965). Describes the role of the nervous and endocrine systems in coordination and integration of body function through animation.

Chapter 11: Neurons, Synapses, and Receptors

HUMAN BODY: SENSE ORGANS (CIF, 19 min., C, 1965). Overview of sensory receptors. Role of various organs in sensory perception--the Meissner, Ruffini, and Paccinian corpuscles, organ of Corti, taste buds, olfactory receptors and rods and cones of the eye. Conversion of stimulus energy into an action potential transmitted to the brain where perception actually occurs is discussed.

THE HUMORAL TRANSMISSION OF THE SYMPATHETIC IMPULSES (ICI, 16 min.,C, 1948). Describes role of sympathin in neurohumoral transmission with the use of animation and experiments on an anaesthetized animal.

THE IONIC BASIS OF THE ACTION POTENTIAL (USNAC, 11 min., C, 1977). Uses computer animation to illustrate the dynamic relationship of three molecular events: channel activity, ion movement, and charge separation at a nerve axon's membrane at rest and during an action potential. Relates ionic activity at the membrane to the oscilloscope tracing.

MOTOR CONDUCTION VELOCITY STUDIES OF THE MEDIAN AND ULNAR NERVES (USNAC, 8 min., C, 1956). After defining the nature of neuropathic lesions, the film demonstrates a way to determine the velocity of motor nerve fibers in the median nerve.

MOVEMENTS OF ORGANELLES IN LIVING NERVE FIBERS (USNAC, 12 min., C, 1947). Time-lapse phase cinemicrography is used to demonstrate the movement of organelles in frog axons.

THE NERVE IMPULSE (EBF, 21 min., C, 1971). History of discoveries of nerve impulse properties. Physiology of nervous system.

THE NERVE IMPULSE - USE OF THE CATHODE RAY OSCILLOSCOPE IN PHYSIOLOGY (IOWA, 19 min., B/W, 1960). Describes the operation and usefulness of the oscilloscope in detection and recording nerve impulses.

PROPERTIES OF ACETYLCHOLINE (ICI, 20 min., C, 1947). Demonstrates the action of acetylcholine on an anaesthetized animal. Examines its action on a sympathetic ganglion.

SENSES OF MAN (IU, 18 min., C, 1965). Through animation describes how external stimuli are converted to nerve impulses.

THE SENSORY WORLD (CRM, 32 min., C, 1971). Through animation, demonstrates how eyes, ears, sense of touch and proprioception systems work.

SYSTEM AND WORKINGS OF A NERVE (IP, 14 min., C). Pulse communication scheme is used to explain the human nervous system.

WORLD OF PULSES (CSUS, Tokyo Cinema Company, Inc., 28 min., C, 1962). Analysis of electrical pulses in the nervous system compared to electronic circuitry. Shows actual photomicrographs of neuronal growth in tissue.

Chapter 12: The Central Nervous System

BODILY SENSATION, PART 1 - CONDUCTING SYSTEMS OF THE CORD AND STEM, PART 2 - THALAMUS AND CORTEX (AVC, 18 min. each, C). Didactic and laboratory approach to human neuroanatomy.

BRAIN AND BEHAVIOR (Out of print, Producer: McGraw-Hill, 22 min., B/W, 1957). Demonstrates two ways to study the function of different brain areas in relation to human behavior: (1) to artificially stimulate different parts of the living brain with an electrode and observe the results, (2) measure, by testing, the change in behavior following injuries to different areas of the brain.

CEREBROSPINAL FLUID (IOWA, 3 min., C, 1971, silent). Illustrates the composition and function of cerebrospinal fluid and factors which influence its pressure within the central nervous system.

CHANGES (CRH, UCLA-Handicapped Students Union, 28 min., C, 1973). Discusses problems and rehabilitation of spinal cord handicapped individuals to stress need for both physical and psychological care; points out barriers--structural, attitudinal, sociological, psychological--facing the handicapped, society's indifference, and some physical rehabilitation technology.

DRUGS AND THE NERVOUS SYSTEM (2nd ed.) (CF, 18 min., C, 1973). Explains in animation how drugs affect the central nervous system and other organs. Discusses stimulants, depressants, hallucinogens and narcotics.

ESSENTIALS OF THE NEUROLOGICAL EXAMINATION (SKF, 50 min., C, 1962). Illustrates how a relatively thorough examination can be performed in the office. Includes the examination of cerebral function, the cranial nerves, cerebellar function, the motor system, the sensory system and deep and superficial reflexes utilizing an actual patient with an aneurysm on the posterior communicating artery.

EXPLORING THE HUMAN BRAIN (BFA, 18 min., C, 1977). Historical development of the understanding of the human brain from the time of Hippocrates. Describes modern methods and equipment which aid the current investigations.

EXPLORING THE HUMAN NERVOUS SYSTEM (CF, 23 min., C, 1963). Anatomy and physiology of the human nervous system. Research in memory and learning. Mostly animated. Compares nervous system of hydra and earthworm to that of humans.

THE HIDDEN UNIVERSE: THE BRAIN, Part 1: 23 min., Part 2: 25 min. (MGHT, 48 min., C, 1977). Covers functions of the brain including motor control, memory and sensory perception, recent research results, including "split brain" studies, biofeedback, and brain malfunction treatment. Highlight is operating room scene where a craniotomy is performed and different parts of the patient's brain are probed as he indicates the sensations this causes.

THE HUMAN BODY - THE BRAIN (CIF, 16 min., C, 1967). Basic functions of the brain described by laboratory demonstrations, x-rays, anatomical specimens and animation.

THE HUMAN BRAIN (EBF, 11 min., B/W, 1955). Briefly describes the development of the brain in man. Illustrates the four basic steps of the nervous-muscular system in the thinking process by using a driver in an emergency situation.

THE HUMAN BRAIN - A DYNAMIC VIEW OF ITS STRUCTURES AND ITS ORGANIZATION (RL, 29 min., C, 1976). Computer visualization of brain structure as a three-dimensional structure. Excellent.

INTRODUCTION TO THE BRAIN (AVC, 18 min., C).

THE MIND OF MAN (IU, 119 min., C, 1971). Comprehensive survey of recent mind research including areas of mind development in children, effects of drugs, dreams, brain structure, chemical changes within the brain, the brain and sexuality, reasoning and the power of the mind in controlling body functions. Includes interviews with Sir John Eccles, Donald Hebb and B. F. Skinner.

MIND OVER BODY (TL, 49 min., C, 1973). This film examines techniques which teach patients to use their minds to ward off or cure illness and looks at new area of research that straddles psychology, physiology and medicine.

MIRACLE OF THE MIND (MGHT, 19 min., 1975). Describes the attempts to understand the nature of the mind by surveying the contributions of science.

THE MUSCLE SPINDLE (JW, 19 min., C, 1970). Describes the proprioceptive activity of the muscle spindle and how it responds to muscle movement and tension. Shows method of removal of muscle spindle and keeping it alive in vitro.

REFLEX ACTION (CCM, 17 min., B/W, 1948, silent).

REFLEXES (CU, 35 min., C, 1946). Animated illustration of proprioceptive and exteroceptive reflexes and reflex arcs including clinical examination.

PERFORMANCE OF SKILLS - THE PYRAMIDAL SYSTEM (AVC, 18 min., C). The role of the pyramidal system in voluntary motor activity is described.

SCIENCE - NEW FRONTIER SERIES: EXPLORING THE HUMAN BRAIN (BFA, 18 min., C, 1977). Describes the evolution of the understanding of human brain functioning from the time of Hippocrates. Examines methods and equipment which are used today to aid in further investigation.

SECRETS OF THE BRAIN (CFI, 15 min., C). Intensive research into the delicate activities of the brain is explored with the hope of finding a method of preventing or treating cerebral palsy. Shows how such methods have been used to treat headaches, muscular spasms and learning disabilities.

SLEEP - THE FANTASTIC THIRD OF YOUR LIFE (MGHT, 2 reels, 26 min. each, C, 1968). Explores the meaning of dreams and reviews the experiments on dream research. Purpose of the experiments is to provide more productive stay-awake hours.

SPLIT BRAIN (IU, 13 min., B/W, 1964). Explanation of the split brain operation and the methods of testing the split brain subjects.

THE SPLIT BRAIN AND CONSCIOUS EXPERIENCE (HR, 17 min., C, 1977). Shows the surgery and the testing procedures on people whose brains have been surgically split to control severe epilepsy.

TO SLEEP - PERCHANCE TO DREAM (HMP, 30 min., B/W, 1967). Experiments conducted at the Sleep Research Laboratory of UCLA determine the relation-ships of dreams to stomach secretions, the amount of time infants spend dreaming, and the effects of dream deprivation. Shows the EEG from a sleeping subject and the rapid eye movements during dreaming.

Chapter 13: Peripheral Nervous System

THE PERIPHERAL NERVOUS SYSTEM (IFB, 19 min., 1977). Demonstrates an infant's reflex action, then explains the components of the spinal reflex arc and the importance of conscious bodily control. Covers the nature of the nerve impulse, examined in micrographs and animation, and the transmission rate of myelinated nerves is introduced. Students in the lab are shown determining the velocity of transmission of nerve impulses in a frog.

SPINAL NERVES (UT, 18 min., C, 1976). The spinal nerves and their relationship to the sympathetic chain illustrated with the use of animation, photomicrography and electron microscopic studies.

Chapter 14: The Autonomic Nervous System

AUTONOMIC NERVOUS SYSTEM (DUKE, 40 min., C, 1953). By the use of diagrams and dissections, shows the double innervation of organs by the two divisions of the ANS.

THE AUTONOMIC NERVOUS SYSTEM (Out of print, Producer: HANDY, 6 min., C, 1967).   Reviews the nervous system and traces the sympathetic and parasympathetic fibers from their origins to their peripheral innervation.

AUTONOMIC NERVOUS SYSTEM (IFB, 17 min., C, 1975). Portrays aspects of the autonomic nervous system, with photographs of gastroscopic and bronchoscopic anatomy, of construction and dilation of the pupils, of the gut during peristalsis, and the control of body temperature.

THE AUTONOMIC NERVOUS SYSTEM - AN OVERVIEW (USNAC, 17 min., C, 1973). A general introduction to the human ANS, explaining the basic functions of the two divisions.

PARASYMPATHETIC AND SYMPATHETIC INNERVATION (2nd Edition) (MD, 40 min., C, 1952). Illustrates the autonomic nervous system through cadaver dissection and animation.

Chapter 15: The Special Senses

COLORS AND HOW WE SEE THEM (AF, 22 min., C, 1977). Shows how and why rods and cones of the retina relay their specific messages to the brain. Describes the Young-Helmholtz three-component theory of color vision.

THE EAR (IP, 14 min., C). Explains the structure and function of the ear, both hearing and equilibrium.

THE EAR AND HEARING (Out of print, Producer: CIBA, 28 min., C, 1965). Presentation of the basic anatomy of the external, middle and internal ear. Demonstrates the effects of hearing loss on speech.

THE EARS (NMAC, 17 min., C, 1970). Demonstration of the physical examination of the pinna, external auditory canal and tympanic membrane. Describes conditions which could cause hearing loss and hearing tests.

THE EARS AND HEARING (EBF, 22 min., C, 1969). Anatomy and physiology of hearing. Computer simulation of basilar membrane movement at different sound frequencies. Process by which sound waves are converted into electrochemical energy and two causes of deafness. Describes how hearing aids or surgery may help conduction deafness.

EMBRYOLOGY OF THE EYE (IEFL, 43 min., C, 1960). Through animation, drawings and models, the development of the eye from ectoderm and mesoderm from the optic cup stage is described.

EYE (DUKE, 21 min., C, 1956). Describes autonomic innervation of iris and ciliary muscle. Discussion of Horner's syndrome (cervical sympathetic nerve trunk paralysis), third nerve paralysis and the resulting deficiencies are provided.

THE EYES AND SEEING (EBF, 20 min., C, 1967). Provides a step by step investigation of how the eye works. Uses close-up studies of the parts of the eye and experiments with human subjects and an alligator in a visual perception laboratory to show the distinction between physical process and the electrochemical process of seeing.

GATEWAYS TO THE MIND (BELL, 52 min., C, 1958). How we receive stimulations through the senses, transmit to the brain and store for future reference. Through live action and animation, shows what happens when we see, hear, taste, smell and feel.

GLAUCOMA (LPI, 15 min., B/W, 1963). Describes acute and chronic forms of glaucoma and the diagnostic procedures necessary to identify the condition. Shows medical and surgical control methods.

HEARING - AUDITORY SYSTEM (AVC, 18 min., C). Part of "Anatomical Basis of Brain Function Series." A professional film.

HOW THE EAR FUNCTIONS (KB, 11 min., B/W, 1940). Provides basic explanation of structure and function of the human ear through the use of animation, photography and sound effects.

THE INCREDIBLE SEEING MACHINE (BL, 26 min., C, 1974). A class of college students explores vision, discovering how the eye works, why optical illusions occur and the use of corrective lenses for visual disorders.

MORE THAN MEETS THE EYE (TL, 30 min., C). Differentiates seeing and perception and demonstrates how experience, memory and prejudice color a person's perception of his/her environment.

SIGHT - VISUAL SYSTEM (AVC, 18 min., C). A professional film on the visual system of the eye.

SIMULATED BASILAR MEMBRANE MOTION (GS, 11 min., B/W, 1966). Computer generated wave motion of the basilar membrane in response to varying pitches. Shows cut off point and how it varies relevant to pitch.

THE STRUCTURE OF THE EYE (CCM, 7 min., C, 1970). Through eyeball dissection and animation, the anatomy and the function of the eye is described.

Chapter 16: The Endocrine System

ADRENALS (DUKE, 18 min., C, 1956). Describes development of the autonomic nervous system and of the adrenal medulla. Explains the relationship of acetylcholine to cholinesterase and shows how epinephrine released by the adrenal medulla effects physiological changes distant from its origin.

ENDOCRINE GLANDS (SEF, 14 min., C, 1973). Description and explanation of function of the endocrine glands and their relationship to bodily health.

HORMONES (USC, 29 min., B/W, 1964). Discusses primarily steroid hormones, especially those related to birth control methods.

PRINCIPLES OF ENDOCRINE ACTIVITY (IU, 16 min., C, 1960). Introduces the endocrine system and the effects of gland secretions. Shows that animals and plants have chemical coordinators regulating growth.

PROSTAGLANDINS: TOMORROW'S PHYSIOLOGY (UP, 22 min., C, 1974). Survey of prostaglandin knowledge as of 1974.

Chapter 17: The Blood

ABO AND LEWIS BLOOD GROUP SYSTEMS (USNAC, 21 min., B/W, 1973). Defines the antigens and antibodies involved in the ABO group system. Describes the importance of the understanding of this information with respect to transfusion, relationship to hemolytic disease of the newborn and accurate laboratory diagnosis.

ARTERIAL AND VENOUS BLOODS (IOWA, 3 min., C, 1971, silent). Compares colors of arterial and venous blood and illustrates the easy reversibility of the oxygen-hemoglobin union.

THE BLOOD (EBF, 16 min., C, 1962). With animation and photomicrography, explains composition of blood, its circulation and its work in maintaining life. Includes blood typing and testing for the Rh factor.

BLOOD (IP, 14 min., C). Blood functions and composition of plasma, corpuscles, hemoglobin and leucocytes. Discusses transfusions.

BLOOD COAGULATION (IOWA, 3 min., C, silent). Blood samples treated with chemicals to hasten or delay coagulation time as compared with normal blood are studied. Illustrates defibrination.

BLOOD COMPONENTS AND THEIR USE (AIF, 45 min., C, 1972). Describes how the various medically necessary blood fractions are prepared and their indications for use.

DETERMINING AMINO ACID SEQUENCES IN PROTEINS, SICKLE CELL ANEMIA (HMC, C, 1977). Discusses sickle cell anemia as an example of a molecular disease caused by a substituted amino acid.

HOW BLOOD CLOTS (BFA, 13 min., C, 1969). Shows how a clot is created through the interaction of plasma proteins and the blood platelets and illustrates the formation of platelets within red bone marrow from the megakaryocyte.

LIFE AND THE STRUCTURE OF HEMOGLOBIN (PAR, 28 min., C, 1976). Computer
   animation is used to describe the structure and, function of hemoglobin.
   Presented by Linus Pauling.

Rh: THE DISEASE AND ITS CONQUEST (ME, 18 min., C, 1974). Describes the
   genetic basis of the Rh problem, the development of antibodies by the Rh
   negative mother and the effects on the fetus. Describes development of anti-
   Rh globulin.

THE SICKLE CELL STORY (MF, 16 min., C, 1977). Portrays both the scientific
   and human aspects of the sickle cell affliction using animation on the
   molecular level and a dramatization of a young black couple faced with the
   genetic possibility of bearing sickle cell children; their predicament introduces
   the topics of prenatal diagnosis and the possibilities of treatment.

THE THEORY OF BLOOD COAGULATION (USNAC, 26 min., B/W, 1969). Describes
   the basic mechanism of blood coagulation and research in the field.

WORK OF THE BLOOD (EBF, 14 min., C, 1957). The structure of blood cells and
   the plasma in a sample of blood is revealed by laboratory analysis. Describes
   the necessity and method of typing blood before giving transfusions.

Chapter 18: The Heart

ACTION OF THE HUMAN HEART VALVES (OSU, 19 min., C, 1956). Demonstrates
   normal function of pulmonary and aortic, and mitral and tricuspid valves. Case
   histories document disorders of these valves and show how these defects have
   been corrected with the technology of the time.

ANGIOGRAPHIC DIAGNOSIS OF CORONARY ARTERY DISEASE (ACR, 27 min.,
   B/W, 1968). Utilizing cinefluorography and animation, describes the diagnosis
   and treatment of partial obstruction of the coronary arteries.

BLOOD CIRCULATION (EBF, 15 min., C). Description of blood and its path
   through the body.

CARDIAC ARREST (ACYD, 17 min., C, 1955). Demonstrates the importance of an
   open airway and adequate blood oxygenation in preventing cardiac arrest and
   ventricular fibrillation using an anaesthetized dog.

CARDIAC ARREST (USNAC, 36 min., C, 1974). Explains function of circulatory
   system, the electrical activity of the normal and abnormal heart and cardiac
   arrest.

CARDIAC ARRHYTHMIAS (SCI, 23 min., C). Shows several cardiac arrhythmias in
   slow motion in dog hearts.

CARDIAC OUTPUT IN MAN (ICI, 38 min., C, 1951). Demonstrates two methods of estimating cardiac output in humans compared to earlier more traumatic methods. Historical approach.

CIRCULATION OF THE BLOOD (ASFT, 9 min., C, 1958). Illustrates the function of the heart and how the elasticity of the arteries controls the circulation of the blood. Through animation, the flow of blood through the body is illustrated.

COMMON HEART DISORDERS AND THEIR CAUSES (MGHT, 19 min., B/W, 1957). Reviews the function of the healthy heart and circulatory system. Describes common heart disorders, giving the history, symptoms and effects of a case of rheumatic fever, the presumed causes of hypertension and the effects of the conditions which may be caused by arteriosclerosis.

CONGESTIVE HEART FAILURE (USNAC, 40 min., B/W, 1966). Discusses the difficulty in diagnosing congestive heart failure, especially that of the left ventricle. Emphasizes the particular signs which should be recognized in the examination.

CONGESTIVE HEART FAILURE - PATHOPHYSIOLOGY AND TREATMENT (SQ, 25 min., C). Describes pathophysiology, diagnosis and treatment of congestive heart failure. Shows that congestive heart failure is rarely of sudden onset, but rather develops insidiously marked by compensatory mechanisms while cardiac reserve is being gradually eroded.

CORONARY COUNTERATTACK (BYU, 21 min., C, 1976). Illustrates multiple factors relating to coronary heart disease, such as high cholesterol, obesity, smoking, and hypertension; emphasizes and illustrates the value of preventative measures and shows recent developments of diagnostic, preventive and therapeutic care.

CPR: TO SAVE A LIFE (EBF, 14 min., C, 1977). Introduces CPR (cardiopulmonary resuscitation) techniques to the general public. Demonstrates emergency procedures to be used on a person whose heart or breathing has stopped.

DISORDERS OF THE HEART BEAT (WLFL, 30 min., C, 1957). Describes the mechanical, electrical and heart sounds (through phonocardiography) of the normal heart and then describes various arrhythmias, such as PVC's, atrial fibrillation and flutter, ventricular fibrillation, conduction defects. Other than the archaic explanation of ventricular fibrillation, this is an excellent film.

DISSECTION OF THE HEART (USNAC, 12 min., C, 1970). Autopsy dissection technique shows removal of the heart, opening the chambers, dissection of the coronary arteries, examination and measurement of the valves and myocardial examination.

HEART (JBL, 6 min., C, 1974). Shows the physical examination of the heart. A professional film.

HEART SOUNDS AND MURMURS-ORIGINS AND CHARACTERISTICS (UW1, 14 min., C, 1963). Comparison of sounds produced by a normal heart with those in hearts with various defects which produce murmurs.

THE HEART (UT, 21 min., C, 1972). The structure, conductive system and circulation through the heart described.

THE HEART-ATTACK (MGHT, 27 min., C, 1972). Presentation of factors such as excess weight, cholesterol levels, high blood pressure, etc., that cause heart disease. Methods of treatment, such as open-heart surgery, are presented.

THE HEART-CARDIOVASCULAR PRESSURE PULSES (SE, 23 min., C). Illustrates experimental methods of recording sphygmograms and discusses their significance in cardiac function. Artificial heart and experimental animals illustrate principles.

THE HEART: COUNTERATTACK (MGHT, 30 min., C, 1972). Provides advice on the prevention of heart attack: control of diet, exercise, smoking and stress.

THE HUMAN HEART (MGHT, 26 min., C, 1968). Walter Cronkite narrates this Twenty First Century TV program which traces the recent medical research in cardiac surgery and the changing opinions which have deterred or encouraged its progress. Interviews with Christiaan Barnard and Michael DeBakey. Discusses interspecies transplants and mechanical heart replacement.

NEW PULSE OF LIFE (PFP, 29 min., C, 1975). Demonstrates how to restore heart beat and breathing in the crucial first minutes after a heart attack.

PATENT DUCTUS ARTERIOSUS (SQ, 25 min., C). Provides general principles of surgical treatment of patent ductus arteriosus from initial examination to post-operative stage in a case study.

PROPERTIES OF CARDIAC MUSCLE (IOWA, 2 min. each, C, 1971, silent).
Part 1.   Heart Responses to Stimuli
Part 2.   Isolation of the Cardiac Pacemaker
Part 3.   Cardiac Muscle contrasted with Skeletal Muscle
Part 4.   Cardiac Response to an Artificial Pacemaker
Part 5.   Sinoatrial Block in the Frog Heart.

REGULATION OF THE HEART BEAT IN THE MAMMAL (IOWA, 4 min., C, 1971, silent). Shows how autonomic control regulates the rate of the heart.

SIR WILLIAM HARVEY (CBC, 30 min., C, 1972). Explores the contributions and life of William Harvey, discoverer of the circulation of the blood.

THE SOUNDS OF THE NORMAL HEART (WAYNE, 17 min., C, 1964). Shows how the various sounds produced by the normal heart are created.

SPORTS FOR LIFE (BARR, 22 min., C, 1976). Attempting to target behavior change towards increased physical fitness, this film emphasizes the concomitant physiological benefits of regular exercise, depicts people from a range of ages, social classes and races participating in and discussing sports, and includes exciting footage of a wide spectrum of sports activities.

WILLIAM HARVEY (UCE, 19 min., C, 1974). Richard Eakins plays the role of Harvey, presenting his ideas regarding the heart and blood vessels and his concepts of the pulmonary and systemic circuits. Experiments and vividly illustrated descriptions support his hypothesis.

WILLIAM HARVEY AND THE CIRCULATION OF BLOOD (RC, 40 min., C, 1955). The story of the development of the concept that the blood moves in a circle including recreations of the experiments which led Harvey to his conclusions.

WORK OF THE HEART (EBF, 19 min., C, 1968). Pictures and description of the heart. Shows how the blood is circulated and discusses factors affecting heart rate. Actual heart operations are presented.

Chapter 19: The Blood Vessels and Lymphatic System

ADRENERGIC RECEPTORS (ICI, 20 min., C, 1966). Describes the vascular pressure changes effected by sympathetic/catecholamine activity and how these effects are blocked by Inderal, a β-blocking agent. Demonstrates methods of determining heart rate, cardiac output, aortic flow and cardiac contractile force. Effects of sympathetic stimulation demonstrated on human volunteers. Clinical applications of beta-blockage described.

AORTIC ARCH AND CAROTID SINUS REFLEXES (IOWA, 4 min., C, 1971). Illustrates Marey's reflex and its control of heart rate by the monitoring of blood pressure within the aortic arch and carotid sinus.

ARTERIAL AND VENOUS PRESSURES (IOWA, 3 min., C, 1971, silent). Illustrates techniques for direct measurement of arterial and venous pressures. The effects of remoteness of the blood from the heart on pressure is illustrated.

ARTERIAL BLOOD PRESSURE REGULATION (JW, 19 min., C, 1970). Uses decerebrate rabbit to show how receptor activity and central integration modify efferent impulses concerned with regulation of arterial blood pressure.

ATHEROSCLEROSIS (MI, 33 min., C, 1974). Discusses the state of knowledge with reference to the treatment of atherosclerosis.

BLEEDING AND SHOCK (USNAC, 51 min., C). Review of vascular anatomy and functions of blood. Describes methods of hemorrhage control to prevent shock.

BLOOD PRESSURE, THEORY AND PROCESS (WAYNE, C). Discusses blood pressure and the clinical effects of high and low blood pressure.

CIRCULATORY RESPONSE TO VASOPRESSOR DRUGS (CIBA, 25 min., C, magnetic sound). Discusses the circulatory system and what happens during shock. The action of several vasopressor agents are shown as well as two clinical instances when angiotension and norepinephrine are used during surgery.

FUNCTIONS OF THE CAROTID SINUS AND AORTIC NERVE (ICI)
Part 1. Pressoreceptors = 38 min.
Part 2. Chemoreceptors = 33 min.
Presents historical exploration of the carotid and aortic bodies and the pressoreceptors. Demonstrates the functions of these structures in actual experimentations along with diagrammatic interpretation.

HUMAN BODY: CIRCULATORY SYSTEM (CIF, 13 min., C, 1956). Describes processes of circulatory system and networks of arteries and veins. Presents cinefluorographic studies of heart, lungs and kidneys.

LYMPH VESSEL MOVEMENT STUDIES IN THE MESENTERY OF THE RAT (Out of print, Producer: CIBA, 7 min., B/W). The movement of lymph in rat mesentery normally and under the influence of drugs is investigated in this laboratory study.

THE LYMPHATIC SYSTEM (ICI, 30 min., C, 1950). Demonstrates, with the use of animation and animal preparations the functional anatomy of the lymphatic system. Historical approach as well as circulation of lymph and lymphocytes described.

RENAL VASCULAR HYPERTENSION - AN APPROACH TO DIAGNOSIS AND TREATMENT (USNAC, 20 min., C). Describes the basis of renal hypertension and then describes the technique of diagnosis to surgery and patient follow-up.

SHOCK (ACYD, 29 min., C, 1965). The value of determining the central venous pressure and cardiac output in the management of shock as part of the diagnostic procedure is emphasized. Treatment is shown.

SHOCK RECOGNITION AND MANAGEMENT (AEF, 17 min., C, 1968). Through animation and live action, illustrates the physiology of shock. Describes how to recognize and manage the shock patient.

VARICOSE VEINS (ASFT, 7 min., C, 1958). Describes structure and function of veins and how they become varicosed. Shows symptoms, treatment and recommendations for victims of varicosities.

Chapter 20: Defense Mechanisms of the Body

ANAPHYLAXIS IN GUINEA PIGS (UCE, 8 min., C, 1961). Demonstrates anaphylactic shock, using a guinea pig sensitized with egg albumen and a control animal injected only with saline solution. Animated sequence shows bronchioles of lungs closed off by smooth muscle contraction.

BODY DEFENSES, PART 1 (EBF, 10 min., B/W, 1956). Presents the mucous membrane and skin as the primary defenses against infection. Relates role of lymphatic system, liver, spleen and circulatory system. Discusses immunization as a factor in strengthening this defense.

FUNDAMENTAL PRINCIPLES OF IMMUNIZATION (WFL, 40 min.).

IMMUNIZATION NOW! (AEF, 20 min., C, 1977). Explains history and need for immunization. Attempts to overcome apathy in childhood immunization.

THE INFLAMMATORY REACTION (ACY, 25 min., C). Reviews the causes of inflammation, the reactions of tissue to trauma as visualized by cinemicrography and animation emphasizing that the inflammatory response is identical regardless of cause.

PHAGOCYTES: THE BODY DEFENDERS (SEF, 10 min., C, 1965). Shows leucocytes undergoing cell division and phagocytosis and illustrates protective role of these cells in maintenance of health.

PHAGOCYTOSIS (UCE, 4 min., C, 1960). With phase contrast cinemicrography, a human leucocyte is shown pursuing and ingesting streptococci. Shows that living and dead cocci may be phagocytized.

SECRET OF THE WHITE CELL (IU, 30 min., B/W, 1966). The method by which leucocytes destroy microorganisms is explored through cinephotomicrography and electron micrographs of the organelles of white cells which release germicidal materials.

WHITE BLOOD CELLS (MGHT, 12 min., C, 1961). Through photomicrography, shows battle of infectious agents invading living tissue and being combated by white blood cells.

Chapter 21: The Respiratory System

BLOOD GASES (USNAC, 27 min., B/W, 1973). Describes the balance between the partial pressures of carbon dioxide and oxygen as alveolar ventilation changes.

BODY FRAMEWORK, PART 4 (ETF, 15 min., B/W, 1962). Structure and function of lungs along with a description of diaphragmatic ventilation of lungs. Importance of healthy lungs is stressed.

CARBON MONOXIDE POISONING-HYPOHEMOGLOBINEMIC HYPOXIA (IOWA, 3 min., C, 1971, silent). Demonstrates symptoms of hypoxia in a guinea pig exposed to carbon monoxide and its recovery.

CHOKING: TO SAVE A LIFE (EBF, 12 min., C, 1977). Realistic presentation of back blows, the Heimlich maneuver and finger probe for asphyxiation emergencies. Dramatizations of choking victims and repetition of rescue procedure. Describes how to avoid choking situations.

HOW THE RESPIRATORY SYSTEM FUNCTIONS (BRAY, 11 min., B/W, 1950). Illustrates structure and function of airways and the exchange of carbon dioxide and oxygen in the lungs. Breathing described through animation. Posture, rest and fresh air emphasized.

HOW TO SAVE A CHOKING VICTIM: THE HEIMLICH MANEUVER (PAR, 11 min., C, 1975). Dr. J. Heimlich describes a simple first-aid method that he has developed which has successfully saved the lives of choking victims.

HUMAN BODY: RESPIRATORY SYSTEM (CIF, 13 min., C, 1961). Describes organs of respiratory system, mechanics of pulmonary ventilation and gas exchange with the blood.

THE HUMAN THROAT (BRAY, 11 min., B/W, 1947). Pharynx and larynx structure described along with mechanism for preventing entrance of food into airway while swallowing.

HYPOXIA (USNAC, 29 min., C, 1975). Hypoxia discussed in a panel led by news correspondent Roy Neal explores different types of hypoxia. Animated sequence.

LARYNX (CDPH, 23 min., B/W, 1951). Shows tumors of vocal cords and discusses treatment techniques.

THE LARYNX (Out of print, Producer: CIBA, 15 min., C, 1961). The anatomy, physiology and diagnosis of laryngeal disorders are described. Depicts examination of larynx with mirror and laryngoscope. Uses endoscopic photography.

LUNGS, PARTS 1 and 2 (NYAM, 60 min. each, B/W, 1963). Visual aids, charts, slides, and live subjects are used by a panel of experts to discuss diseases of the lung.

MEDICAL ASPECTS OF DIVING (NF, 29 min. each, C, 1962).
   Part 1.   The Mechanical Effects of Pressure. Explains how swimmers and divers can prevent harm by avoiding development of unequal pressures.
   Part 2.   Effects of Elevated Partial Pressures of Gases. Explains how the body is affected by the partial pressure and relative concentrations of the gases that you breathe.

NEW BREATH OF LIFE (PFP, 24 min., C, 1978). Teaches emergency maneuvers in case of choking or when breathing has stopped because of trauma, presenting diagrams and models to supplement live action; demonstrates in enough detail to allow viewer to practice and then apply techniques.

THE NOSE (USNAC, 20 min., C, 1969). Physical examination of the nose, normal and abnormal. Demonstrates various abnormalities and meaning of nasal discharges.

THE NOSE - STRUCTURE AND FUNCTION (EBF, 11 min., C, 1954). Through animation and microphotography the breathing and smelling functions of the nose are demonstrated.

PRINCIPLES OF RESPIRATORY MECHANICS, PARTS 1 and 2 (AMA, 21 min. each, C, 54). Compares breathing activity in a normal subject with those with various functional and organic disorders. Points out the value of mechanical measurements and the principles of elastic and resistive properties of the lungs.

RESPIRATION (PMR, 20 min., B/W, 1969). A BBC-TV film describing external respiration and lung structure. Physiology of reflex breathing mechanism and role of alveolar surface tension discussed.

RESPIRATION (UEVA, 12 min., B/W, 1953). Describes internal and external respiration, showing role of chest activity in breathing. Lung and air passage structure illustrated along with explanation of how oxygen is distributed throughout the body and its necessity.

RESPIRATION AND CIRCULATION (USNAC, 26 min., C, 1961). Explains coordinated functioning of respiratory and circulatory systems and their control through chemoreceptors. Relates to the physiological problems of flight.

RESPIRATION IN MAN (EBF, 26 min., C, 1969). Overview of important concepts of respiration, structure and functions of respiratory system.

SMOKING/EMPHYSEMA: A FIGHT FOR BREATH (MGHT, 12 min., C, 1975). Uses real scenes and animation to show respiratory system, diseased and healthy lungs, and microscopic activity of respiratory structures. Relates emphysema to smoking.

SPIROMETRY: THE EARLY DETECTION OF CHRONIC OBSTRUCTIVE PULMONARY DISEASE (ALA, 25 min., C, 1968). Shows function of normal respiratory system and the effects of impaired ventilatory function.

THE TOBACCO PROBLEM--WHAT DO YOU THINK (EBF, 17 min., C, 1971). Describes the dangers of smoking by discussion and demonstrations by laboratory scientists. Looks at the stated reasons for smoking.

TOO TOUGH TO CARE (ALA, 18 min., C, 1964). Tells, with satire and jests, the story of a fictitious tobacco company's efforts to find new smokers. Points out misleading claims made in ads and spoofs the image of the smoker generated by advertising.

Chapters 22, 23: The Digestive System

ALIMENTARY TRACT (EBF, 11 min.).

BODY DEFENSES, PART 2 (EBF, 10 min., B/W, 1956). Shows mechanical actions of esophagus, stomach and intestine: peristalsis, antiperistalsis, segmentation, stomach hunger contractions and hypermotility described; shows villi in their absorbing movements.

BODY DEFENSES, PART 3 (EBF, 10 min., B/W, 1956). Describes digestive
     processes occurring in mouth, stomach and small intestine. Explains both
     mechanical and chemical aspects of digestion.

THE DIGESTIVE SYSTEM (EBF, 17 min., C, 1965). Digestive system explained by
     animation and practical laboratory tests. Includes remarkable scenes of
     digestive sequences filmed by x-ray motion picture photography.

GOOD SENSE ABOUT YOUR STOMACH (TL, 15 min., C, 1973). Through animation
     and dramatization illustrates digestion. Provides practical methods for
     controlling quantity and quality of food and relaxation necessary for good
     digestion.

HUMAN BODY: THE CHEMISTRY OF DIGESTION (CIF, 16 min., C, 1968).
     Through x-rays, animation and laboratory demonstrations, the major steps in
     the digestion of the three major foodstuffs is illustrated.

HUMAN BODY: DIGESTIVE SYSTEM (CIF, 14 min., C, 1958). Through animation
     and x-rays, shows the functions of the digestive organs and how complex
     foodstuffs are broken down into assimilable molecules.

HUMAN GASTRIC FUNCTION (ASF, 18 min., C). Documents function of the
     stomach with Tom, an irascible old man who had swallowed lye as a youngster.
     He was provided with a gastric fistula which provided an excellent opportunity
     to witness stomach function. Many of the experiments performed are
     presented in this film. It is quite interesting, though now old.

INTRA-ORAL AND PHARYNGEAL STRUCTURES AND THEIR MOVEMENTS
     (USNAC, 18 min., C, 1950). Pictures the losses in activity or the tongue, soft
     palate and epiglottis as the result of the surgical removal of a cancerous
     maxillary sinus.

LET'S EAT FOOD (MGHT, 35 min., C, 1976). Presents the American diet, its
     relationship to diseases such as dental caries, heart disease, and obesity, and
     supplies suggestions for diet alternatives. Includes commentaries by noted
     nutritionists Joan Gussow, Dr. Jean Mayer and Dr. William Connor.

THE MECHANICS OF SWALLOWING (WARDS, 3 min., B/W, 1966). X-rays are used
     to visualize the motions of the jaw, tongue and larynx while chewing and
     swallowing. X-ray views of baby and adult drinking.

PROBING THE SECRETS OF THE STOMACH (UJ, 18 min., C, 1969).
     Cinemicrography is used to study the structure of the stomach. Hydrochloric
     acid and pepsinogen secretion is illustrated along with the effects of gastrin
     and acetylcholine on gastric secretion.

REGULATION OF PANCREATIC SECRETION (USNAC, 28 min., B/W. 1970).
     Describes etiology and pathogenesis of acute pancreatitis and the influence of
     the condition on secretory function.

ROENTGEN ANATOMY OF THE NORMAL ALIMENTARY CANAL (AMA, 27 min., B/W, 1966). Cinefluorographic presentation of functioning anatomy of the digestive tract from mouth to rectum.

THE SMALL INTESTINE - MOTILITY AND VASOMOTOR ACTIVITY (UIO, 18 min., C, 1966). Mobility of the intestine is explored through animation and demonstration in an anaesthetized rabbit. Demonstrates changes in vascularization as the result of vasomotor stimulation.

WILLIAM BEAUMONT (UCE, 23 min., C, 1976). Richard Eakin plays the role of Beaumont, presenting a dynamic lecture about his experience in medical school and in the army where he found the injured man whose gunshot wound served as his "physiological laboratory" for the study of gastric digestion. A concluding contrast is made between past and present medical data.

Chapter 24: Metabolism, Nutrition, and Temperature Regulation

BODY DEFENSES, PART 4 (EBF, 10 min., B/W, 1956). Discusses dietary requirements for three classes of foodstuffs, as well as minerals, vitamins and water. Through animation, sugar and fat processing (digestion, absorption and assimilation) is described.

A CHEMICAL FEAST (BNF, 11 min., C, 1973). Shows how dyes, gums, emulsifiers and other additives often substitute for real nutritional value in many processed foods in American supermarkets. Describes how "creative playfoods" often cost more per pound than steak.

CHOLESTEROL--EAT YOUR HEART OUT (SEF, 14 min., C, 1975). Promotes awareness of dietary cholesterol, its effects on serum cholesterol level and possible benefits of low-cholesterol diets in reducing atherosclerotic incidence and severity particularly to a young audience which can relate to the teenage actors used in the film; intended to encourage early attention to dietary habits.

THE CURIOUS CASE OF VITAMIN E (UCE, 29 min., B/W, 1975). Vitamin E, which is said to prevent old age, stimulate the sex drive, and protect the body from the hazards of smoking and smog, is examined. Includes a discussion by scientists who have conducted research on this vitamin.

A DECLARATION: "THE RIGHT TO EAT!" (SRS, 20 min., C, 1976). Lorne Greene presents the moral dilemma of the unequal distribution of the world's protein resources and gives the Hunger Foundation's nine-point program to give everyone the Right to Eat. Suggests that Americans can make a significant impact on the world hunger situation by changing their personal food consumption patterns; emphasizes the substitution of textured vegetable proteins for meat, animal proteins. Addresses by notables such as Daniel E. Shaughnessy and Hubert Humphrey.

DIET AND ATHEROSCLEROTIC DISEASE - ARE THEY RELATED? (USNAC, 29 min., 1969). Dr. Stamler presents statistical data to implicate poor diet as the cause of atherosclerosis. Dr. Altschule describes his feelings as to the inadequacy of the studies carried out to date.

EAT, DRINK AND BE WARY (CF, 20 min., C, 1975). Discusses problems associated with the increased reliance on manufactured and processed foods. Examines the issues of additives, food processing and the changes taking place in the American diet.

ENERGY RELEASE FROM FOOD (Out of print, Producer: Upjohn Co., 26 min., C, 1946). Roles of nicotinic acid, thiamine and riboflavin in the energy exchanges of carbohydrates described in animation. Selected clinical deficiencies created by insufficient B-complex are illustrated.

FOR TOMORROW WE SHALL DIET (CF, 24 min., C, 1976) Changing eating habits to lose weight. Discussion of dangers of fad diets, relation of calories to energy output, proper nutrition and exercise.

HOW THE BODY USES ENERGY (MGHT, 15 min., 1963). Explains why energy is required for life function. Discusses digestion, oxidation of food products and energy production.

HUMAN BODY: NUTRITION AND METABOLISM (CIF, 14 min., C, 1962). Contrasts basal and activity metabolism levels and requirements in terms of calories. Role of foodstuffs in energy production emphasized.

NUTRITION - ENERGY, CARBOHYDRATES, FATS AND PROTEINS (WAYNE, Part 1: 29 min., Part 2: 26 min. B/W, 1974). Identifies factors which determine the caloric value of food and which affect body weight; discusses various reducing diets. Describes major nutrients and their sources.

NUTRITION-FUELING THE HUMAN MACHINE (BFA, 18 min., C, 1978). Lists facts oi nutrition that affect our physical health and well-being. Encourages a balanced and varied diet, discusses the possible dangers of food additives and sugar, and emphasizes the importance of exercise to avoid obesity.

NUTRITION IS. . . (WGC, 26 min., C, 1976). The basics of nutrition and the relevancy of health are summarized.

NUTRITION SERIES (BARR, C, 1976).
The All-American Meal (11 min.) Stresses two concepts: the need to eat at home and nutritional aspects of fast food. Home meals build family as well as dietary fiber.
You Are What You Eat (10 min.) Eating the correct combination of proteins carbohydrates, fats, vitamins, and minerals and their utilization in the body are shown through animation.
The Consumer and the Supermarket (15 min.) Introduces the four food groups, notes their common forms values and costs, mentions advertising, store layout, packaging and labeling.

REGULATING BODY TEMPERATURE (EBF, 22 min., C, 1972). Demonstrates how physiological and behavioral mechanisms regulate body temperature in humans as compared to other mammals.

Chapter 25: The Urinary System

ARTIFICIAL KIDNEY (LPI, 12 min., B/W, 1963). An auto accident causes the failure of a child's natural kidney. Describes and shows an artificial kidney (1963 vintage) in use comparing it to the natural kidney.

DIURESIS (CIBA, 48 min., C, magnetic sound). Comprehensive study of the kidney including its anatomy, physiology and function, formation of urine, and hairpin countercurrent principle. Discusses kidney malfunctioning and disease and mechanisms of action and usefulness of diuretics.

THE DYNAMIC KIDNEY (EL, 29 min., C, 1976). A description of kidney function with the use of animation.

ELIMINATION (UEVA, 13 min., B/W, 1950). Through animation, elimination through the skin, lungs, kidney and gut is described. Gives detailed descriptions.

EMBRYOLOGY OF THE URINARY SYSTEM (BURTNA, 17 min., C, 1975). Development of human kidney, ureter, bladder and urethra in males and females is described with the use of animated diagrams.

FUNCTIONAL ANATOMY OF THE HUMAN KIDNEY (AMA, 25 min., C, 1967). Light and electron micrographs used to identify and describe parts of the kidney and nephrons. Describes mechanisms of formation of urine, including that of the counter current multiplier system.

THE HUMAN BODY - EXCRETORY SYSTEM (CIF, 13 min., C, 1960). Structure and function of the excretory system and its relation to the regulation of blood chemistry.

THE HUMAN KIDNEY - TUBULES, BLOOD VESSELS, AND VASCULAR TUBULAR RELATIONS (HP, 20 min., C, 1975).

KIDNEY FUNCTION IN DISEASE (EL, 46 min., C). Discusses renal pressor system and changes in structure and function in kidneys in hypertension, glomerulonephritis, myelonephritis, renal calculus, crush syndrome, etc. Discusses water metabolism, electrolytes, acids and bases with reference to kidney function.

KIDNEY FUNCTION IN HEALTH (EL, 38 min., C). Follows urine formation from glomerulus through kidney tubules. Describes renal function tests designed to measure filtration.

THE KIDNEYS, PARTS 1 and 2 (NYAM, 60 min. each, B/W, 1963). Panel discussion using visual aids, charts and slides to describe kidney disease.

KIDNEYS, URETERS AND BLADDERS (BRAY, 11 min., B/W, 1947). Animated
   film describes organs of urinary system and urine formation.

PHYSIOLOGY OF THE KIDNEY (ICI, 46 min., C, 1948). Describes anatomy of
   kidney and its circulation. Describes glomerular filtration and tubular
   secretion and the effects of blood pressure, adrenaline and chloroform
   overdose.

RENAL HYPERTENSION/BILATERAL NEPHRECTOMY KIDNEY
   TRANSPLANTATION (UP, 23 min., C). Film features three noted kidney
   authorities including Dr. Harry Goldblatt.

VASCULAR AND TUBULAR ORGANIZATION OF THE KIDNEY (AHP, 20 min.,
   C, 1972). Explores vascular and tubular systems by injection of red latex and
   white silicone solutions through photomicrographs.

WORK OF THE KIDNEYS (EBF, 20 min., 1972). Structure and function of the
   renal system. Demonstrates formation of urine, regulation of blood
   composition, functioning of the bladder. Explains relation of blood pressure to
   urine flow and rate of secretion as affected by sugar water and temperature.
   Emphasizes the importance of the kidneys as regulators of the internal
   environment of the body.

Chapter 26: Fluid and Electrolyte Balance and Acid-Base Regulation

ACID-BASE BALANCE (WAYNE, 31 min., B/W, 1974). Reviews need for maintenance
   of normal pH and osmotic pressure. Demonstrates the effects of disturbances
   in electrolytes, pH, buffers on the development of acidosis and alkalosis.

BASIC CONCEPTS OF BODY FLUIDS AND THE EFFECT OF MAJOR IONS
   (WAYNE, 30 min., B/W, 1971). Describes the three fluid compartments and
   their components. Use of a frog heart in a kymograph experiment reveals the
   effects of changing the chemical balance on living cells.

CHEMICAL BALANCE THROUGH RESPIRATION (AFI, 23 min., C). Illustrates
   respiratory physiology in regard to gas exchange and the maintenance of a
   stable hydrogen ion concentration through breathing and the buffer action of
   the blood. Respiratory and metabolic acidosis, respiratory and metabolic
   alkalosis are all shown as well as the effects of uneven ventilation.

DYNAMICS OF FLUID EXCHANGE (WAYNE, 30 min., B/W, 1974). Describes
   osmotic and hydrostatic pressure and their relationship to edema. Discusses
   the role of the kidney in the maintenance of fluid and electrolyte balance.

THE PATIENT WHO CANNOT DRINK (MC, 19 min., C, 1965). Reviews facts of
   fluid balance. Explains role of kidney in maintaining electrolyte balance.

VITAL FLUID (PMR, 20 min., B/W, 1969). Describes the composition and
   homeostatic maintenance of intracellular fluid.

Chapters 27, 28: Reproduction

ACHIEVING SEXUAL MATURITY (JW, 21 min., C, 1973). Deals with the sexual anatomy, physiology, and behavior of both sexes from conception through adulthood. Uses explicit, live photography of nude males and females to identify the external genitalia and graphics and diagrams to show internal anatomy. Describes hormones involved in both males and females with respect to menstrual cycle and sperm production and sex hormone control. Masturbation discussed and photographed in both sexes.

ANATOMY AND PHYSIOLOGY OF THE HUMAN REPRODUCTIVE ORGANS (USNAC, 34 min., C, 1973). Discusses and illustrates the internal female and male reproductive system. Shows how structure is specialized for production and transport of gametes. Describes hormonal changes during menstrual cycle and pregnancy and relates pelvic structure to parturition.

BABYMAKERS (MGHT, 4 min., C). Discusses experimentation in artificial insemination in both animal and human populations. Controversies surrounding egg-embryo manipulation, surrogate mothers and artificial insemination are discussed in the film.

THE BEGINNING OF LIFE (PFP, 26 min., C, 1970). Uses microphotography of up to 200,000 magnification to show sperm and egg cell formation in healthy living mammals. Shows fertilization, cell division, and growth to first heartbeat.

THE BEGINNING OF LIFE (BNF, 30 min., C, 1969). Stages of growth of the human embryo inside the womb. Swedish Broadcasting Corporation.

BIOGRAPHY OF THE UNBORN (EBF, 17 min., B/W, 1956). Photomicrography, animation, live photography and x-rays record the development of the embryo and the fetus in the womb.

BIRTH CONTROL: THE CHOICES (CF, 25 min., C, 1976) Methods of birth control -- their uses, limitations, and side effects. Includes discussions of vasectomy, tubal ligation, and abortion.

BIRTH CONTROL--FIVE EFFECTIVE METHODS (SF, 10 min., C, 1976). Graphic description of the methods of application, mechanisms of action (via animation), relative effectiveness (via bar graphs) and pros and cons of the diaphragm, IUD, contraceptive foam, pill, and condom. Describes the ovulation cycle and the rhythm method.

BIRTH OF A BABY (CHM, 28 min., B/W, 1967). Shows the passage of baby through the vagina. Illustrates breech births, and birth of twins. Discusses prematurity, miscarriage and abortion.

BIRTH OF A FAMILY (PE, 24 min., C, 1976). Focuses on preparation for childbirth and is accompanied by a text, lesson plan, and glossary of terms. Follows a couple actively involved in conditioning for a natural birth, and pictures the birth from labor to emergence of child and placenta, at a Chicago hospital. Includes the couple's postpartum testimonial to the effectiveness of their training.

BIRTH WITHOUT VIOLENCE (NYF, 21 min., C, 1975). Reviews Frederick Leboyer's method of child delivery which aims to minimize the birth trauma and provide a peaceful transition for the newborn from the womb to life.

THE BODY HUMAN: THE MIRACLE MONTHS (TC, 50 min., C). Captures for the first time on film the earliest phases of human development, including the only film ever taken of human ovulation, of a living 40-day embryo, and of the world of a 12-week fetus as its hands and feet are taking shape. Glimpses of development through thermography and ultrasound pictures (e.g., baby sucking its thumb inside the womb).

CHILD IS BORN (BFA, 22 min., C, 1971). Traces the beautiful relationship between wife and husband from the moment they decide to have children to the homecoming of their first baby. Shows the development of the embryo and fetus as well as the actual birth process.

CONTRACEPTION (JW, 23 min., C, 1973). Explains the need for contraceptives, the varieties available, their principles of operation, history and advantages and disadvantages.

DIAGNOSIS OF HIDDEN CONGENITAL ANOMOLIES (MF, 10 min.). The film points out the necessity and ease of examination for hidden congenital anomalies within a few minutes after birth. Illustrated with procedures for detection of choanal atresia, cleft palate and diaphragmatic hernia.

ENDOCRINOLOGIC ASPECTS OF AMENORRHEA (USNAC, 34 min., B/W, 1969). Describes causes of secondary amenorrhea and diagnostic testing for the condition.

FEMALE GENITALIA (JBL, 10 min., C, 1974). Demonstrates the physical examination of the vulva in the physician's office.

FERTILIZATION AND EARLY CLEAVAGE OF THE RABBIT EGG (UF, 8 min., C). Documents the development of the egg from the entry of the sperm to the morula stage.

THE FERTILIZATION PROCESS (PROBLEMS AND PHASES) (MGHT, 15 min., C, 1967). Describes the phases of fertilization: release of sperm, approach, penetration, fusion and activation. Explores chemotoxic orientation of sperm with respect to egg and the work of pincers in parthenogenesis.

THE FIRST DAYS OF LIFE (MGHT, 16 min., C, 1971). Shows natural childbirth, reviews the development of the fetus and traces some aspects of the development of the baby up to the age of three.

FROM GENERATION TO GENERATION (MGHT, 30 min., C, 1959). Tells the story of human reproduction as an integral part of the universal pattern of nature, explaining that childbearing is an emotional and spiritual experience as well as a physical one.

GENTLE BIRTH (POLY, 15 min., C, 1976). Features an actual hospital delivery by the Leboyer method of birth, including the father's presence in the delivery room, the dimming of lights and immersion of the baby in warm water reminiscent of the womb to alleviate the shock of birth. Does not explain specifically how the Leboyer method differs from standard delivery, but neither does it make highly emotional and unsupported claims concerning its effectiveness. Integration of the Leboyer method with Lamaze in the modern hospital setting is emphasized.

HEREDITY AND ENVIRONMENT (MGHT, 27 min., C, 1978) Shows how the physical and mental development of an infant is affected by environmental forces.

HEREDITY AND ENVIRONMENT (MGHT, 27 min., C, 1978). Describes the environmental factors which can affect the physical and mental growth of the infant.

HIGH-SPEED CINEMATOGRAPHY OF HUMAN SPERMATOZOA (NYU, 17 min., C). Ultraslow motion pictures of human spermatozoa showing their movement patterns in a free environment. Shows abnormal sperm and photographs of detailed sperm anatomy through the electron microscope.

HORMONE CONTROLS IN HUMAN REPRODUCTION (MGHT, 16 min., C. 1967). Correlates ovarian and menstrual cycles through the action of the various hormones. Traces the development of a human embryo.

THE HUMAN BODY: REPRODUCTIVE SYSTEM (CIF, 14 min., C, 1959). Similarities and differences in the reproductive anatomy of male and female described with animation and photomicrography. Shows how these organs are involved in the creation of a new life.

HUMAN REPRODUCTION (MGHT, 21 min., C, 1966). Human reproduction and process of birth factually presented. Models and animation show anatomy embryology and body mechanics in the delivery process. Stresses biological normalcy or reproduction and emphasizes clear, objective familiarity with these facts.

HUMAN REPRODUCTION (CHM, 28 min., B/W, 1967). Describes human growth and development including sex hormones and their role in ovulation and menstruation. Discusses spermatogenesis and fertilization.

MALE GENITALIA (JBL, 10 min., C, 1974). Demonstrates examination of the male genitalia in the doctor's office.

THE MALE SEX HORMONE (SC, 24 min.).

THE MENSTRUAL CYCLE (EL, 12 min., C). Hormonal and histological changes in the ovaries and uterus in the normal menstrual cycle are described by animation.

MIGRATION OF THE PRIMORDIAL SEX CELLS INTO THE GENITAL RIDGE (EAT, 18 min., C., 1974). Describes the embryological development of the female reproductive system with particular emphasis on the migration of the primordial sex cells into the structure which will form the adult ovary.

THE MILK EJECTION REFLEX (JW, 21 min., C, 1972). Shows the mechanism of milk secretion and its "let-down" by stimulation of the nerve fibers leading to the posterior lobe of the pituitary.

THE MIRACLE OF LIFE (PFP, 15 min., C, 1975). Live action film shot through microscopic photography and showing processes of fertilization, cell division, and growth to first heartbeat.

A NORMAL BIRTH (MAP, 11 min., B/W, 1951). Delineates the actual birth of a child, from the time the mother enters the delivery room until she leaves it.

OVULATION AND EGG TRANSPORT IN MAMMALS (UW, 14 min., C, 1972). Illustrates ovulation, oviduct pick up of the ovum, and follows the egg to the site of fertilization.

PHYSIOLOGY OF CONCEPTION (UN, 29 min., B/W, 1965). Discussion of diagnosis of ovulation timing and the physiology of sperm migration, fertilization and early development of the zygote. Includes coverage of female reproductive physiology and anatomy as well as a discussion of infertility.

PRENATAL DEVELOPMENT (CRM, 23 min., C, 1974). Describes the biology of the developing infant. Presents the latest theories and facts about the influence of the environment on the fetus within the uterus.

PRENATAL DIAGNOSIS BY AMNIOCENTESIS (MF, 26 min., C, 1972). Portrays two genetic counseling situations in which amniocentesis is suggested - one concerning the risk of biochemical defect, the other concerning the risk of a chromosomal defect. Shows the actual technique used to withdraw amniotic fluid and illustrates the complex laboratory procedures used to evaluate the fluid.

RECENT ADVANCES IN REPRODUCTIVE PHYSIOLOGY (UNL, 29 min., B/W, 1965). Presentation of current research in endocrine, biochemical, genetic and clinical pathological aspects of human reproductive physiology.

REPRODUCTION IN THE SEA URCHIN (CIF, 13 min., C, 1965). Describes life cycle of the sea urchin with emphasis on the events surrounding fertilization: gamete release, fertilization, early cleavage.

SPERM MATURATION IN THE MALE REPRODUCTIVE TRACT: DEVELOPMENT OF MOTILITY (WASH, 13 min., C, 1968). Shows gradual attainment of motility as spermatozoa pass from seminiferous tubules through the ductuli efferentes, and epididymis, and finally enter the ductus deferens.

STUDIES IN HUMAN FERTILITY (OP, 30 min., B/W, 1955). Describes the anatomy and physiology of the male and female reproductive systems, and gives a detailed description of contraceptive methods and techniques.

TWELVE RECENT ADVANCES IN REPRODUCTIVE PHYSIOLOGY (UN, 29 min., B/W, 1965). The advances in reproductive physiology covered in this film involve biochemistry, endocrinology, genetics and clinical pathology.

VASECTOMY (CF, 17 min., C, 1972). Through personal experience recounting, explores vasectomy as a method of contraception. Follows the discussion of a couple about the operation with their doctor and the operation itself.

VASECTOMY (NEIF, 8 min., B/W). Through live action and animation shows method by which doctors should proceed with vasectomy operations.

VD: A NEWER FOCUS (AEF, 20 min., C, 1976). Problems, myths, and facts of VD presented in a combination of real and dramatic situations. Includes discussion of attitudes which prevent treatment. Describes symptoms of the various venereal diseases.

VENEREAL DISEASES: THE HIDDEN EPIDEMIC (EBF, 27 min.). Presents a brief history of venereal disease, examines the symptoms of syphillis and gonorrhea and the microorganisms which cause diseases, and emphasizes the importance and methods of treatment.

WHEN LIFE BEGINS (MGHT, 12 min., C, 1971). The sequential growth and development of the human fetus from ovulation to birth is accurately presented. The relationship between the developing fetus and its umbilical cord, placenta and intrauterine environment of amniotic fluid and surrounding membranes is chronologically traced. Film concludes with a natural childbirth without drugs or instruments.

X-RAY, ULTRASOUND AND THERMOGRAPHY IN DIAGNOSIS (UP, 22 min., C, 1969). Shows diagnostic potential of x-rays, ultrasound and thermography for physicians. Reveals a twelve week fetus moving during ultrasound viewing. Early detection of breast cancer is shown to be possible by thermography.

YOUR PELVIC AND BREAST EXAMINATION (PE, 12 min., C, 1975) Film designed to relieve the anxiety of routine but important medical examinations. The viewer observes how a young woman performs a self-examination of her breasts; clear detailed photography shows the cervix, instruments used for pelvic examination, and how a pap smear and gonorrhea culture are taken.

Distributor Code Key

ACR	American College of Radiology, 20 N. Wacker Dr. Chicago, IL 60606.
ACY	American Cyanamid Co., Berdan Ave., Wayne, NJ 07470.
ACYD	American Cyanamid Co., Davis & Geck Division, 1 Casper St., CT 06810.
AEF	American Educational Films, 132 Lasky Dr., Beverly Hills, CA 90212.
AF	Academy Films, P.O. Box 1023, Venice, CA 90291.
AFI	Association Films, 799 Stevenson St., San Francisco, CA 94103.
AHP	Alfred Higgins Productions, 9100 Sunset Blvd., Los Angeles, CA 90069.
AIF	Australian Instructional Films, 39 Pitt St., Sydney, Australia.
ALA	American Lung Association, Public Relations Dept., 1740 Broadway, New York, NY 10019.
AMA	American Medical Association, 535 N. Dearborn, Chicago, IL 60610.
ASF	Association Sterling Films, 866 Third Ave., New York, NY 10022.
ASFA	American Science Film Association.
ASFT	Association-Sterling, 8615 Directors Row, Dallas, TX 75240.
AVC	A-V Corp., 2518 N. Boulevard, Houston, TX 77066.
BARR	Barr Films, 3490 E. Foothill Blvd. Pasadena, CA 91107.
BELL	Bell Telephone Companies, Information Dept., 195 Broadway, Room 07-1106, New York, NY 10007.
BFA	BFA Educational Media, 2211 Michigan Ave., P.O. Box 1785, Santa Monica, CA 90404.
BFI	British Film Institute, 72 Dean St., London W1V 5HB, England.
BL	Bausch and Lomb, Inc., Film Distribution Service, 635 St. Paul St., Rochester, NY 14602.
BNF	Benchmark Films, Inc., 145 Scarborough Road, Briarcliff Manor, NY 10510.
BRAY	Bray Studios, Inc., 630 Ninth Avenue, New York, NY 10036.
BURTNA	Burton, Alexis L.
BYU	Brigham Young University, DMDP Media Business Services, W 164 Stadium, Provo, UT 84602.
CBC	Canadian Broadcasting Corp., P.O. Box 500 Terminal A, Toronto 5, Ontario, Canada.
CCM	CCM Films, MacMillan Films, Inc., 34 MacQuesten Parkway S, Mt. Vernon, NY 10550.
CDPH	California Department of Public Health, 2151 Berkeley Way, Berkeley, CA 94704.
CF	Churchill Films, 662 N. Robertson Blvd. Los Angeles, CA 90069.
CFI	Counselor Films, Inc., 1740 Cherry St., Philadelphia, PA 19103.
CHM	Cleveland Health Museum, 8911 Euclid Ave., Cleveland, OH 44106.
CIBA	Ciba Pharmaceutical Co., Medical Communications Dept., 556 Morris Ave., Summit, NJ 07901.
CIF	Coronet Instructional Films, 65 E. South Water St., Chicago, IL 60601.
CRH	Craig Rehabilitation Hospital.
CRM	CRM Educational Films, 1011 Camino Del Mar, Del Mar, CA 92104.
CSUS	University Media Services, California State University, Sacramento, 6000 J St., Sacramento, CA 95819.

CU	Center for Mass Communication of Columbia University Press, 562 West 113th St., New York, NY 10025.
DUKE	Duke University School of Medicine, Durham, NC 27702.
EAT	Eaton Labs, Division of Norwich Pharmacal Co., 17 Eaton Ave., Norwich, NY 13815.
EBF	Encyclopedia Britannica Educational Corp., 425 N. Michigan Ave., Chicago, IL 60611.
EL	Eli Lilly and Co., Medical Division, Indianapolis, IN 46206.
ETF	Eastman Teaching Films, 343 State St., Rochester, NY 14650.
GS	Garden State/Novo, Inc., 630 Ninth Ave., New York, NY 10036.
HMC	Houghton Mifflin Co., One Beacon St., Boston, MA 02107.
HMP	Harold Mayer Productions, 155 West 72nd St., New York, NY 10023.
HP	Hoechst Pharmaceuticals.
HR	Harper and Row Publishers, 10 E. 53rd St., New York, NY 10022.
ICI	Imperial Chemical Industries, Inc., P.O. Box 1274, 151 South St., Stamford, CT 06904.
IEFL	International Eye Film Library, 264 Danforth Ave., Jersey City, NJ 07305.
IFB	International Film Bureau, 332 S. Michigan Ave., Chicago, IL 60604.
IOWA	Film Productions Unit, Iowa State University, Ames, IA 50010.
IP	Iwanami Productions, Inc., 22-2 Kanda Misakicho, Chiyoda-Ku, Tokyo, Japan.
IU	Indiana University, Audio-Visual Center, Bloomington, IN 47401.
JBL	J. B. Lippincott, East Philadelphia Square, Philadelphia, PA 19105.
JJ	Johnson and Johnson, New Brunswick, NJ 08903.
JTC	John Tracy Clinic, 806 W. Adams Blvd., Los Angeles, CA 90007.
JW	John Wiley and Sons, Inc., Pub., 605 3rd Ave., New York, NY 10016.
KB	Knowledge Builders, Lowell Avenue and Cherry Lane, Floral Park, NY 11001.
LPI	Lawren Productions, Inc., P.O. Box 666, Mendocino, CA 95460.
MAP	Medical Arts Production.
MC	Mayo Clinic, Section of Photography, Rochester, MN 55902.
MD	National Foundation - March of Dimes, 1275 Mamaroneck Ave., White Plains, NY 10605.
MF	Milner-Fenwick, Inc., 2125 Greenspring Dr., Timonium, MD 21093.
MG	Media Guild, Box 881, Solana Beach, CA 92075.
MGHT	McGraw-Hill Films, 110 15th St., Del Mar, CA 92014.
MI	Medcom, Inc., 1633 Broadway, New York, NY 10019.
MP	Montage Productions, 2401 East Douglas, Wichita, KS 67211.
NASA	National Aeronautics and Space Administration, Code FAD, Washington, DC 20546.
NEIF	National Education and Information Films, Ltd., National House, Tullock Rd., Apollo Bunder, Bombay 1, India.
NF	Norwood Films, P.O. Box 1894, Wheaton Post Office, Silver Spring, MD 20902.
NGS	National Geographic Society, 17th and M Sts. NW., Washington, DC 20036.
NMAC	National Medical Audiovisual Center, 2111 Plaster Bridge Rd., Atlanta, GA 30324.
NYAM	New York Academy of Medicine, 2 East 103rd St., New York, NY 10029.

NYF	New Yorker Films, 16 West 61st St., New York, NY 10023.
NYU	New York University, Film Library, 26 Washington Place, New York, NY 10030.
OP	Ortho Pharmaceutical Corp., Film Division, Raritan, NJ 08869.
OSU	Ohio State University, Motion Picture Division, Film Distribution, 156 West 19th Ave., Columbus, OH 43210.
PAR	Paramount Communications, 5451 Marathon St., Hollywood, CA 90038.
PE	Perennial Education, Inc., P.O. Box 855 Ravinia, 477 Roger Williams, Highland Park, IL 60035.
PFP	Pyramid Films, P.O. Box 1048, Santa Monica, CA 90406.
PHASE	Phase Films.
PMR	Peter M. Robeck and Co. Inc., Dist. by Time-Life Films Inc., 20 Biekman Place, New York, NY 10022.
POLY	Polymoph Films, Inc., 331 Newbury St., Boston, MA 02115.
RC	Royal College of Physicians of London.
RL	Roche Labs.
SB	Senses Bureau, University of California at San Diego., P.O. Box 109, La Jolla, CA 92037.
SC	Schering Corporation, Galloping Hill Rd., Kenilworth, NJ 07033.
SCI	Scientificom Audio Visual Distribution Center, 708 Dearborn St., Chicago, IL 60610.
SE	Searle Company, G.D.
SEF	Sterling Educational Films, 241 E. 34th St., New York, NY 10016.
SKF	Smith, Kline and French Laboratories, Smith, Kline and French Services Dept., 1530 Spring Garden Street, Philadelphia, PA 19101.
SQ	E. R. Squibb and Sons.
SRBO	Sports and Recreation Bureau of Ontario, Toronto, Ontario, Canada.
SRS	SRS Productions, 4224 Ellenita Ave., Tarzana, CA 91356.
TC	Trainex Corporation, 12601 Industry St., Garden Grove, CA 92641.
TF	Tribune Films, 38 West 32nd St., New York, NY 10001.
TL	Time-Life Films, Inc., 43 W. 16th St., New York, NY 10001.
UCE	University of California Extension Media Center, 2223 Fulton St., Berkeley, CA 94720.
UEVA	Universal Education and Visual Arts, 100 Universal City Plaza, Universal City, CA 91608.
UF	University of Frauenklinik, Berlin, Germany.
UIO	University of Iowa, AV Center, C-5 East Hall, Iowa City, IA 52240.
UJ	Unijapan Films, 9-13 Ginza 5-Chone, Chuo-ku, Tokyo, 104 Japan.
UN	University of Nebraska, Instructional Media Center, University of Nebraska Extension Division, Lincoln, NE 68508.
UNL	United Learning, 6633 W. Howard St., Niles, IL 60648.
UP	Upjohn Co., Film Library, 7000 Portage Rd., Kalamazoo, MI 94001.
USC	University of Southern California, Department of Cinema, University Park, Los Angeles, CA 90007.
USNAC	U.S. National Audiovisual Center, General Services Administration, Washington, DC 20409.
UT	University of Texas Medical Branch, Galveston, TX 77550.
UWM	University of Washington, School of Medicine, Seattle, WA 98105.
UW	University of Washington Press, 1416 N.E. 41st St., Seattle, WA 98105.
WARDS	Ward's Natural History Establishment, P.O. Box 1712, Rochester, NY 14603.

WASH   Washburn, 28 Union, Worcester, MA. 01608.
WAYNE   Wayne State University, CIT Productions Center, 5035 Woodward,
           Detroit, MI 48202.
WFL   Wellcome Film Library, Wellcome Building, 183-193 Euston Road,
           London, N.W.l, England.
WGC   West Glen Communications, 565 5th Ave., New York, NY 10017.
WLFL   Wyeth Laboratories Film Library, P.O. Box 8299, Philadelphia, PA
           19101.

BIBLIOGRAPHY

1. **Anatomical Orientation**

   Basmajian, J. V. Surface Anatomy: An Instruction Manual. Baltimore: Williams & Wilkins Co., 1977.

   Basmajian, J. V. Grant's Method of Anatomy. 9th ed. Baltimore: Williams & Wilkins Co., 1975.

   Donnelly, J. E. "The Living Anatomy Technique." Journal of Physical Education and Recreation, April 1980.

   Genant, H. K., et al. "Computed Tomography of the Musculoskeletal System." Journal of Bone and Joint Surgery, vol. 62-a(7), 1980.

   Gray, H. Anatomy of the Human Body. Edited by C. M. Goss, 29th ed. Philadelphia: Lea & Febiger, 1973.

   Hamilton, W. J. Surface and Radiological Anatomy for Students and General Practitioners. 5th ed. Baltimore: Williams & Wilkins Co. 1971.

   Hollingshead, W. H. Textbook of Anatomy. 3rd ed. New York: Harper & Row, 1974.

   Katch, F. I. "Estimation of Body Fat From Skinfolds and Surface Area." Human Biology, vol. 51, Spring 1979.

   Leldey, R. S. "Computerized Transaxial X-ray Tomography of the Human Body." Science, 186:4160, October 18, 1974.

   McKears, D. W., and R. H. Owen. Surface Anatomy for Radiographers. Chicago: New Book Medical Pubs., 1979.

   "Modern Men of Parts." Time, vol.110, No. 11, March 18, 1974.

   Peterson, R. R. Cross-Sectional Approach to Anatomy. Chicago: Year Book Medical Pubs., 1980.

   Riegler, H. F., and A. P. Peppard. Surface Anatomy for Coaches and Athletic Trainers. Springfield, Ill.: C. C. Thomas, 1979.

   Royce, J. Surface Anatomy. Philadelphia: F. A. Davis, 1965.

2. **The "Chemical and Physical Basis of Life**

   Baker, J. and G. E. Allen. Matter, Energy, and Life. 4th ed. Reading, Mass.: Addison-Wesley Publishing Co., 1981.

   Doty, P. "Proteins." Scientific American, September, 1957.

   Eyre, D. R. "Collagen: Molecular Diversity in the Body's Protein Scaffold." Science, 207:1315-1322.

   Kamen, M. "A Universal Molecule of Living Matter." Scientific American, August, 1958.

   McIver, R. T. "Chemical Reactions Without Solvation." Scientific American, November, 1980.

   Masterton, W. L., et al. Chemical Principles. 5th ed. Philadelphia: Saunders College Publishing, 1981.

   McGilvery, R. W. Biochemistry: A Functional Approach. 2nd ed. Philadelphia: W. B. Saunders Co., 1979.

   Mortimer, C. E. Chemistry: A Conceptual Approach. 4th ed. New York: D. Van Nostrad Co., 1979.

   Phillips, D. C. "The Three-Dimensional Structure of an Enzyme Molecule." Scientific American, November, 1966.

Sackheim, G. I., and R. M. Schultz. Chemistry for the Health Sciences. 3rd ed. New York: Macmillan Publishing Co., Inc., 1977.

Sharon, N. "Carbohydrates." Scientific American, November 1980.

3.   The Cell

Andreoli, T. E., et al. (eds.). Membrane Physiology. New York: Plenum Press, 1980.

Baker, J. J. W., and G. E. Allen. Matter, Energy, and Life. 3rd ed. Reading. Mass., Addison-Wesley Publishing Co., 1975.

Baumann, G. "Cell Membrane Ready for Action: Steps Toward Building a Living Cell." Chemistry, vol. 51, December 1978.

Baumann, G. "Opening the Membrane Channel: Steps Toward Building a Living Cell." Chemistry, vol. 52, January 1979.

Baxter, J. D. "Recombinant DNA and Medical Progress." Hospital Practice, February 1980.

Chambon, P. "Split Genes." Scientific American, May 1981.

Cohen, S. N. "The Manipulation of Genes." Scientific American, July 1975.

Cohen, S. N., and J. Shapiro. "Transposable Genetic Elements." Scientific American, February 1980.

Cudmore, L. L. L. The Center of Life: A Natural History of the Cell. New York: Quadrangle/New York Times Book Co., 1977.

Davidson, E. H. (ed.). Gene Activity in Early Development. 2nd ed. New York: Academic Press, 1976.

Dawkins, R. The Selfish Gene. London: Oxford University Press, 1976.

DeRobertis, E. D. P., et al. Cell Biology. 7th ed. Philadelphia: Saunders College Publishing, 1980.

Dustin, P. "Microtubules." Scientific American, August 1980.

Gilbert, W. and L. Villa-Komaroff. "Useful Proteins from Recombinant Bacteria." Scientific American, April 1980.

Hayflick, L. "Cell Biology of Human Aging." Scientific American, January 1980.

Hinkle, P. C. and R. E. McCarty. "How Cells Make ATP." Scientific American, (1) :58, 1980.

Klug, et al. "The Nucleosome." Scientific American, February 1981.

Kotyk, A. and K. Janacek. Cell Membrane Transport. 2nd ed. New York: Plenum Publishing Corp., 1975.

Lake, J. A. "The Ribosome." Scientific American, August 1981.

Lawrence, W., and J. J. Terz (eds.). Cancer Management. New York: Grune & Stratton, 1977.

Lodish, H. F., and J. E. Rothman. "The Assembly of Cell Membranes." Scientific American, 240(1):48, 1979.

McElroy, W. D., and C. P. Swanson. Modern Cell Biology. 2nd ed. Englewood Cliffs, N. J: Prentice-Hall, 1976.

McKusick, V. A., and F. H. Ruddle. "The Status of the Gene Map of the Human Chromosomes. Science, April 22, 1977.

Metcalfe, J. C. (ed.). Biochemistry of Cell Walls and Membranes II. Baltimore: University Park Press, 1978.

Metzler, D. E. Biochemistry: The Chemical Reactions of Living Cells. New York: Academic Press, 1977.

Novick R. P., "Plasmids." Scientific American, December 1980.

Porter, K. "The Ground Substance of the Living Cell." Scientific American, March 1981.

Rost, R. L., and E. M. Gifford (eds.). Mechanisms and Control of Cell Division. Stroudsburg, Pa.: Dowden, Hutchinson & Ross, 1977.

Rubinstein, E. "Diseases Caused by Impaired Communication Among Cells." Scientific American, March 1980.

Russell, T. R., et al. (eds.). From Gene to Protein: Information Transfer in Normal and Abnormal Cells. New York: Academic Press, 1979.

Schimke, R. T. "Gene Amplification and Drug Resistance." Scientific American. November 1980.

Schuster, P., et al. "The Origin of Genetic Information." Scientific American, April 1971.

Strechler, B. L. Time, Cells, and Aging. 2nd ed. New York: Academic Press, 1977.

Swanson, D. P., and P. I. Webster. The Cell. 4th ed. Englewood Cliffs N. J.: Prentice-Hall. 1977.

Watson, J. D. The Double Helix. New York: Atheneum Publishers, 1968.

Watson, J. D. Molecular Biology of the Gene. 3rd ed. New York: The Benjamin Co. 1976.

Watson, J.D. Molecular Biology of the Gene. 4th ed. Menlo Park, CA. The Benjamin/Cummings Publishing Co. Inc., 1987.

Weissman, G., and R. Claiborne (eds.). Cell Membranes: Biochemistry Cell Biology, and Pathology. New York: Hospital Practice Publishing, 1976.

"What Tells a Cell How Old it is?" Family Health, vol. 12, March 1980.

Yunis, J. J. (ed.). Molecular Structure of Human Chromosomes. New York: Academic Press, 1977.

4. **Tissues**

Bloom, W., and D. Fawcett. A Textbook of Histology. 10th ed. Philadelphia: W. B. Saunders Co., 1975.

Bulfer, J. M., and C. E. Allen. "Fat Cells and Obesity." Bioscience, vol. 29, December 1979.

Chamlet-Campbell, J., et al. "Smooth Muscle Cell in Culture." Physiological Reviews, vol. 59, Body. January 1979.

Clark, W. E. The Tissues of the Body. 6th ed. Philadelphia: J. B. Lippincott Co.. 1979.

Hughes, Graham, R. V. Connective Tissue Diseases. 2nd ed. St. Louis: The C. V. Mosby Co., 1981.

Motta, Pietro, et al. Microanatomy of Cell and Tissue Surfaces: An Atlas of Scanning Electron Microscopy. Philadelphia: Lea & Febiger, 1977.

Staehelin, L. A., and B. E. Hull. "Junctions between Living Cells." Scientific American, 238(5):140, 1978.

Tseng, H. Atlas of Ultrastructure. New York: Appleton-Century-Crofts, 1980.

5. **The Integumentary System**

"Black Skin Problems." American Journal of Nursing, June 1979.

Breathneck, A. S. Atlas of the Ultrastructure of the Human Skin. London: Churchill Livingstone, 1971.

Brenton, M. "Dandruff." Family Health, December 1978.

Daniels, F., Jr. "Saving Sun Worshippers From Their God." Medical Opinion, July 1976.

Enna, C. D., et al. "The Histomorphology of the Elastic Tissue System in the Skin of the Human Hand." Hand, July 1979.

Flandermeyer, K. L. "That Ol' Devil Acne." Family Health, February 1980.

Kaminester, L. H. "Sunlight, Skin Cancer, and Sunscreens." Journal of the American Medical Association, June 30, 1975.

Kaminester, L. H. "Warts: Another Look at a Nuisance." Consultant, September 1976.

Langer, K. "On the Anatomy and Physiology of the Skin: Conclusions by Professor K. Langer." British Journal of Plastic Surgery, October 1978.

Levine, N. "Sun Worship: Reducing the Ritual's Danger and Damage." Modern Medicine, June 1980.

Mier, P. D., and D. W. K. Cotton. The Molecular Biology of the Skin. Oxford: Blackwell Scientific Publications, 1976.

Mihm, M. C., Jr. "The Structure of Normal Skin and the Morphology of Atopic Eczema." Journal of Investigative Dermatology, September 1976.

Montagna, W. and P. F. Paraktial. The Structure and Function of the Skin. 3rd ed. New York: Academic Press, 1974.

"New Dressing Aids in Burn Healing." Journal of the American Medical Association, July 26, 1976.

Peterson. H. D. "Burn Sepsis-Management of the Severe Burn." Forum on Infection, September 1976.

Pillsbury, D. M. and C. L. Heaton. A Manual of Dermatology. 2nd ed. Philadelphia: W. B. Saunders Co., 1980.

Rao, D. B. "Management of Dermal and Decubitus Ulcers." Drug Therapy, October 1976.

Sanders, B. B., Jr., and G. S. Stretcher. "Warts--Diagnosis and Treatment." Journal of the American Medical Association, June 28, 1976.

"Skin Cancer: 3,000,000 new cases in 1976." Journal of the American Medical Association, April 26, 1976.

Spearman, R. I. The Integument: A Textbook of Skin Biology. Cambridge University Press, 1973.

Uhler, D. M. "Common Skin Changes in the Elderly." American Journal of Nursing, August 1978.

Wurtman, R. J. "The Effects of Light on the Human Body." Scientific American, July 1975.

6. **The Skeletal System**

Beresford, A. "The Aging Intervertebral Disc." Physical Therapy, August 1979.

Bourne, G. W. The Biochemistry and Physiology of Bone. 2nd ed. New York: Academic Press, 1972.

Burkhardt, R. Bone Marrow and Bone Tissue: Color Atlas of Clinical Histopathology. New York: Springer-Verlag, 1971.

Cumming. G. R. "Skeletal Age." Modern Athlete and Coach. January 1979.

Dahlin, D. C. Bone Tumors. 3rd ed. Springfield. Ill.: C. C. Thomas, 1978.

DeLuca, H. F. "The Vitamin D Hormonal System: Implications for Bone Diseases." Hospital Practice, April 1980.

Elliott, J. "Electrical Stimulation of Bone Growth Wins Clinical Acceptance." Chapter 7 - Articulations. Journal of American Association, 243, April 1980.

Garn, S. "Bone Loss and Aging." Physiology and Pathology of Human Aging. New York: Academic Press, 1975.

Harris, W. H., and R. P. Heaney. Skeletal Renewal and Metabolic Bone Disease. Boston: Little, Brown & Co., 1970.

Klein, Karl K. "Developmental Asymmetries of the Weight-Bearing Skeleton and its Implications in Knee Stress and Knee Injury." Athletic Training, Summer 1978.

Langone, J. "Replacing Damaged Bones." Discover, April 1981.

Mack, R. W. "Height, Skeletal Maturation and Adiposity in Adolescents with High Relative Weight at One Year of Age." Annals of Human Biology, January - February 1979.

Marx, J. L. "Osteoporosis: New Help for Thinning Bones." Science 207:628-630, 1980.

Teitz, Carol C. "Problems Associated with Tibial Fractures With Intact Fibulae." Journal of Bone and Joint Surgery, 1980.

Tronzo, R. G. "Bone: Self-Repairing, Self-Renewing." Consultant, April 1972.

7.   **Articulations**

Bingham, R. "Rheumatoid Disease: Has One Investigator Found its Cause and its Cure?" Modern Medicine, February 15, 1976.

Blau, S. P. "How to Select Therapy in Rheumatoid Arthritis." Consultant, September 1976.

Bones, Joints and Muscles of the Human Body: A Programmed Text for Physical Therapy Aides. New York: Macmillan, 1970.

Distefano, V. J., "Functional Anatomy and Biomechanics of the Wrist and Elbow Joints." Athletic Training, Fall 1979.

Ehrlich, G. E. "Learning How to Fight Rheumatoid Arthritis." Medical Tribune, May 12, 1976.

Fowler, P. "Ligamentous Anatomy and Physical Examination." American Journal of Sports Medicine, November-December, 1977.

Hollander, J. L., and D. J. McCarty, Jr. Arthritis and Allied Conditions. 8th ed. Philadelphia: Lea & Febiger, 1972.

Kelley, W., et al. Textbook of Rheumatology. Philadelphia: W. B. Saunders Co. 1980.

Kelsey, J., et al. Musculoskeletal Disorders: Their Frequency of Occurrence and Their Impact on the Population of the United States. New York: Neale Watson Academic Publications, 1978.

O'Donoghue, D. H. Treatment of Injuries to Athletes. Philadelphia: W. B. Saunders Co., 1976.

Pope, M. H., et al. "The Role of the Musculature in Injuries to the Medical Collateral Ligament." Journal of Bone and Joint Surgery, 1979.

Protzman, R. R. "Anterior Instability of the Shoulder." Journal of Bone and Joint Surgery, 1980.

Rasch, P., and R. Burke. Kinesiology and Applied Anatomy: The Science of Human Movement. 6th ed. Philadelphia: Lea & Febiger, 1978.

Rocke, A. F., and H. Wainer. Skeletal Maturity: The Knee Joint as a Biological Indicator. New York: Plenum Publishing Corp., 1975.

Root, L., and T. Siegal. "Osteotomy of the Hip in Children: Posterior Approach." Journal of Bone and Joint Surgery, 1980.

Rudd, E. M. "Living with Arthritis." Drug Therapy, March 1979.

Rushfeld, P. D. "Influence of Cartilage Geometry on the Pressure Distribution in the Human Hip Joint. Science, vol. 204, April 27, 1979.

Sampson, P. "Replacement of Some Arthritic Joints is Possible - But Still a Major Job." Journal of the American Medical Association, March 29, 1976.

Sokoloff, L. (ed.) Joints and Synovial Fluid. Vol. 1. New York: Academic Press, 1978.

Sonstegard, D. A., and L. S. Matthews. "The Surgical Replacement of the Human Knee Joint." Scientific American, January 1978.

Thompson, G. R. "Diagnosis and Treatment of Septic Arthritis." Hospital Medicine, June 1976.

Zihlman, A., and D. Carmer. "Human Locomotion." Natural History, Vol. 85, January 1976.

8, 9 The Muscular System

Basmajian, J. V., and M. A. MacConaill. Muscles and Movements: A Basis for Human Kinesiology. Baltimore: Williams & Wilkins Co., 1977.

Bourne, G. H. (ed.). The Structure and Function of Muscle. 2nd ed. New York: Academic Press, 1973.

Casey, J. "In Praise of Women's Muscles." Esquire, May 1980.

Cohen, C. "The Protein Switch of Muscle Contraction." Scientific American, November 1975.

Cole, R. P. "Myoglobin Function in Exercising Skeletal Muscle." Science, 30, April 1982.

Daniel, E. E., and S. Sarna. "The Generation and Conduction of Activity in Smooth Muscle." Annual Rev. Pharmacol. Toxicol., 18:145, 1978.

Hartschorne, D. J., and A. Gorecka. Biochemistry of the Contractile Proteins of Smooth Muscle. In: Bohr, D. F., et al. (eds.), Handbook of Physiology. Sec. 2, Vol. II, Baltimore: Williams & Wilkins Co., 1980, p. 93.

Homsher, E., and C. J. Kean. "Skeletal Muscle Energetics and Metabolism." Annu. Rev. Physiol., 40:93, 1978.

Huddart, H., and S. Hunt. Visceral Muscle. New York: Halstead Press, 1975.

Huxley, H. E. "The Mechanism of Muscular Contraction." Science, 164:1356-1366, 1969.

Margaria, R. "The Sources of Muscular Energy." Scientific American, March 1972.

Mauro, A., and R. Bischoff (eds.). Muscle Regeneration. New York: Raven Press, 1979.

Murphy, R. A. "Filament Organization and Contractile Function in Vertebrate Smooth Muscle." Annu. Rev. Physiol., 41:737. 1979.

Murray, J. M., and A. Weber. "The Cooperative Action of Muscle Proteins." Scientific American, February 1974.

Peachey, L. D. Muscle and Motility. New York: McGraw-Hill Book Co., 1974.

Rasch, P. J., and R. K. Burke. Kinesiology and Applied Anatomy: The Science of Human Movement. Philadelphia: Lea & Febiger, 1978.

Sugi, H., and G. H. Pollack (eds.). Cross-Bridge Mechanism in Muscle Contraction. Baltimore: University Park Press, 1979.

Tonomuna, Y. Muscle Protein, Muscle Contraction and Cation Transport. Baltimore: University Park Press, 1972.

Uehara, Y. Muscle and Its Innervation. Baltimore: Williams & Wilkins Co., 1976.

10. The Nervous System: Its Organization, Development, and Components

Barr, M. L. The Human Nervous System. 3rd ed. New York: Harper and Row, 1979.

Bennett, T. L. The Sensory World: An Introduction to Sensation and Perception. Monterey, Calif.: Brooks/Cole Publishing Co., 1978.

Carpenter, M. B. Human Neuroanatomy. 7th ed. Baltimore: Williams & Wilkins Co. 1976.

Carpenter, M. B. Core Text of Neuroanatomy. 2nd ed. Baltimore: Williams & Wilkins Co. 1978.

Coren, S., et al. Sensation and Perception. New York: Academic Press, 1979.

Cotman, C. W. (ed). Neuronal Plasticity. New York: Raven Press, 1978.

Dunkerley, G. B. A Basic Atlas of the Human Nervous System. Philadelphia: Davis Co. 1976.

Gardner, E. D. Fundamentals of Neurology. 6th ed. Philadelphia: W. B. Saunders Co., 1975.

Goldstein, E. B. Sensation and Perception. Belmont, Calif.: Wadsworth Publishing Co., 1980.

Haymaker, W. and R. D. Adams. Histology and Histopathology of the Nervous System. Springfield, Ill.: C. C. Thomas, 1980.

Kenshalo, D. R. (ed.). Sensory Function of the Skin of Humans. New York: Plenum Press 1979.

Morell, P. and Norton, W. T. "Myelin." Scientific American, May 1980.

Noback, C. and R. DeMarest. The Human Nervous System: Basic Principles of Neurobiology. 2nd ed. New York: McGraw-Hill Book Co. 1975.

Peele, T. L. The Neuroanatomic Basis for Clinical Neurology. 3rd ed. New York: McGraw-Hill Book Co., 1976.

11. Neurons, Synapses, and Receptors

Agayo, A. J., and G. Karpati (eds.). Current Topics in Nerve and Muscle Research. New York: Elsevier/North-Holland, 1979.

Axelrod, J. "Neurotransmitters." Scientific American, June 1974.

Bolton, T. B. "Mechanisms of Action of Transmitters and Other Substances on Smooth Muscle." Physiol. Rev., 59:606, 1979.

Ceccarelli, B., and W. P. Hurlbut. "Vesicle Hypothesis of the Release of Quanta of Acetylcholine." Physiol. Rev., 60:396, 1980.

Daftary, A. V., and R. I. Wang. "Dopamine - the Neglected Neurotransmitter." Drug Therapy, June 1976.

DeWied, D. "Hormonal Influences on Motivation, Learning, and Memory Processes." Hospital Practice, January 1976.

Fambrough, D. M. "Control of Acetylcholine Receptors in Skeletal Muscle." Physiol. Rev., 59:165, 1979.

Gage, P. W. "Generation of End-Plate Potentials." Physiol. Rev., 56:177, 1976.

Guroff, G. Molecular Neurobiology. New York: Marcel Dekker, 1979.

Katz, B. Nerve, Muscle, and Synapse. New York: McGraw-Hill Book Co., 1966.

Keynes, R. D. "Ion Channels in the Nerve-Cell Membrane." Scientific American, 240(3):126, 1979.

Langer, S. Z., et al. (eds.). Presynaptic Receptors. New York: Pergamon Press, 1979.

Lasek, R. "Axonal Transport and the Use of Intracellular Markers in Neuroanatomical Investigations." Federation Proc., 34:1603-1611, 1975.

McEwen, B. S. "Interactions Between Hormones and Nerve Tissues." Scientific American, July 1976.

Otsuka, M., and Hall, Z. W. (eds.) Neurobiology of Chemical Transmission. New York: John Wiley & Sons, 1978.

"Pain." SciQuest, May 1979.

Schwartz, J. H. "The Transport of Substances in Nerve Cells." Scientific American, April 1980.

Stephenson, W. K. Concepts of Neurophysiology. New York: John Wiley & Sons, 1980.

Stevens, C. F. "The Neuron." Scientific American, 241(3):54, 1979.

Swchwartz, J. H. "Axonal Transport: Components, Mechanisms, and Specificity." Annu. Rev. Neurosci. 2:476, 1979.

Wallick, E. T., et al. "Biochemical Mechanisms of the Sodium Pump." Annu. Rev. Physiol., 41:397, 1979.

12. The Central Nervous System

Asanuma, H. "Recent Developments in the Study of The Columnar Arrangements of Neurons Within the Motor Cortex." Physiological Review, 1975.

Bekhtereva, N. P. The Neurophysiological Aspects of Human Mental Activity. New York: Oxford University Press, 1978.

Cook, A. W. "Electrical Stimulation in Multiple Sclerosis." Hospital Practice, April 1976.

Cooper, I. S. Cerebellar Stimulation in Man. New York: Raven Press, 1978.

Dalessio, D. J. "Headache: Clinical Guide to Identifying the Cause." Hospital Medicine, September 1979.

Diamond, S. J. Neuropsychology: A Textbook of Systems and Psychological Functions of the Human Brain. Boston: Butterworths, 1979.

Dole, V. P. "Addictive Behavior." Scientific American, December 1980.

Eccles, J. C. The Understanding of the Brain. New York: McGraw-Hill Book Co., 1973.

Ehrlich, Y. H. (ed.). Modulators Mediators, and Specificers in Brain Function. New York: Plenum Press, 1979.

Evarts, E. V. "Brain Mechanisms of Movement. Scientific American, 241(3):164, 1979.

Galaburda, A. M., et al. "Right-Left Asymmetries in the Brain." Science, February 24, 1978.

Gotto, A. M., et al. (eds.). Brain Peptides: A New Endocrinology. New York: Elsevier/North-Holland, 1979.

Hobson, J. A., and M. A. B. Brazier (eds.). The Reticular Formation Revisited. New York: Raven Press, 1980.

Holden, C. "Pain, Dying, and the Health Care System." Science, 203:934, 1979.

"Know Your Own Brain." Economist, December 22, 1979.

Llinas, R. R. "The Cortex of the Cerebellum. Scientific American, January 1975.

Loken, M. K. and M. Frick. "Scanning the Brain: Radionuclide Scintigraphy." Modern Medicine, May 1, 1976.

Massion, J., and K. Sasaki (eds.). Cerebro-Cerebellar Interactions. New York: Elsevier/North-Holland, 1979.

Mechner, F. "Patient Assessment: Neurological Examination." Part 3, American Journal of Nursing, April 1976.

"Multiple Sclerosis: Explorations of Cause." Journal of the American Medical Association, June 14, 1976.

Nathanson, J. A, and P. Greengard. "Second Messenger in the Brain." Scientific American, August 1977.

Pearson, K. "The Control of Walking." Scientific American, December 1976.

Pepeu, G., Hormonal (eds.). Receptors for Neurotransmitters and Peptide Hormones. New York: Raven Press, 1980.

Pinsker, H. M., and W. D. Willis (eds.). Information Processing in the Nervous System. New York: Raven Press, 1980.

Ruch, T. C., and H. D. Patton. Physiology and Biophysics, Vol. 1: The Brain and Neural Function. Philadelphia: W. B. Saunders Co., 1979.

Stellar, S., and Y. Bhandari. "Results of Present-Day Therapy of Parkinson's Disease." Journal of the Medical Society of New Jersey, April 1976.

Ter-Pogossian, M. M., et al. "Positron-Emission Tomography." Scientific American, October 1980.

"The Brain." Scientific American, September 1979. (Entire issue is devoted to the brain and human nervous system.)

Uttley, A. M. Information Transmission in the Nervous System. New York: Academic Press, 1979.

Wallace, P. "Neurochemistry: Unraveling the Mechanism of Memory." Science, vol. 190, pgs. 1076-1078, 1975.

13. The Peripheral Nervous System

Asbury and Johnson. Pathology of Peripheral Nerve. Philadelphia: W. B. Saunders Co., 1978.

Brushart, T. M., and M. M. Mesulam. "Alterations in Connections Between Muscle and Anterior Horn Motor Neurons After Peripheral Nerve Repair." Science, May 9, 1980.

DeAngelis, A. M. Surgical Anatomy of Peripheral Nerves. Mount Kisco, N. Y.: Future Pub., 1973.

Hargens, A. R., et al. "Peripheral Nerve-Conduction Block by High Muscle-Compartment Pressure." Journal of Bone and Joint Surgery, 1979.

Hubbard, J. I. Peripheral Nervous System. New York: Plenum Press, 1974.

Hunter, L. Y., et al. "The Saphenous Nerve: Its Course and Importance in Medial Arthotomy." American Journal of Sports Medicine, July-August 1979.

Landon, D. H. Peripheral Nerve. New York: Chapman & Hall, Methuen, 1976.

Levi-Montalcini, R. and P. Calissano. "The Nerve-Growth Factor," Scientific American, June 1979.

Omer, G., and M. Spinner. Management of Peripheral Nerve Problems. Philadelphia: W. B. Saunders Co., 1980.

Schade, J. P. Peripheral Nervous System. New York: Elsevier/North Holland, 1973.

Spinner, M. Injuries to the Major Branches of the Peripheral Nerves of the Forearm. Philadelphia: W. B. Saunders, 1978.

Sunderland, S. Nerves and Nerve Injuries. London: Churchill Livingstone, 1972.

Vital, C. and J. M. Vallat. Ultrastructural Study of the Human Diseased Peripheral Nerve. New York: Masson Pub., 1980.

Weller, R. O., and J. C. Navarro. Pathology of Peripheral Nerves: A Practical Approach. Woburn, Mass.: Butterworths, 1977.

14. The Autonomic Nervous System

Appenzeller, O. The Autonomic Nervous System. 2nd ed. New York: North-Holland/Elsevier, 1976.

Barabasz, A. F. "Biofeedback, Mediated Biofeedback and Hypnosis in Peripheral Vasodilation Training." American Journal of Clinical Hypnosis, July 1978.

Basmajian, J. V. "Biofeedback: The Clinical Tool Behind the Catchword." Modern Medicine, October 1, 1976.

Bhagat, B. D. Mode of Action of Autonomic Drugs. Flushing, N. Y.: Graceway Publishing Co., 1979.

"Biofeedback Training for Childbirth." Medical World News, June 30, 1975.

DeQuattro, V., et al. Anatomy and Biochemistry of the Sympathetic Nervous System. In DeGroot, L. J., et al. (eds.) Endocrinology, Vol. 2, New York: Grune & Stratton., 1979, page 1241.

Gaardner, K. R., and P. S. Montgomery. Clinical Biofeedback: A Procedural Manual. Baltimore: Williams & Wilkins Co., 1977.

Gabella, G. Structure of the Autonomic Nervous System. New York: Chapman & Hall, Methuen, 1976.

Garrity, M. K. "Biofeedback Demonstration." Physics Teacher, September 1978.

Jacobson, A. M. "Behavioral Treatment for Raynaud's Disease: A Comparative

Study with Long-Term Follow-Up." American Journal of Psychiatry, June 1979.

Patton, D. M. (ed.). The Release of Catecholamines from Adrenergic Neurons. New York: Pergamon Press, 1979.

Schneider, R. D. "Re-examination of the Relationship Between Locus of Control and Voluntary Heart Rate Change." Journal of General Psychology, July 1978.

"Tic Douloureaux: An End to the Agony." Family Health, February 1979.

Tueck, S. (ed.). The Cholinergic Synapse. New York: Elsevier/North-Holland, 1979.

Van Toller, C. The Nervous Body: An Introduction to the Autonomic Nervous System and Behavior. New York: John Wiley & Sons, 1979.

Zimkina, A. M., et al. "Interaction Between Systems as the Basic Principle of Organization of Activity of the Nervous System." Human Physiology, May-June 1979.

15. The Organs of the Special Senses

Allen, E. W. Essentials of Ophthalmic Optics. New York: Oxford University Press, 1979.

Armington, J., et al. Visual Psychophysics. New York: Academic Press, 1978.

Bettner, J. "Protecting Your Eyes from Sports Accidents." Business Week, October 15, 1979.

Carterette, E. C., and M. P. Friedman. Hearing. New York: Academic Press, 1978.

Durrant, J. D., and J. H. Lovrinic. Bases of Hearing Science. Baltimore: Williams & Wilkins Co., 1977.

Engen, T. "Why the Aroma Lingers On." Psychology Today, May 1980.

Evans, E. F., and J. P. Wilson (eds.). Psychophysics and Physiology of Hearing. New York: Academic Press, 1977.

Favreau, O. E., and M. C. Corballis. "Negative After-Effects in Visual Perception." Scientific American, December 1976.

Freese, A. S. The Miracle of Vision. New York: Harper & Row, 1977.

Gerwin, K. S., and A. Glorig. A Detection of Hearing Loss and Ear Disease in Children. Springfield, Ill.: C. C. Thomas, 1974.

Glickstein, M., and A. R. Gibson. "Visual Cells in the Pons of the Brain." Scientific American, November 1976.

Green, D. M. An Introduction to Hearing. New York: Halsted Press, 1976.

Hubel, D. H., and T. N. Wiesel. "Brain Mechanisms of Vision." Scientific American, vol. 241(3) pg. 150, 1979.

Jan, J. E., R. D. Freeman, and E. P. Scott. Visual Impairment in Children and Adolescents. New York: Grune & Stratton, 1977.

Johansson, G. "Visual Motion Perception." Scientific American, June 1975.

Kaneko, A. "Physiology of the Retina." Annu. Rev. Neurosci., vol. 2, pg. 169, 1979.

Kupchella, C. E. Sights and Sounds: The Very Special Senses. Indianapolis: Bobbs-Merrill Co., 1975.

Land, E. H. "The Retinex Theory of Color Vision." Scientific American, vol. 237(6), pg. 108, 1977.

Ming, A. L. S. Color Atlas of Ophthalmology. Pacific Palisades, Calif.: Houghton Mifflin Professional Pubs., Medical Division, 1979.

Moulton, D. G. "Spatial Patterning of Responses to Odors in the Peripheral Olfactory System." Physiological Reviews 1976.

Ohloff, G., and A.F. Thomas,(eds.). Gustation and Olfaction. New York: Academic Press, 1971.

Oldendorf, W. "See Inside Your Eye." Science Digest, January-February 1981.

Parker, D. E. "The Vestibular Apparatus." Scientific American, November 1980.

Records, R. E. Physiology of the Human Eye and Visual System. Hagerstown, Md.: Harper & Row, 1979.

Regan, D., et al. "The Visual Perception of Motion in Depth." Scientific American, July 1979.

Ross, J. "The Resources of Binocular Perception." Scientific American, March 1976.

Ruston, W. A. H. "Visual Pigments and Color Blindness." Scientific American, March 1975.

Sekuler, R., and E. Levinson. "The Perception of Moving Targets." Scientific American, vol. 236(1), pg. 60, 1977.

Sewell, W. F., et al. "Detection of an Auditory Nerve-Activating Substance. Science, vol. 202, pgs. 910-912, 1978.

Stiles, W. S. Mechanisms of Color Vision. New York: Academic Press, 1978.

"Structure of the Cochlea." Physics Today, August 1980.

Van Heyningen, R. "What Happens to the Human Lens in Cataract?" Scientific American, December 1975.

Yannuzzi, L. A., et al. (eds.). The Macula: A Comprehensive Text and Atlas. Baltimore: Williams & Wilkins Co., 1978.

16. The Endocrine System

Austin, C. R., and R. V. Short (eds.). Mechanisms of Hormone Action. New York: Cambridge University Press, 1979.

Bargmann, W., et al. (eds.). Neurosecretion and Neuroendocrine Activity: Evolution, Structure, and Function. New York: Springer-Verlag, 1978.

Baxter, J. D., and G. G. Rousseau. Glucocorticoid Hormone Action. New York: Springer-Verlag, 1979.

Brownstein, M. J., et al. "Synthesis, Transport, and Release of Posterior Pituitary Hormones." Science 207:373-378, 1980.

Cryer, P. E. Diagnostic Endocrinology. London: Oxford University Press, 1976.

DeGroot, L. J. (ed.). Endocrinology. New York: Grune & Stratton, 1979.

Esmann, V. (ed.). Regulatory Mechanisms of Carbohydrate Metabolism. New York: Pargammon Press, 1978.

Field, J. (ed.) "Endocrinology." In Handbook of Physiology. Baltimore: Williams & Wilkins Co., 1972.

Fitzgerald, P. J. and A. B. Morrison (eds.). The Pancreas. Baltimore: Williams & Wilkins Co., 1980.

Frieden, E. H. Chemical Endocrinology. New York: Academic Press, 1976.

Frohman, L. A. "Neurotransmitters as Regulators of Endocrine Function." Hospital Practice, April 1975.

Grave, G. D. (ed.). Thyroid Hormones and Brain Development. New York: Raven Press, 1977.

Hershman, J. M. (ed.). Endocrine Pathophysiology: A Patient Oriented Approach. Philadelphia: Lea & Febiger, 1977.

James, V. H. (ed.). The Adrenal Gland. New York: Raven Press, 1979.

Li, C. H. (ed.). Thyroid Hormones. New York: Academic Press, 1978.

Manx, J. "New Theory of Hormones Proposed." Science 215:1383, 1981.

McEwen, B. S. "The Brain as a Target Organ of Endocrine Hormones." Hospital Practice, May 1975.

Notkins, A. L. "The Causes of Diabetes." Scientific American, November 1979.

O'Malley, B. W., and W. T. Schrader. "The Receptors of Steroid Hormones." Scientific American, February 1976.

Richenberg, H. V. (ed.). Biochemistry and Mode of Action of Hormones II. Baltimore: University Park Press, 1978.

Tepperman, J. Metabolic and Endocrine Physiology: An Introductory Text. Chicago: Year Book Medical Publishers, 1980.

Turner, C. D. General Endocrinology. 6th ed. Philadelphia: W. B. Saunders Co., 1976.

Villee, D. Human Endocrinology. Philadelphia: W. B. Saunders Co., 1975.

Wurtman, R. J. "The Pineal as a Neuroendocrine Transducer." Hospital Practice, January 1980.

17. The Circulatory System: The Blood

Beutler, E. Hemolytic Anemia in Disorders of Red Cell Metabolism. New York: Plenum Press, 1978.

Cooper, H. A., et al. "The Platelet: Membrane and Surface Reactions." Annu. Rev. Physiol., 38:501, 1976.

Doolittle, R. F. "Fibrinogen and Fibrin." Scientific American, December 1981.

Ellory, C., and V. L. Lew (eds.). Membrane Transport in Red Cells. New York: Academic Press 1977.

Escobar, M. R., and H. Friedman,(eds.). Macrophages and Lymphocytes: Nature, Functions and Interaction. New York: Plenum Press, 1979.

Friedman, H., et al. (eds.). The Reticuloendothelial System. New York: Plenum Press, 1979.

Gowans, J. L. Blood Cells and Vessel Walls: Functional Interactions. Princeton, N. J.: Excerpta Medica, 1980.

Hirsch, J., et al. Concepts in Hemostasis and Thrombosis. New York: Churchill Livingstone, 1979.

Katz, A. J. "Transfusion Therapy: Its Role in the Anemias." Hospital Practice, June 1980.

Kelemen, E., et al. Atlas of Human Hemopoietic Development. New York: Springer-Verlag, 1979.

Maugh, T. H. "A New Understanding of Sickle Cell Emerges." Science, 211:265, 1981.

Murano, G. and R. L. Bick (eds.). Basic Concepts of Hemostasis and Thrombosis. Boca Raton, Fla.: CRC Press, 1980.

Platt, W. R. Color Atlas and Textbook of Hematology. Philadelphia: J. B. Lippincott Co. 1973.

Schmidt-Nielsen, K. "Countercurrent Systems in Animals." Scientific American, May 1981.

Simone, J. V. "Childhood Leukemia: The Changing Prognosis." Hospital Practice, July 1974.

Spivak, J. L. (ed.). Fundamentals of Clinical Hematology. Hagerstown, Md.: Harper & Row, 1980.

Wallerstein, R. O. "Role of the Laboratory in the Diagnosis of Anemia." Journal of the American Medical Association, August 2, 1976.

Zucker, M. "The Functioning of Blood Platelets." Scientific American, June 1981.

Zucker-Franklin, D., et al. Atlas of Blood Cells--Function and Pathology. Philadelphia: Lea & Febiger, 1980.

18. The Circulatory System: The Heart

Angelor, G., et al. "Physical Development and Body Structure of Children with Congenital Heart Disease." Human Biology, 1980.

"Arrhythmias." Hospital Practice, May 1980.

Baan, J., et al. (eds.). Cardiac Dynamics. Boston: M. Nijhoff, 1979.

Benditt, E. P. "The Origin of Atherosclerosis." Scientific American, February 1977.

Berne, R. M., and M. N. Levy. Cardiovascular Physiology. 3rd ed. St. Louis: C. V. Mosby Co., 1977.

Berne, R. M., et al. (eds.). Handbook of Physiology. Sec. 2, Vol. 1. Baltimore: Williams & Wilkins Co., 1979.

Bierenbaum, M. L. "A Review of Some Epidemiological Factors in Coronary Heart Disease." Journal of the Medical Society of New Jersey, December 1975.

Burch, G. E., and T. Winsor. A Primer of Electrocardiography, 6th ed. Philadelphia: Lea & Febiger, 1972.

De Bakey, M., and A. Gotto. The Living Heart. New York: David McKay Co., 1977.

De Motu, C. (C. D. Leake, translator). Anatomical Studies on the Motion of the Heart and Blood. Springfield, Ill.: C. C. Thomas, 1978.

Effler, D. B. "Myocardial Revascularization." Journal of the American Medical Association, February 1976.

Felner, J. M., and R. C. Schlant. Echocardiography: A Teaching Atlas. New York: Grune & Stratton, 1976.

Grace, W. J. "Guide to the Management of the Complications of Acute Myocardial Infarction." Hospital Medicine, November 1975.

Grady, D. "The Plug-In Artificial Heart." Discover, April 1981.

Griffith, L. S. C. "The Proper Use of Coronary Arteriography." Modern Medicine, February 15, 1975.

Harris, C. C. A Primer of Cardiac Arrhythmia: A Self-Instructional Program. St. Louis: C. V. Mosby, 1979.

Hurst, J. W. (ed.). The Heart: Arteries and Veins. 3rd ed. New York: McGraw-Hill Book Co., 1974.

Jarvik, R. K. "The Total Artificial Heart." Scientific American, January 1981.

Jones, P. Cardiac Pacing. New York: Appleton-Century-Crofts, 1980.

Katz, A. M. Physiology of the Heart. New York: Raven Press, 1977.

Kezdi, P. You and Your Heart. New York: Atheneum Press, 1977.

Laragh, J. H. "Hypertension." Drug Therapy, January 1980.

Lear, M. W. Heartsounds. New York: Simon & Schuster, 1979.

Mason, S. J., and S. Wolfson. "Angina Pectoris," Hospital Medicine, February 1980.

Perloff, J. K. "When to Think of Bypass Surgery for Angina." Consultant, October 1976.

Pollack, G. H. "Cardiac Pacemaking: An Obligatory Role of Catecholamines?" Science, May 13, 1977.

Ritchie, J. L., G. W. Hamilton, and F. J. T. Wackers (eds.). Thallium-201 Myocardial Imaging. New York: Raven Press, 1978.

Shepherd, J. T., and P. M. Vanhoutle (eds.). The Human Cardiovascular System: Facts and Concepts. New York: Raven Press, 1979.

Simpson, A. G., and A. F. Morris. "Athletic Heart Syndrome: Some Recent Observations (Cardiovascular Effects)." American Corrective Therapy Journal, March-April 1979.

Sonnenblick, E. H., and M. Lesch (eds.). Exercise and Heart Disease. New York: Grune & Stratton, 1977.

Stallones, R. "The Rise and Fall of Ischemic Heart Disease." Scientific American, November 1980.

"Stadards and Guidelines for CPR and ECC." Journal of American Medical Association, 244, August 1980.

Tarazi, R. C. "The Heart in Hypertension: Its Load and Its Role." Hospital Practice, December 1975.

Thompson, D. A. Cardiovascular Assessment: Guide for Nurses and Health Professionals. St. Louis: C. V. Mosby Co., 1981.

Walker, W. J. "Curbing Risk Factors Has Helped Reduce U.S. Coronary Deaths." Modern Medicine, June 1, 1976.

Watanabe, Y., and L. S. Dreifus. Cardiac Arrhythmias: Electrophysiologic Basis for Clinical Interpretation. New York: Grune & Stratton, 1977.

19. The Circulatory System: Blood Vessels and Lymphatics

Baron, H. C. "Valvular Incompetence and Varicose Veins." Hospital Medicine, April 1976.

Benditt, E. P. "The Origin of Atherosclerosis." Scientific American, February 1977.

Bergman, S. G. "The Atherosclerosis Problem." Clinical Pediatrics, January 1975.

Bevan, J. A., et al. (eds.). Vascular Neuroeffector Mechanisms. New York: Raven Press, 1980.

Bohr, D. F., et al. (eds.). Handbook of Physiology. Vol. 2. Baltimore: Williams & Wilkins Co., 1980.

Caro, C. G., et al. The Mechanics of the Circulation. New York: Oxford University Press, 1979.

Elues, M. W. The Lymphocytes. Chicago: Year Book Medical Publishers, 1972.

Fox, E. L. Sports Physiology. Philadelphia: W. B. Saunders Co., 1979.

"Hypertension: Getting-and-Keeping-It Under Control." Medical World News, March 22, 1976.

James, D. G. (ed.). Circulation of the Blood. Baltimore: University Park Press, 1978.

Kaley, G., and B. M. Altura (eds.). Microcirculation. Vol. 2. Baltimore: University Park Press, 1977.

Keatinge, W. R., and M. C. Harman. Local Mechanisms Controlling Blood Vessels. New York: Academic Press, 1979.

Kirchheim, H. R. "Systemic Arterial Baroreceptor Reflexes." Physiol. Rev. 56:100, 1976.

Laird, W. P. "Childhood and Diet as Related to Atherosclerosis." Clinical Pediatrics May 1975.

Merrill, J. P. "Curable Hypertension." Hospital Medicine, April 1976.

Rushmer, R. F. Structure and Function of the Cardiovascular System. 2nd ed. Philadelphia: W. B. Saunders Co., 1976.

Shepard, J. T. and P. M. Vanhoutte. The Human Cardiovascular System: Facts and Concepts. New York: Raven Press, 1979.

Sparks, H. V., and F. L. Belloni. "The Peripheral Circulation: Local Regulation. Annu. Rev. Phsiol., 40:67, 1978.

Vanhoutte, P. M., and I. Leusen (eds.). Mechanics of Vasodilation. New York: S. Karger, 1978.

Wolf, S., et al. (eds.). Structure and Function of the Circulation. New York: Plenum Press, 1979.

20. Defense Mechanisms

Allison, A. C., et al. Inflammation. New York: Springer-Verlag, 1978.

Amos, D. B., et al. (eds.). Immune Mechanisms and Disease. New York: Academic Press, 1979.

Benacerraf, B., and Unanue, E. R. Textbook of Immunology. Baltimore: Williams & Wilkins Co., 1979.

Burke, D. C. "The Status of Interferon." Scientific American, 236(4):42, 1977.

Cunningham, B.A. "The Structure and Function of Histocompatibility Antigens." Scientific American, 234(4):96, 1977.

Escobar, M. R., and H. Friedman,(eds.). Macrophages and Lymphocytes: Nature, Functions, and Interactions. Mew York: Plenum Press, 1979.

Friedman, H., et al. The Reticuloendothelial System. New York: Plenum Press, 1979.

Gadebusch, H. J. (ed.). Phagocytosis and Cellular Immunity. West Palm Beach, Florida: CRC Press, 1979.

Houck, J. C. (ed.). Chemical Messengers of the Inflammatory Process. New York: Elsevier/North-Holland, 1980.

Hudson, L., and F. Hoy. Practical Immunology. 2nd ed. St. Louis: C. V. Mosby Co., 1981.

Kaley, G., and B. M. Altura (eds.). Microcirculation. Vol. 3. Baltimore: University Press, 1977.

Kaplan, J. G. (ed.). The Molecular Basis of Immune Cell Function. New York: Elsevier/North-Holland, 1979.

Klebanoff, S. J., and R. A. Clark. The Neutrophil: Function and Clinical Disorders. New York: Elsevier/North-Holland, 1978.

Lisiewicz, J. Human Neutrophils. Bowie, Md.: Charles Press Publishers, 1979.

Marx, J. L. "Antibodies: Getting Their Genes Together." Science, 212, 1981.

Marx, J. L. "Natural Killer Cells Help Defend the Body." Science, November 1980.

Marx, J. L. "T-Cells Scrutinized at Rudesheim Meeting." Science, 214:983, 1981.

McConnell, I., et al. The Immune System-A Course on the Molecular and Cellular Basis of Immunity. 2nd ed. St. Louis: C. V. Mosby Co., 1981.

Milstein, C. "Monoclonal Antibodies." Scientific American, October 1980.

Mims, C. A. Pathogenesis of Infectious Disease. New York: Grune & Stratton, 1976.

Mohn, J. F., et al. (eds.). Human Blood Groups. New York: S. Karger, 1977.

Page, H. "Our Cells Learn Sensitivities--and Transfer Them." Modern Medicine, March 15, 1976.

Paul, W. E., and B. Benacerraf. "Functional Specificity of Thymus-Dependent Lymphocytes." Science, March 25, 1977.

Quastel, M. R. (ed.) Cell Biology and Immunology of Leukocyte Function. New York: Academic Press, 1979.

Raff, M. C. "Cell-Surface Immunology." Scientific American, May, 1976.

Rose, N. R. "Autoimmune Diseases." Scientific American, February 1981.

Rose, N. R., and F. Milgrom (eds.). Principles of Immunology. New York: Macmillian Publishing Co., Inc., 1979.

Schwartz, L. M., and W. Miles (eds.). Blood Bank Technology. 2nd ed. Baltimore: Williams & Wilkins Co., 1977

Stobo, J. D. "Basic Mechanisms of Immunity." Hospital Medicine, July 1980.

"The Leukotrines in Allergy and Inflammation." Science, 215: 1383, 1982.

Van Furth, R. (ed.). Mononuclear Phagocytes: Functional Aspects of Mononuclear Phagocytes. Boston: M. Nijhoff, 1979.

21. The Respiratory System

Bartels, H. and R. Bvaumann. "Respiratory Function of Hemoglobin." Int. Rev. Physiol, 14:107, 1977.

Bouhuys, A. Physiology of Breathing: A Textbook for Medical Students. New York: Grune & Stratton, 1977.

Bradley, G. W. "Control of the Breathing Pattern." Int. Rev. Physiol., 14:185, 1977.

Bryant, C. The Biology of Respiration. Baltimore: University Park Press, 1979.

Cohen, D., et al. "Smoking Impaires Long-Term Dust Clearance From the Lung." Science, 204:514, 1979.

Control Mechanisms in Breathing. New York: Pergamon Press, 1980.

Ellis, P. D. and D. M. Billings. Cardiopulmonary Resuscitation: Procedures for Basic and Advanced Life Support. St. Louis: C. V. Mosby, 1979.

Fink, B. R. The Human Larynx: A Functional Study. New York: Raven Press, 1975.

Fishman, A. P. Assessment of Pulmonary Function. New York: McGraw-Hill, 1980.

Fraser, R. G., and J. A. Pare. Organ Physiology: Structure and Function of the Lung. 2nd ed. Philadelphia: W. B. Saunders Co., 1977.

Garvey, J. "Infant Respiratory Distress Syndrome." American Journal of Nursing, April 1975.

Gerrick, D. J. Mechanics of Breathing. Lorain, Ohio: Dayton Labs, 1978.

Guz, A. "Regulation of Respiration in Man." Annu. Rev. Physiol., 37:303, 1975.

Heimlich, H. J. "A Life-Saving Maneuver to Prevent Food Choking." Journal of the American Medical Association, October 27, 1975.

Kettel, L. J. "Acute Respiratory Acidosis." Hospital Medicine, February 1976.

Loeschche, H. H. Central Nervous Chemoreceptors. In MTP International Review of Science. Physiology, Vol. 2. Baltimore: University Park Press, 1974, p. 167.

Manfredi, F. "Acid-Base Problems Originating in the Lungs." Medical Opinion, March 1976.

Naeye, R. L. "Sudden Infant Death." Scientific American, April 1980.

Norman, P. S. "Asthma." Drug Therapy, January 1980.

Perutz, M. F. "Hemoglobin Structure and Respiratory Transport." Scientific American, 139(6):92, 1978.

Riechel, J. "Pulmonary Emphysema." Hospital Medicine, February 1980.

Slonin, N. B., and L. H. Hamilton. Respiratory Physiology. 3rd ed. St. Louis: C. V. Mosby Co., 1976.

Straus, M. J. (ed.). Lung Cancer: Clinical Diagnosis and Treatment. New York: Grune & Stratton, 1977.

Sundbeg, J. "The Acoustics of the Singing Voice." Scientific American, March 1977.

Tisi, G. M., Pulmonary Physiology in Clinical Medicine. Baltimore: Williams & Wilkins Co., 1980.

Von Euler, C. and H. Langercrantz (eds.). Central Nervous System Control Mechanisms in Breathing. New York: Pergamon Press, 1980.

Wagner, P. D., "Diffusion and Chemical Reaction in Pulmonary Gas Exchange." Physiol. Rev., 57:257, 1977.

Wanner, A., and K. R. Ratzan. "Chronic Bronchitis and Emphysema." In Forum on Infection. New York: Biomedical Information Corp., 1975.

West, J. B. Pulmonary Pathophysiology: The Essentials. Baltimore: Williams & Wilkins Co., 1977.

Williams, M. H., Jr. "Answers to Questions on Respiratory Emergencies. Hospital Medicine, October 1976.

22. The Digestive System Anatomy

Atanassova, E., and M. Papasova. "Gastrointestinal Motility." Int. Rev. Physiol, 12:35, 1977.

Cohen, S., et al. Gastrointestinal Motility. In: R. K. Crane. (ed.). International Review of Physiology: Gastrointestinal Physiology III, vol. 19. Baltimore: University Park Press, 1979, p. 107.

Heimlich, H. J. "A Life-Saving Manuever to Prevent Food Choking." J. Amer. Med. Assoc., October 27, 1975.

Hurwitz, A. L., et al. Disorders of Esophageal Motility. Philadelphia: W. B. Saunders Co., 1979.

Levin, B. "Early Detection of Colon Cancer." Consultant, July 1976.

Levitt, M. D., and J. H. Bond. "Flatulence." Annu. Rev. Med., 31:127, 1980.

Luy, M. L. M. "The Roughage Rage: Would More Fiber in Our Diet Improve Health?" Modern Medicine, April 1, 1976.

Mandel, I. D. "Dental Caries." American Scientist, 67:680-688, 1979.

Moog, F. "The Lining of the Small Intestine." Scientific American, November 1981.

Van Der Reis, L. (ed.). The Esophagus. New York: S. Karger, 1978.

White, T. T., and H. Sarles, and J. P. Benhamou. Liver, Bile Ducts, and Pancreas. New York: Grune & Stratton, 1977.

23. Digestive System Physiology

Binder, H. J. (ed.). Mechanisms of Intestinal Secretion. New York: A. R. Liss, 1979.

Crane, R. K. (ed.). International Review of Physiology: Gastrointestinal Physiology III. Vol. 19. Baltimore: University Park Press, 1979.

Davenport, H. W. A Digest of Digestion. 2nd ed. Chicago: Year Book Medical Publishers, 1978.

Davenport, H. W. Physiology of the Digestive Tract: An Introductory Text. 4th ed. Chicago: Year Book Medical Publishers, 1977.

Davenport, H. W. "Why the Stomach Does Not Digest Itself." Scientific American, January 1972.

Gall, E. A., and F. K. Mostofi (eds.). The Liver. Huntington, New York: R. E. Krieger Co., 1980.

Glass, G. B. (ed.). Gastrointestinal Hormones. New York: Raven Press, 1980.

Goldstein, F. "Tracking and Treating Gallstones." Modern Medicine, pgs. 15-30, May 1979.

Graham, D. Y. "Update on Peptic Ulcer Therapy." Consultant, February 1976.

Jenkins, G. N. The Physiology and Biochemistry of the Mouth. Philadelphia: J. B. Lippincott Co., 1978.

Johnson, L. R. "Gastrointestinal Hormones: Physiological Implications." Symp. Federation Proc., 36:1929-1951, 1977.

Kappas, A., and A. P. Alvares. "How the Liver Metabolizes Foreign Substarces." Scientific American, June 1975.

Kassel, B., and Kay, J. "Zymogens of Proteolytic Enzymes." Science, 180:1022-1027, 1973.

McCarty, D. J. "The Management of Gout." Hospital Practice, September 1979.

Neurath, H. "Protein Digesting Enzymes." Scientific American, December 1974.

Roth, J. "Insulin Receptors in Diabetes." Hospital Practice, May 1980.

Sernka, T. J., and E. D. Jacobsen. Gastrointestinal Physiology: The Essentials. Baltimore: Williams & Wilkins Co., 1979.

Sharon, N. "Carbohydrates." Scientific American, November 1980.

24. Metabolism

Alexander, G. Cold Thermogenesis. In: Robertshaw, D. (ed.). International Review of Physiology: Environmental Physiology III. Vol. 20. Baltimore: University Park Press, 1979, p. 43.1.

Alfin-Slater, R. B., and D. Kritcheysky (eds.). Nutrition and the Adult: Macronutrients. New York: Plenum Press, 1979.

Bender, D. A. Amino Acid Metabolism. New York: John Wiley & Sons, 1978.

Coleman, J. E. Metabolic Interrelationships Between Carbohydrates, Lipids and Proteins. In: Bondy, P. K., and L. E. Rosenberg (eds.). Metabolic Control and Disease. 8th ed. Philadelphia: W. B. Saunders Co., 1980, p. 161.

Cudlipp, E. Vitamins. New York: Grosset & Dunlap, 1978.

Dickerson, E. "Cytochrome C and the Evolution of Energy Metabolism." Scientific American, March 1980.

Friedman, H. C. (ed.). Enzymes. Stroudsburg, Pa.: Dowden, Hutchinson & Ross. 1980.

Hardy, R. N. Temperature and Animal Life. Baltimore: University Park Press, 1979.

Heller, H. C., et al. "The Thermostat of Vertebrate Animals." Scientific American, 239(9):102, 1979.

Herman, R. H., et al. (eds.). Metabolic Control in Mammals. New York: Plenum Press, 1979.

Hunt, S. M., et al. Nutrition: Principles and Clinical Practice. New York: John Wiley & Sons, 1980.

Kharasch, N. (ed.). Trace Elements in Health and Disease. New York: Raven Press, 1979.

Langone, J. "The Girth of a Nation." Discover, February 1981.

Lieber, C. S. "The Metabolism of Alcohol." Scientific American, March 1976.

McCarty, R. E. "How Cells Make ATP." Scientific American, 238:(3):104, 1978.

McKean, K. "Exercise." Discover, November 1981.

Mayer, J. Human Nutrition: Its Physiological, Medical, and Social Aspects. Springfield, Ill., Charles C. Thomas, 1979.

Robinson, A. M., and D. H. Williamson. "Physiological Roles of Ketone Bodies as Substrates and Signals in Mammalian Tissues." Physiol. Rev., 60:143, 1980.

Shoden, R. J., and W. S. Griffin. Fundamentals of Clinical Nutrition. New York: McGraw-Hill, 1980.

Sloan, A. W. Man in Extreme Environment. Springfield, Ill.: Charles C. Thomas, 1979.

Thompson, C. I. Controls of Eating. New York: Spectrum Publications, 1979.

Waterlow, J. C., et al. Protein Turnover in Mammalian Tissues and in the Whole Body. New York: Elsevier/North-Holland, 1978.

Young, V. R., and N. S. Scrimshaw. "The Physiology of Starvation." Scientific American, October 1971.

25. The Urinary System

Barajas, L. "Anatomy of the Juxta-Glomerular Apparatus." American Journal of Physiology, November 1979.

Bauman, J. W., and F. P. Chinard. Renal Function: Physiological and Medical Aspects. St. Louis: C. V. Mosby Co., 1975.

Bissada, N. K., and A. E. Finkbeiner. Lower Urinary Tract Function and Dysfunction. New York: Appleton-Century-Crofts. 1978.

Brenner, B. M., and F. C. Rector (eds.). The Kidney. 2nd ed. Two volumes. Philadelphia: W. B. Saunders Co. 1981

Brenner, B. M., and J. H. Stein (eds.). Hormonal Function and the Kidney. New York: Churchill Livingstone, 1979.

"Chronic Kidney Disease: Why You Need a Special Diet." Drug Therapy, June 1976.

Churg, J. (ed.). The Kidney. Baltimore: Williams & Wilkins, 1979.

Deetjen, P., et al. Physiology of the Kidney and of Water Balance. New York: Springer-Verlag, 1974.

Dennis, V. W., et al. "Renal Handling of Phosphate and Calcium." Annu. Rev. Physiol., 41:257, 1979.

Diamond L. H., and J. E. Balow (eds.). Nephrology Reviews, New York: John Wiley & Sons, 1980.

"Dramatic Advances Against Kidney Diseases." Medical Tribune, June 23 1976.

Ehrlich, E. N. "Adrenocortical Regulation of Salt and Water Metabolism." In:

DeGroot, L. J., et al. (eds.) Endocrinology. Vol 3. New York: Grune & Stratton, 1979, p. 1883.

Fisher, J. W. (ed.). Kidney Hormones. New York: Academic Press, 1970.

Guder. W. G. and J. Schmidt (eds.). Biochemical Nephrology Bern, H., Huber, 1978.

Harvey, R. J. The Kidneys and the Internal Environment. New York: Halsted Press, 1976.

Merrill, J. P. "The Artificial Kidney." Scientific American, July 1961.

Roberts, J. A. "A Guide to the Urologic Examination." Hospital Medicine, February 1976.

"Urinary Frequency: Simple Technique Causes Prostate Problems. " R. N., August 1980.

Valtin, H. Renal Functions: Mechanisms Preserving Fluid and Soluble Balance in Health. Boston: Little, Brown & Co., 1973.

Vander, A. J. Renal Physiology. New York: McGraw-Hill Book Co., 1975.

Weller, J. M. (ed.). Fundamentals of Nephrology. Hagerstown, Md.: Harper & Row, 1979.

26. <u>Fluid and Electrolyte Regulation and Acid-Base Balance</u>

Brenner, B. M., and J. J. Stein (eds.). Acid-Base and Potassium Homeostasis New York: Churchill Livingstone, 1978.

Gilles, R. (ed.). Mechanisms of Osmoregulation. New York: John Wiley & Sons, 1979.

Goldberger, E. Primer of Water, Electrolyte, and Acid-Base Syndromes. 6th ed. Philadelphia: Lea & Febiger, 1980.

Grollman, A. "Body Fluids and Electrolytes." Consultant, May 1976.

Guyton, A. C., et al. Circulatory Physiology II: Dynamics and Control of the Body Fluids. Philadelphia: W. B. Saunders Co., 1975.

Jones, N. L. Blood Gases and Acid-Base Physiology. New York: B. C. Decker, 1981.

Kettle, L. J. "Acute Respiratory Acidosis." Hospital Medicine, February 1976.

Manfredi, F. "Acid-Base Problems Originating in the Lungs." Medical Opinion, March 1976.

Maxwell, M. H., and C. R. Kleeman (eds.). Clinical Disorders of Fluid and Electrolyte Metabolism. 3rd ed. New York: McGraw-Hill, 1979.

Quintero, J. A. Acid-Base Balance: A Manual for Clinicians. St. Louis: W. H. Green. 1979.

Rector, F. C., and M. G. Cogan. "The Renal Acidoses." Hospital Practice, April 1980.

Smith, K. Fluids and Electrolytes: A Conceptual Approach. New York: Churchill Livingstone, 1980.

27. Reproductive System

Abrams, J. "The Pap Test in 1980." J. Med. Soc. of New Jersey, 77:11, October 1980.

Baachus, H. Essentials of Gynecologic and Obstetric Endocrinology. Baltimore: University Park Press, 1975.

Buckner, W. P. Medical Readings on Human Sexuality. Stanford, California: Medical Reachings Inc., 1978.

Channing, C. P., and J. M. Marsh (eds.). Ovarian Follicular and Corpus Luteum Function. New York: Plenum Press, 1979.

Cohen, R. J. "Breast Cancer: Guide to Early Diagnosis." Hospital Medicine, June 1980.

Coppen, A., and M. Bishop. "Hysterectomy, Hormones, and Behavior: A Prospective Study." Lancet, January 17, 1981.

Crile, G., Jr. "Let's Stop Scaring Women Away From Mammography." Modern Medicine, October 15, 1976.

Davidson, M. "Hormones and Sexual Behavior in the Male." Hospital Practice, September 1975.

DeGroot, L. J., et al. (eds.). Endocrinology. Vols. 1, 2, 3. New York: Grune & Stratton, 1979.

Dekretser, B. D. (ed.). The Testis. New York: Raven Press, 1981.

DiBenedetto, R. J., et al. "The Physiology of the Uterine Tube." Journal of the Medical Society of New Jersey, July 1975.

Droegemuller, W., and R. Bessler. "Effectiveness and Risks of Contraception." Annu. Rev. Med., 31:329, 1980.

Eckert, C. "How to Evaluate and Manage Breast Lumps." Journal of the American Medical Association, November 24, 1975.

Epel, D. "The Program of Fertilization." Scientific American, 237(5):128, 1977.

Greep, R. O. (ed.). International Review of Physiology: Reproductive Physiology III. Vol. 22. Baltimore: University Park Press, 1980.

Grobstein, C. "External Human Fertilization." Scientific American, June 1979.

Grumbach, M. M. "The Neuroendocrinology of Puberty." Hospital Practice, March 1980.

Hafez, E. S., et al. (eds.). Accessory Glands of the Male Reproductive Tract. Ann Arbor, Mich.: Ann Arbor Science, 1979.

Holmes, K. K. "Sexually Transmitted Diseases: An Overview and Perspectives on the Next Decade." Infectious Diseases, April 1980.

Howell, N. "Human Reproduction Reconsidered." Nature, November 23, 1978.

Huffman, S. L. "Postpartum Amenorrhea: How Is It Affected by Maternal Nutritional Status?" Science, June 9, 1978.

Kolata, G. B. "Infertility: Promising New Treatments." Science, October 13, 1978.

"Mammography May Be Better Than You Think." Hospital Practice, December 1975.

McGuire, W. L. "Steroid Receptors and Breast Cancer." Hospital Practice, April 1980.

McNab, W. L. "Other Venereal Infections." American Journal of Nursing, July 1979.

Michael, R. P. "Hormones and Sexual Behavior in the Female." Hospital Practice, December 1975.

Peck, D. R., and R. M. Lowman. "Mammography." Journal of the American Medical Association, October 18, 1976.

Peters, H., and K. P. McNatty. The Ovary: A Correlation of Structure and Function in Mammals. Berkeley: University of California Press, 1980.

Piver, M. S. "Chemotherapy for Gynecologic Malignancies." Drug Therapy, October 1976.

Purifoy, F. E., et al. "Steroid Hormones and Aging: Free Testosterone, Testosterone and Androstenedione in Normal Females Aged 20-87 Years." Human Biology, 1980.

Resnick, M. I. "How to Find and Stage Prostatic Carcinoma." Modern Medicine, October 15, 1976.

Segal, S. J. "The Physiology of Human Reproduction." Scientific American, September 1974.

Shearman, R. P. (ed.). Human Reproductive Physiology. Oxford: Blackwell, 1972.

Shocket, B. R. "Medical Aspects of Sexual Dysfunction." Drug Therapy, June 1976.

Speroff, L., et al. Clinical Gynecologic Endocrinology and Infertility. Baltimore: Williams & Wilkins Co., 1978.

Steinberger, A., and E. Steinberger (eds.). Testicular Development, Structure, and Function. New York: Raven Press, 1980.

"Three Methods of Tubal Occlusion--And All Look Good." Modern Medicine, September 15, 1976.

Tobiason, S. J. "Benign Prostatic Hypertrophy." American Journal of Nursing, February 1979.

Wallach, E. E., and R. D. Kempers. Modern Trends in Infertility and Conception Control. Baltimore: Williams & Wilkins Co., 1979.

Weideger, P. Menstruation and Menopause: The Physiology and Psychology, the Myth and the Reality. Revised ed. New York: Delta/Dell, 1977.

Yen, S. S., and R. B. Jaffe (eds.). Reproductive Endocrinology. Philadelphia: W. B. Saunders Co., 1978.

28. Pregnancy and Early Embryonic Development

Alexander, A. R., and G. J. Everidge. "Pregnant But With No Uterus." American Journal of Nursing," February 1980.

Beaconsfield, P., et al. "The Placenta. " Scientific American, August 1980.

Burnett, L. S., et al. "An Evaluation of Abortion: Techniques and Protocols. " Hospital Practice, August 1975.

Centerwall, W. R. and J. L. Murdoch. "Human Chromosome Analysis." American Family Physician, April 1975.

Chamberlin, G., and A. Wilkinson (eds.). Placenta Transfer. Baltimore: University Park Press 1979.

"Conception." Discover, October 1982.

DeGroot, L. J., et al. (eds.). Endocrinology. Vols. 1, 2, 3. New York: Grune & Stratton, 1979.

Ehrhardt, A. A., et al. "Effects of Prenatal Sex Hormones on Gender Related Behavior." Science, 211:1312-1318, 1981.

Epel, D. "The Program of Fertilization." Scientific American, November 1977.

Fehr, P. E. "Guidelines for Prescribing in Pregnancy." Modern Medicine, June 15, 1976.

Friedmann, T. "Prenatal Diagnosis of Genetic Disease." Scientific American, November 1971.

Fuchs, F. "Genetic Amniocentesis." Scientific American, June 1979.

Greep, P. O. (ed.). International Review of Physiology: Reproductive Physiology III. Vol. 22. Baltimore: University Park Press, 1980.

Grobstein, C. "External Human Fertilization." Scientific American, June 1979.

Hafez, E. S. (ed.). Human Reproduction: Conception and Contraception. Hagerstown, Md.: Harper & Row, 1979.

Haller, J. O., and M. Schneider. Pediatric Ultrasound. Chicago: Year Book Medical Publishers, 1980.

Haseltine, F. P., and S. Ohno. "Mechanisms of Gonadal Differentiation." Science, 211:1272, 1981.

Hobbins, J. C., and F. Winsberg. Ultrasonography in Obstetrics and Gynecology. Baltimore: Williams & Wilkins Co., 1977.

Jones, M. D. "Oxygen Delivery to the Brain Before and After Birth." Science, April 1982.

Jones, O. W. "Where We Stand Today in Prenatal Diagnosis." Medical Opinion, May 1976.

Konner, M. "She and He." Science, 3(7). September 1982.

Langer, A., et al. "Amniotic Fluid Analysis in Prenatal Diagnosis." Journal of the Medical Society of New Jersey, July 1975.

Lindley, M. "Life and Death Before Birth." Nature, August 23, 1979.

Marx, J. L. "Dysmenorrhea." Science, 205:175-176, 1979.

Masters, W. H., and V. E. Johnson. Human Sexual Response. Boston: Little, Brown & Co., 1966.

Michael, R. P. "Hormones and Sexual Behavior in the Female." Hospital Practice, December 1975.

"Midtrimester Amniocentesis for Prenatal Diagnosis." Journal of the American Medical Association, September 27, 1976.

Miller, H. C., and T. A. Merritt. Fetal Growth in Humans. Chicago: Year Book Medical Publishers, 1979.

Nilsson, L., et al. A Child is Born. Revised ed. New York: Delacorte/Lawrence, 1977.

Page, E. W., et al. Human Reproduction: Essentials of the Core Content of Reproductive Obstetrics, Gynecology, and Prenatal Medicine. 3rd ed. Philadelphia: W. B. Saunders Co., 1981.

Patten, B. M., and B. M. Carlson. Foundations of Embryology. 3rd ed. New York: McGraw-Hill Book Co., 1974.

Rowley, P. T. "Genetic Screening: Whose Responsibility? Journal of the American Medical Association, July 26, 1976.

Rugh, R., and L. B. Shettles. From Conception to Birth. New York: Harper & Row, 1971.

Seligmann, J. "New Science of Birth." Newsweek, November 15, 1976.

Smith, D. W. "The Fetal Alcohol Syndrome." Hospital Practice, October 1979.

Thornburn, G. D., and J. R. Challis. "Endocrine Control of Parturition." Physiol. Rev., 59:863, 1979.

Vorherr, H. (ed.). Human Lactation. New York: Grune & Stratton, 1979.

Wilson, J. D., et al. "The Hormonal Control of Sexual Development." Science, 211:1278, March 1981.

Yen, S. S., and R. B. Jaffe (eds.). Reproductive Endocrinology. Philadelphia: W. B. Saunders Co., 1978.

Yunek, M. J., and R. M. Lojek. "Fetal and Maternal Monitoring: Intrapartal Fetal Monitoring." American Journal of Nursing, December 1978.

d   1.   Which one of the following listed is the second most complex
         structural level of organization?
         a)   system
         b)   cell
         c)   organ
         d)   tissue

a   2.   The embryonic tissue that forms brain, spinal cord, and nerves is
         known as:
         a)   ectoderm
         b)   mesoderm
         c)   endoderm
         d)   somatoderm

d   3.   The tissue that lines body cavities, ducts, and vessels is known as
         _____tissue.
         a)   muscular
         b)   nervous
         c)   connective
         d)   epithelial

a   4.   Which one of the following organ systems controls and integrates body
         activities?
         a)   endocrine
         b)   respiratory
         c)   integumentary
         d)   circulatory
         e)   urinary

b   5.   If the body is erect with the feet together, the upper limbs hanging
         down at the sides with the palms of the hands facing forward and the
         fingers extended, the body is in the _____ position.
         a)   fetal
         b)   anatomical
         c)   prone
         d)   supine

e   6.   Which one of the following is not paired correctly as opposites?
         a)   medial-lateral
         b)   proximal-distal
         c)   external-internal
         d)   superior-inferior
         e)   all of the above are correctly paired as opposites

d   7.   The portion of the body between the neck and the abdomen is the
_____.
    a)   lumbar region
    b)   sacral region
    c)   axilla
    d)   thorax

d   8.   The region just inferior to the umbilical is known as the _____
region.
    a)   hypochondriac
    b)   iliac
    c)   epigastric
    d)   hypogastric

c   9.   The plane that divides the body lengthwise into anterior and posterior
portions is known as a _____ plane.
    a)   midsagittal
    b)   sagittal
    c)   frontal
    d)   horizontal

a   10.   Which one of the following does not lie in the ventral cavity?
    a)   spinal cord
    b)   mediastinum
    c)   stomach
    d)   bladder

d   11.   Which one of these spaces contains all the others?
    a)   pleural cavity
    b)   mediastinum
    c)   pericardial cavity
    d)   thoracic cavity

e   12.   Which one of the following is not found in the pelvic cavity?
    a)   ovary
    b)   uterus
    c)   urinary bladder
    d)   rectum
    e)   all of the above are found in the pelvic cavity

c   13.   The membrane that lines the thoracic cavity is known as:
    a)   visceral pleura
    b)   visceral peritoneum
    c)   parietal pleura
    d)   visceral pleura
    e)   none of the above is a correct answer

d   14.   Which one of the following statements is not correct?
            a)   mesenteries are made of a double layer of peritoneum
            b)   mesenteries suspend organs in the abdominal cavity
            c)   mesenteries provide a pathway for vessels and nerves to travel to
                 and from organs
            d)   organs covered by mesenteries are said to be retroperitoneal

d   15.   Maintaining optimum internal operating conditions for the body's cells
            in the face of changing environmental conditions is known as:
            a)   hemostasis
            b)   negative feedback
            c)   positive feedback
            d)   none of the above is correct

c   16.   Which one of the following utilizes sound waves to obtain images of
            the body?
            a)   dynamic spatial reconstructor
            b)   computed tomography
            c)   ultrasonography
            d)   nuclear magnetic resonance

b   17.   A body part that is situated above another specific body part is said
            to be ____to that part.
            a)   inferior
            b)   superior
            c)   anterior
            d)   medial

NAME_____

SECTION _____

c 1. Which one of the following listed is the third most complex structural
       level of organization?
       a) system
       b) cell
       c) organ
       d) tissue

b 2. The embryonic tissue that gives rise to most of the connective tissue
       is:
       a) ectoderm
       b) mesoderm
       c) endoderm
       d) somatoderm

c 3. Which of the following is not an example of connective tissue?
       a) bone
       b) fascia
       c) cardiac
       d) tendon

d 4. Which one of the following organ systems controls and integrates
       body activities?
       a) digestive
       b) reproductive
       c) skeletal
       d) nervous
       e) lymphatic

c 5. If the body is lying horizontally with the face downward, it is in the
       _____position.
       a) fetal
       b) anatomical
       c) prone
       d) supine

a 6. Which one of the following is not paired correctly as opposites?
       a) anterior-ventral
       b) medial-lateral
       c) proximal-distal
       d) superior-inferior
       e) external-internal

b   7.   The portion of the back between the thorax and the pelvis is known as the _____ region.
     a)   cervical
     b)   lumbar
     c)   sacral
     d)   plantar

a   8.   The region just lateral to the hypogastric is known as the _____ region.
     a)   iliac
     b)   hypogastric
     c)   lumbar
     d)   hypochondriac

b   9.   The plane that divides the body into equal left and right halves is known as a _____ plane.
     a)   coronal
     b)   midsagittal
     c)   sagittal
     d)   horizontal

e   10.   Which one of the following does not lie in the dorsal cavity?
     a)   meninges
     b)   cerebrospinal fluid
     c)   brain
     d)   spinal cord
     e)   all of the above lie in the dorsal cavity

e   11.   Which one of these spaces contains all the others?
     a)   thoracic cavity
     b)   abdominal cavity
     c)   pleural cavity
     d)   pelvic cavity
     e)   ventral cavity

e   12.   Which one of the following is not found in the abdominal cavity?
     a)   small intestine
     b)   spleen
     c)   liver
     d)   pancreas
     e)   all of the above are found in the pelvic cavity

e   13.   The membrane that covers the heart is known as the _____.
     a)   parietal pleura
     b)   meninges
     c)   parietal pericardium
     d)   mediastrinum
     e)   visceral pericardium

d   14. Which one of the following statements is incorrect?
a)   mesenteries are used to provide a pathway for vessels and nerves traveling to and from organs
b)   the lungs are covered by a mesentery
c)   mesenteries suspend organs, thereby providing them with support
d)   all of the above are correct statements

d   15. Maintaining optimum internal operating conditions for the body's cells in the face of changing environmental conditions is known as:
a)   negative feedback
b)   hemostasis
c)   positive feedback
d)   none of the above is correct

d   16. Which of the following use x-rays to obtain images of the body?
a)   nuclear magnetic resonance
b)   dynamic spatial reconstructor
c)   computed tomography
d)   both b and c use x-rays.

a   17. The membrane that lines the abdominal cavity is known as the:
a)   parietal peritoneum
b)   visceral pleura
c)   parietal pleura
d)   visceral peritoneum

NAME_____

SECTION _____

e   1.   Which one of the following listed is the smallest in size?
   a)   molecule
   b)   atom
   c)   element
   d)   proton
   e)   electron

c   2.   In an atom the second energy level can hold a maximum of _____ electrons.
   a)   3
   b)   6
   c)   8
   d)   18
   e)   32

b   3.   The number of protons in an atom is known as the:
   a)   atomic weight
   b)   atomic number
   c)   mass number
   d)   valence number

a,c   4.   Isotopes of a particular chemical element have the same number of: (two answers possible)
   a)   protons
   b)   neutrons
   c)   electrons

d   5.   If something has the same mass as an electron but can be positively or negatively charged, it could be a:
   a)   gamma ray
   b)   alpha particle
   c)   x-ray
   d)   beta particle

a   6.   If two atoms share two pairs of electrons equally, the bond is known as a/an:
   a)   nonpolar covalent bond
   b)   hydrogen bond
   c)   polar covalent bond
   d)   ionic bond

c   7.   Which one of the following listed is third largest in size?
   a)   glucose
   b)   fructose
   c)   sucrose
   d)   glycogen

a   8.   Which one of the following is not characteristic of lipids?
     a)   contain as many oxygen atoms as carbon atoms
     b)   contain carbon, hydrogen, and oxygen atoms
     c)   are insoluble in water
     d)   are soluble in organic solvents

a   9.   Proteins are made up of groups of smaller subunit molecules known as:
     a)   amino acids
     b)   fatty acids
     c)   nucleic acids
     d)   strong acids

a   10.   The sequence of amino acids in a polypeptide chain constitutes the
     _____ structure of a protein.
     a)   primary
     b)   secondary
     c)   tertiary
     d)   quaternary

c   11.   Nucleotides of DNA may contain all but which one of the following substances?
     a)   purine
     b)   pyrimidine
     c)   ribose
     d)   phosphate

d   12.   Which one of the following bases is not found in DNA?
     a)   adenine
     b)   guanine
     c)   cytosine
     d)   uracil

a   13.   Which one of the following bases will bond with guanine?
     a)   cytosine
     b)   thymine
     c)   uracil
     d)   adenine

d   14.   Which one of the following substances is the immediate source of energy for organismal activity?
     a)   glucose
     b)   ADP
     c)   phosphate
     d)   none of the above is correct

d   15.   Which one of the following is a correct statement about metabolism?
a)   anabolic reactions are break-down reactions
b)   catabolic reactions are synthesis reactions
c)   at normal body temperature metabolic reactions occur fast enough
d)   metabolic reactions produce heat, which helps maintain a constant body temperature

c   16.   Enzymes increase the rate of a chemical reaction by:
a)   increasing the temperature of reactants
b)   reacting with inhibitor molecules
c)   lowering the activation energy necessary for the reaction to take place
d)   forming enzyme-product complexes

a   17.   Which one of the following is not true of competitive inhibition?
a)   the inhibitor molecule will react with the enzyme molecule to form a product
b)   the inhibitor molecule and the normal substrate molecule are similar in structure
c)   the inhibitor molecule can combine reversibly with the enzyme molecule
d)   substrate and inhibitor concentrations are important in this kind of inhibition

d   18.   Which one of the following is not true of a true solution?
a)   the solute particles do not settle out
b)   a light path is not visible if a beam of light is passed through the solution
c)   the dispersed particles within a true solution are very small in size
d)   the substance present in the greatest amount is called the solute

c   19.   Which one of the following is not one of the properties of water?
a)   high latent heat of vaporization
b)   high specific heat
c)   nonpolar liquid
d)   capable of reacting with salts to form hydrated ions

c   20.   A 10% glucose solution is made by placing 10g of glucose in:
a)   100 moles of water
b)   100 ml. of water
c)   enough water to make 100 ml. of solution
d)   none of the above is correct

b   21.   Which one of the following is not true of colloids?
- a)   a light path is visible if a beam of light is passed through the colloid
- b)   cooked egg white is an example of a colloid
- c)   the dispersed particles do not readily settle out
- d)   all of the above are true of colloids

d   22.   If the pH shifts from 7.5 to 5.5, there is a:
- a)   10 fold increase in (OH+) ions
- b)   10 fold decrease in (OH+) ions
- c)   100 fold increase in (OH+) ions
- d)   100 fold increase in H+ ions
- e)   100 fold decrease in H+ ions

b   23.   Which one of the following is not a true statement about diffusion?
- a)   the diffusing particles move along precise pathways
- b)   the higher the temperature, the faster the rate of diffusion
- c)   at equilibrium there is no longer any movement of diffusing particles, because they are equally distributed
- d)   all of the above are true statements about diffusion.

e   24.   Dialysis is the:
- a)   movement of ions and other small particles from high to low concentration
- b)   movement of ions and other small particles from low to high concentration
- c)   movement of water from low to high concentration
- d)   movement of water from high to low concentration
- e)   none of the above is correct

d   25.   Which one of the following is not an example of bulk flow?
- a)   flow of blood within blood vessels
- b)   flow of air into the lungs
- c)   flow of urine from the bladder
- d)   flow of substances across capillary walls

d   1.   Which one of the following listed is the second smallest in size?
          a)   molecule
          b)   atom
          c)   element
          d)   proton
          e)   electron

d   2.   In an atom, the third energy level can hold a maximum of _____ electrons.
          a)   2
          b)   32
          c)   64
          d)   none of the above is correct

d   3.   The number of protons and neutrons in an atom is known as the:
          a)   valence number
          b)   atomic number
          c)   atomic mass
          d)   mass number

b   4.   Isotopes of a particular chemical element have different numbers of:
          a)   protons
          b)   neutrons
          c)   electrons

c   5.   If something has no detectable mass or electric charge but has lots of energy, it could be a/an:
          a)   alpha particle
          b)   beta particle
          c)   gamma ray
          d)   negatron

b   6.   The transfer of electrons from one atom to another leads to the formation of the kind of bond known as:
          a)   nonpolar covalent
          b)   ionic
          c)   polar covalent
          d)   hydrogen

b   7.   Which one of the following is a disaccharide?
          a)   glucose
          b)   sucrose
          c)   starch
          d)   glycogen

a   8.   Which one of the following is not a characteristic of lipids?
       a)   insoluble in organic solvents
       b)   contain only a few oxygen atoms
       c)   contain carbon, hydrogen, and oxygen atoms
       d)   serve as energy sources

c   9.   Which substance is composed of carbon, hydrogen, and oxygen, with the hydrogen and oxygen frequently present in the same 2:1 ratio as they are in water?
       a)   stearic acid
       b)   cholesterol
       c)   glucose
       d)   adenine
       e)   none of the above

a   10.   Dipeptides are formed by joining the:
       a)   amino group of one amino acid with the carboxyl group of another amino acid
       b)   amino group of one amino acid with the amino group of another amino acid
       c)   amino group of one amino acid with the carboxyl group of a fatty acid
       d)   amino group of one amino acid with the carboxyl group of a nucleic acid

b   11.   Hydrogen bonding is particularly important in determining the _____ structure of a protein.
       a)   primary
       b)   secondary
       c)   tertiary
       d)   quaternary

b   12.   Nucleotides of RNA may contain all but which one of the following substances?
       a)   phosphate
       b)   deoxyribose
       c)   pyrimidine
       d)   purine

d   13.   Which one of the following bases is not found in RNA?
       a)   adenine
       b)   guanine
       c)   cytosine
       d)   thymine

b,d 14. Which one of the following bases will bond with adenine (more than
one answer possible)?
a) cytosine
b) uracil
c) guanine
d) thymine

b 15. Which one of the following is not a component of ATP?
a) phosphate
b) deoxyribose
c) adenine
d) all of the above are found in ATP

c 16. Which term includes all of the others?
a) catabolic reactions
b) anabolic reactions
c) metabolic reactions
d) synthesis reactions
e) breakdown reactions

d 17. Which one of the following is not true of enzymes?
a) they are commonly proteins
b) they can generally be denatured by heat or changes in acidity
c) they may be produced in inactive forms
d) they are used up during the reactions they catalyze

b 18. Which one of the following is true of competitive inhibition?
a) the inhibitor combines irreversibly with the active site on the
enzyme
b) the bond between the inhibitor and enzyme molecule occurs at a
site other than the active site of the enzyme
c) substrate and inhibitor concentrations are not important in this
kind of inhibition
d) none of the above are true of competitive inhibition

b 19. Which of the following is not correct of true solutions?
a) the substance present in the greatest amount is called the solvent
b) a light path is visible if a beam of light is passed through the
solution
c) the particles dispersed with a true solution are very small
d) a solution may be a mixture of two different liquids

a 20. Which one of the following is not one of the properties of water?
a) low specific heat
b) high latent heat of vaporization
c) polar liquid at room temperature
d) capable of reacting with salts to form hydrated ions

d   21. A 10% solution of ethyl alcohol is made by placing 10 ml. of ethyl alcohol in:
a)   1 mole of water
b)   100 ml. of water
c)   100 grams of water
d)   none of the above is correct

a   22. Which one of the following listed has the dispersed particles listed in size correctly from largest to smallest?
a)   suspension, colloid, solution
b)   solution, suspension, colloid
c)   colloid, solution, suspension
d)   suspension, solution, colloid
e)   solution, colloid, suspension

c   23. If the pH shifts from 4.5 to 2.5, there is a:
a)   10 fold increase in H+ ions
b)   10 fold decrease in OH- ions
c)   100 fold increase in H+ ions
d)   100 fold increase in OH- ions

e   24. Osmosis is the movement of solvent + from:
a)   high to low solvent + concentration
b)   low to high solvent + concentration
c)   high solute concentration to low solute concentration
d)   low solute concentration to high solute concentration
e)   none of the above is correct

c   25. The flow of blood within the blood vessels is an example of:
a)   filtration
b)   dialysis
c)   bulk flow
d)   osmosis
e)   none of the above.

b   1.   The ability to respond to stimuli is known as:
         a)   contractility
         b)   irritability
         c)   conductivity
         d)   metabolism

b   2.   Most of the cell organelles are located in the:
         a)   nucleus
         b)   cytoplasm
         c)   plasma membrane
         d)   extracellular fluid
         e)   none of the above

b   3.   The membrane channels:
         a)   are open all the time
         b)   allow the passage of certain polar molecules
         c)   allow the passage of only nonpolar molecules
         d)   do not allow the passage of ions
         e)   none of the above

d   4.   Specificity means that:
         a)   only a certain amount of substance can cross the cell membrane
              at a given time
         b)   no more than two different kinds of molecules can bind with a
              particular carrier molecule
         c)   a specific amount of energy is required to move the molecule to
              be transported
         d)   none of the above is correct

d   5.   Which one of the following is not true of facilitated diffusion?
         a)   shows the characteristic of specificity
         b)   shows the characteristic of competition
         c)   shows the characteristic of saturation
         d)   requires energy
         e)   can move substances from high to low concentration of those
              substances

e   6.   The accumulation of solid particles in a vesicle is known as:
         a)   exocytosis
         b)   pinocytosis
         c)   facilitated diffusion
         d)   mediated transport
         e)   none of the above is correct

c   7.   If a cell is placed in an isotonic solution, the net movement of water
         is:
         a)   into the cell
         b)   out of the cell
         c)   there will be no net movement of water into or out of the cell
         d)   none of the above is correct

c   8.   If a red blood cell is placed in a test tube of distilled water, the red
         blood cell would be _____ to the distilled water.
         a)   hypotonic
         b)   isotonic
         c)   hypertonic
         d)   cannot be determined

c   9.   The synthesis of RNA from a DNA template is known as:
         a)   transduction
         b)   translation
         c)   transcription
         d)   transference

d   10. If a small part of one side of a DNA molecule has the base sequence
         a-t-c-g-t-a, then the corresponding RNA base sequence would be:
         a)   t-a-g-c-a-t
         b)   t-a-c-g-t-a
         c)   u-a-g-g-a-t
         d)   u-a-g-c-a-u

c   11. The sites of protein synthesis within the cell are the:
         a)   lysosomes
         b)   peroxisomes
         c)   ribosomes
         d)   mitochondria

d   12. Which one of the following is not true of the endoplasmic reticulum?
         a)   play a role in the production of steroids
         b)   play a role in the production of fatty acids
         c)   almost always interconnected with the nuclear envelope
         d)   produce centrioles
         e)   all of the above are true

e   13. The organelle that contains enzymes that break down proteins and
         lipids is the:
         a)   Golgi apparatus
         b)   peroxisomes
         c)   ribosomes
         d)   mitochondria
         e)   none of the above is correct

d 14. The organelle that contains enzymes that produce hydrogen peroxide is
        known as the:
        a) lysosome
        b) mitochondria
        c) endoplasmic reticulum
        d) none of the above is correct

c 15. If an organelle is made up of nine groups of tubules in a circle with
        three tubules per group, then the organelle could be a _____.
        a) cilium
        b) flagellum
        c) centriole
        d) mitochondrion

d 16. Which one of the following listed is not generally considered to be an
        inclusion?
        a) hemoglobin
        b) melanin
        c) glycogen
        d) chondrotin sulfate

b 17. In normal mitosis, which one of the stages listed occurs fourthly?
        a) anaphase
        b) telophase
        c) prophase
        d) metaphase

a 18. The stage in normal mitosis where the nucleoli are disappearing and
        the chromosomes are appearing is known as:
        a) prophase
        b) telophase
        c) interphase
        d) anaphase
        e) metaphase

b 19. Mitosis with cytokinesis would usually yield:
        a) haploid cells
        b) diploid cells
        c) multinucleate cells
        d) cannot occur in nature

b 20. Crossing over could take place during:
        a) anaphase II
        b) metaphase I
        c) interphase
        d) prophase II
        e) telophase II

a   21.   If the normal diploid chromosome number is 12 for a particular kind of animal, then during interkinesis of meiosis a cell of that organism would contain _____chromosomes.
- a)   6
- b)   12
- c)   18
- d)   24

e   22.   Changes in the genetic information within a cell are known as:
- a)   sports
- b)   nucleotides
- c)   genes
- d)   chromosomes
- e)   none of the above is correct

d   23.   Which of the following help form the cytoskeleton of the cell?
- a)   microtubules
- b)   intermediate filaments
- c)   microfilaments
- d)   all of the above help form the cytoskeleton

a   24.   During which phase of interphase does DNA synthesis occur?
- a)   S phase
- b) G_1 phase
- c) G_2 phase
- d)   M phase

d   1.   The ability to process foods, obtain energy, and synthesize products is
          known as:
          a)   conductivity
          b)   contractility
          c)   irritability
          d)   metabolism

d   2.   In general, the lipid layer of the cell membrane permits the passage
          of:
          a)   small polar molecules
          b)   most ions
          c)   essentially all molecules in the cytoplasm
          d)   nonpolar substances

c   3.   Membrane channels
          a)   are always open
          b)   are the primary means by which large nonpolar molecules enter
               and leave cells
          c)   provide one route by which ions such as sodium ions or potassium
               ions can enter or leave cells
          d)   are present in membranes surrounding lysosomes and mitochondria
               but are not present in the plasma membrane
          e)   none of the above

b   4.   Saturation means that:
          a)   the carrier molecules are made of saturated fats
          b)   only a certain number of molecules may be transported across the
               cell membrane at a time
          c)   more than one kind of molecule can bind with a particular
               carrier molecule
          d)   energy is required to move material across the plasma membrane
          e)   none of the above is correct

c   5.   Which one of the following is not true of active transport?
          a)   shows the characteristic of saturation
          b)   shows the characteristic of competition
          c)   uses no energy
          d)   shows the characteristic of specificity
          e)   can move substances against concentration gradients

d   6.   The accumulation of extracellular fluid within a vesicle is known as:
          a)   exocytosis
          b)   osmosis
          c)   mediated transport
          d)   pinocytosis
          e)   facilitated diffusion

a 7. If a cell is placed in a hypotonic solution, the net movement of water is:
  a) into the cell
  b) out of the cell
  c) there will be no net movement of water either into or out of the cell
  d) none of the above is correct

a 8. If a red blood cell is placed in a test tube of 10% saline solution, the red blood cell would be _____ to the saline solution.
  a) hypotonic
  b) hypertonic
  c) isotonic
  d) cannot be determined

c 9. The synthesis of a protein using the instructions provided by messenger RNA is known as:
  a) transduction
  b) transference
  c) translation
  d) transcription

d 10. If a small part of one side of a DNA molecule has the base sequence a-t-c-g-t-a, then the corresponding RNA base sequence would be
  a) a-u-g-c-t-t
  b) t-a-g-c-u-a
  c) a-t-c-g-u-t
  d) none of the above is correct

d 11. During the synthesis of proteins, messenger RNA becomes associated with:
  a) lysosomes
  b) mitochondria
  c) nucleolus
  d) ribosomes
  e) DNA

e 12. Lysosomes
  a) originate from peroxisomes
  b) contain DNA
  c) are the major cell organelle involved in ATP production
  d) are present in the nucleus of a cell but not in the cytoplasm
  e) none of the above

c 13. The protein actin is a primary constituent of
    a) microtubules
    b) intermediate filaments
    c) microfilaments
    d) endoplasmic filaments
    c) none of the above

e 14. The organelles that are the powerhouses of the cell are the:
    a) basal bodies
    b) peroxisomes
    c) lysosomes
    d) endoplasmic reticulum
    e) mitochondria

d 15. If an organelle is made up of nine groups of tubules in a circle, with two tubules per group and two additional tubules in the center of the circle, then the organelle could be a _____.
    a) centriole
    b) basal body
    c) microfilament
    d) cilium

d 16. Which one of the following listed can occur as an extracellular substance?
    a) collagen
    b) hyaluronic acid
    c) elastin
    d) all of the above

d 17. In normal mitosis, which one of the stages listed occurs thirdly?
    a) metaphase
    b) telophase
    c) prophase
    d) anaphase

d 18. The stage in normal mitosis where the chromosomes are migrating toward the poles is known as:
    a) prophase
    b) telophase
    c) interphase
    d) anaphase
    e) metaphase

c 19. Mitosis without cytokinesis would usually yield:
    a) haploid cells
    b) diploid cells
    c) multinucleate cells
    d) cannot occur in nature

c   20. DNA synthesis occurs during which portion of interphase?
         a)   interkinesis
         b) G_1 phase
         c)   S phase
         d) G_2 phase
         e)   none of the above

a   21. If the normal diploid chromosome number for a particular kind of
         animal is 16, then at the conclusion of telophase of the second
         division sequence of meiosis a cell of that organism would contain
         _____ chromosomes.
         a)   8
         b)   16
         c)   24
         d)   32
         e)   40

a   22. In a female human, a normal somatic cell would contain
         a)   44 autosomes
         b)   no nucleus
         c)   two y chromosomes
         d)   one sex chromosome
         e)   none of the above

c   23. Which one of the following are hollow, cylindrical, unbranched
         structures that are a part of the cytoskeleton of a cell?
         a)   microfilaments
         b)   microtrabecular lattice
         c)   microtubules
         d)   intermediate filaments

b   24. During which phase of interphase does RNA synthesis occur?
         a)   S phase
         b) G_1 phase
         c) G_2 phase
         d)   M phase

d   1.   Which one of the following is not a function of epithelial tissues?
           a)   absorption
           b)   excretion
           c)   secretion
           d)   all of the above are functions of epithelial tissues

c   2.   Epithelial cells may have their free surfaces thrown into many small folds called:
           a)   cilia
           b)   junctional complexes
           c)   microvilli
           d)   gap junctions

d   3.   If an epithelium has all cells touching the basement membrane but not all cells reaching the free surface, then one would call that tissue a _____epithelium.
           a)   simple
           b)   striated
           c)   transitional
           d)   pseudostratified
           e)   stratified

b   4.   If an epithelium has cells that are taller than they are wide, then the shape of the cells is said to be:
           a)   cuboidal
           b)   columnar
           c)   squamous
           d)   pyramidal

c   5.   Simple squamous epithelium is not found in the:
           a)   alveoli
           b)   kidneys
           c)   lining of the trachea
           d)   blood vessels

d   6.   Simple squamous epithelium could act to aid in:
           a)   absorption or protection
           b)   secretion or absorption
           c)   protection or filtration
           d)   filtration or diffusion

b   7.   Stratified squamous epithelium is not found in the:
           a)   anus
           b)   stomach
           c)   mouth
           d)   vagina

a   8.   Which one of the following is not a characteristic of mucous membranes?
a)   line body cavities
b)   are absorptive and secretory
c)   can have a lamina propria
d)   can have a muscularis mucosa

c   9.   Which one of the following is not characteristic of glands that produce hormones?
a)   multicellular
b)   lack ducts
c)   are called exocrine glands
d)   all of the above are characteristics of the glands that produce hormones

b   10. The mode of secretion of a gland where the secretion accumulates within a cell and then the cell ruptures, liberating the secretion, is known as:
a)   apocrine
b)   holocrine
c)   merocrine
d)   endocrine

a   11. Which one of the following is not a characteristic of connective tissues?
a)   cells close together
b)   fibers
c)   matrix
d)   have a blood supply

d   12. Which one of the following is not a characteristic of collagen fibers?
a)   strong
b)   white
c)   inelastic
d)   branching

c   13. If a tissue has all 3 kinds of fibers in an unorganized arrangement, the tissue could be:
a)   adipose
b)   elastic connective tissue
c)   areolar connective tissue
d)   cartilage

a   14. If a tissue has many fine collagen fibers that are not visible with a
          light microscope, and lacunae containing chondrocytes, then the tissue
          would be:
          a)   hyaline cartilage
          b)   elastic cartilage
          c)   fibrocartilage
          d)   reticular cartilage

b   15. Which one of the following is not found in both bone and cartilage?
          a)   lacunae
          b)   blood supply
          c)   collagen fibers
          d)   matrix

e   16. Which one of the following is not an example of a serous membrane?
          a)   pleura
          b)   peritoneum
          c)   pericardium
          d)   mesentery
          e)   all of the above are examples of serous membranes

a   17. Which one of the following is not a location of stratified columnar
          epithelium in the body?
          a)   small intestine
          b)   male urethra
          c)   epiglottis
          d)   anal canal
          e)   parts of the pharynx

b   18. The tissue that lines the chambers of the heart and blood vessels is
          called:
          a)   mesothelium
          b)   endothelium
          c)   epithelium
          d)   pericardium

d   19. Which one of the following is not a characteristic of connective
          tissues?
          a)   cells
          b)   fibers
          c)   matrix
          d)   avascular

e   20. The dermis of the skin is primarily made up of:
          a)   areolar connective tissue
          b)   dense regular connective tissue
          c)   adipose tissue
          d)   elastic connective tissue
          e)   none of the above is correct

b   21.   If a tissue has a blood supply, lacunae, and canaliculi, that tissue is probably:
- a) elastic cartilage
- b) bone
- c) fibrocartilage
- d) hyaline cartilage

d   22.   If a tissue has intercalated disks, striations, and one nucleus per cell, then that tissue could be:
- a) skeletal muscle
- b) smooth muscle
- c) visceral muscle
- d) cardiac muscle

e   23.   When a wound in the skin is healing, the new epithelial cells arise from:
- a) fibrin
- b) fibroblasts
- c) subcutaneous tissue
- d) dermis
- e) none of the above is correct

a   24.   Adjacent cells of muscle tissue and nerve tissue, as well as those of epithelial tissues, are held together by:
- a) gap junctions
- b) intermediate junctions
- c) tight junctions
- d) all of the above

d   25.   The moist epithelial membranes that line the digestive, respiratory, urinary, and reproductive tracts are called:
- a) stratified squamous epithelium
- b) serous membranes
- c) elastic connective tissue
- d) mucous membranes

b   1.   Which one of the following is not a characteristic of epithelial
           tissues?
           a)   many cells close together
           b)   much intracellular material
           c)   presence of a basement membrane
           d)   line body cavities and passageways

d   2.   Which one of the following four terms is not a synonym for the other
           three?
           a)   striated border
           b)   brush border
           c)   microvilli
           d)   cilia

a   3.   If an epithelium is composed of many layers with only the deepest in
           contact with the basement membrane, then one would call that tissue
           a _____ epithelium.
           a)   stratified
           b)   simple
           c)   pseudostratified
           d)   striated

c   4.   If an epithelium has cells that are thin and flat, then the shape of the
           cells is said to be:
           a)   cuboidal
           b)   columnar
           c)   squamous
           d)   pyramidal

d   5.   Simple columnar epithelium could act to aid in:
           a)   protection
           b)   filtration
           c)   diffusion
           d)   absorption

b   6.   Simple columnar epithelium is not found in the:
           a)   ducts of glans
           b)   esophagus
           c)   stomach
           d)   uterus

c   7.   Stratified squamous epithelium is not found in the:
           a)   mouth
           b)   esophagus
           c)   trachea
           d)   vagina

d   8.   Which one of the following layers of a mucous membrane would be the deepest?
- a) lamina propria
- b) basement membrane
- c) epithelium
- d) muscularis mucosa

c   9.   Which one of the following is not a characteristic of glands that produce secretions such as sweat or digestive enzymes?
- a) are multicellular
- b) have ducts
- c) are called endocrine glands
- d) all of the above are characteristic of such glands

a   10.   The mode of secretion of a gland where a bit of the secretory cell pinches off with the secretion is known as:
- a) apocrine
- b) holocrine
- c) merocrine
- d) endocrine

c   11.   Which one of the following is not a function of connective tissues?
- a) anchor
- b) support
- c) absorb
- d) provide a framework

a   12.   Which one of the following is not a characteristic of elastic fibers?
- a) unbranched
- b) long
- c) thin
- d) often form networks

c   13.   If a tissue has all 3 kinds of fibers, including an abundancy of collagen fibers that form a closely interwoven network, then the tissue could be:
- a) areolar connective tissue
- b) elastic connective tissue
- c) dense irregular connective tissue
- d) dense regular connective tissue

c   14.   If a tissue has collagen fibers in thick bundles that are visible with the light microscope, and lacunae containing chondrocytes, then the tissue could be:
- a) hyaline cartilage
- b) elastic cartilage
- c) fibrocartilage
- d) reticular cartilage

d   15.   Which one of the following is not found in both bone and cartilage?
a)   lacunae
b)   collagen fibers
c)   matrix
d)   canaliculi

d   16.   The pleura are an example of:
a)   mucous membranes
b)   mucous acini
c)   serous acini
d)   serous membranes
e)   none of the above is correct

b   17.   Which one of the following is not a location of simple squamous epithelium?
a)   lining of blood vessels
b)   ducts of glands
c)   the alveoli
d)   Bowman's capsule

c   18.   The tissue that forms the surface layer of serous membranes is known as:
a)   endothelium
b)   epithelium
c)   mesothelium
d)   perithelium

d   19.   One of the most common kinds of cells found in connective tissue is:
a)   goblet cells
b)   mast cells
c)   neurons
d)   fibroblasts

c   20.   The kind of connective tissue found in fascia is:
a)   dense white irregular
b)   elastic
c)   dense white regular
d)   areolar

c   21.   If a tissue has lacunae but no canaliculi and lots of fibers of 2 different kinds, then the tissue is probably:
a)   hyaline cartilage
b)   fibrocartilage
c)   elastic cartilage
d)   can't tell

a   22.   If a tissue has striations and several flattened peripheral nuclei per
         cell, then the tissue could be:
         a)   skeletal muscle
         b)   smooth muscle
         c)   cardiac muscle
         d)   visceral muscle

c   23.   When a wound in the skin is healing, the new connective tissue that
         is formed comes from the:
         a)   epidermis
         b)   fibrin
         c)   fibroblasts
         d)   dermis

d   24.   Which of the following are included in a junctional complex?
         a)   desmosome
         b)   tight junction
         c)   intermediate junction
         d)   all of the above

b   25.   The thin epithelial membranes that line the ventral body cavities are
         called:
         a)   stratified squamous epithelium
         b)   serous membranes
         c)   dense regular connective tissue
         d)   mucous membranes

d 1. Which one of the following listed is pierced thirdly by the point of a pin entering the body from the outside?
   a) stratum lucidum
   b) stratum corneum
   c) stratum germinativum
   d) stratum granulosum

c 2. Which one of the following listed is not a characteristic or a function of the stratum corneum?
   a) protects the body
   b) helps to limit the loss of body water
   c) is made up of cells that are cuboidal in shape
   d) has the cytoplasm of its cells replaced with keratin

d 3. The bluish skin characteristic of cyanosis is due to the presence of:
   a) carotene
   b) melanin
   c) dilation of the capillaries in the skin
   d) unoxygenated hemoglobin

a 4. Which one of the following is not true of the dermis?
   a) its outer portion is called the reticular layer
   b) it contains elastic, reticular, and collagen fibers
   c) it contains specialized glands
   d) it contains specialized sense organs

b 5. Which one of the following is not a downgrowth of the epidermis into the dermis?
   a) sweat glands
   b) blood vessels
   c) hair follicles
   d) sebaceous glands

a 6. Which one of the following is not true of merocrine sweat glands?
   a) part of the cytoplasm of the secreting cell is included in the secretion
   b) they are found mostly all over the body except the nipples, hips, and portions of the skin of the genital organs
   c) they are coiled tubular glands
   d) they are stimulated by sympathetic nerves

b 7. Which one of the following is not true of sebaceous glands?
   a) their secretion is usually oily
   b) their secretion usually empties directly onto the surface of the skin
   c) their secretion contains substances toxic to bacteria
   d) they are usually simple alveolar types of glands

a   8.   Which one of the following structures listed would be pierced thirdly
         by the point of a pin entering a hair at right angles to its long axis?
         a)   medulla
         b)   cuticle
         c)   cortex
         d)   sebum

d   9.   The structure that contracts to cause "goose pimples" is the:
         a)   dermal papillae
         b)   roots
         c)   hair follicle
         d)   arrector pili

a   10. Which one of the following is not found at the proximal end of the
         finger nail?
         a)   hyponychium
         b)   lunula
         c)   nail matrix
         d)   eponychium

b   11. Which one of the following is not a function of the skin?
         a)   protection
         b)   production of vitamin C
         c)   regulation of body temperature
         d)   excretion

d   12. The organism that causes athlete's foot is a:
         a)   protozoa
         b)   virus
         c)   bacterium
         d)   fungus

b   13. Which one of the following is not true of acne?
         a)   it is most common during puberty
         b)   it is caused by a fungus
         c)   the causative organism stimulates excessive secretion of sebaceous
              glands
         d)   symptoms are pimples, blackheads, and dandruff

a   14. Which one of the following conditions is not caused by a virus?
         a)   moles
         b)   shingles
         c)   fever blisters
         d)   warts

c 15. A burn where there is damage to both the epidermis and dermis so that they can only regenerate from the edges of the wound would be classified as a _____ burn.
a) first degree
b) second degree
c) third degree
d) cannot be accurately determined

b 16. Which one of the following does not contain any blood vessels?
a) hypodermis
b) epidermis
c) reticular layer of the dermis
d) papillary layer of the dermis

b 17. The product of ceruminous glands is:
a) sweat
b) wax
c) sebum
d) oil

d 18. If a layer of the skin consists of flattened, closely packed cells with indistinct cell boundaries and inclusions of a substance known as eleiden, then the layer must be the:
a) stratum corneum
b) stratum granulosum
c) stratum basale
d) none of the above is correct

d 19. The pH of the skin tends to be:
a) less than 3.0
b) 6.8-7.4
c) greater than 7.4
d) none of the above is correct

b 20. Which one of the following is not a consequence of aging?
a) decrease in hair follicle activity
b) increased sebum production
c) reduced sweating
d) atrophy of melanocytes

c   1.   Which one of the following listed is pierced thirdly by the point of a
          pin entering the body from the outside?
          a)   stratum lucidum
          b)   hypodermis
          c)   dermis
          d)   stratum granulosum

b   2.   Which one of the following is not true of the stratum germinativum?
          a)   rests upon a basement membrane
          b)   has granules of keratohyalin in its cells
          c)   has cells linked by desmosomes
          d)   furnishes replacement cells for those lost from the more
                superficial layers

a   3.   Blushing is due to:
          a)   dilation of capillaries in the epidermis
          b)   an increase in the amount of melanin
          c)   an increase in the amount of carotene
          d)   a decrease in the amount of carotene

b   4.   Which one of the following is not true of the hypodermis?
          a)   it is composed of areolar connective tissue
          b)   it is part of the skin
          c)   it contains blood vessels
          d)   it contains nerve endings and fat cells

c   5.   Which set of the following glands is the most widely distributed?
          a)   ciliary and meibomian
          b)   ceruminous and sebaceous
          c)   sweat and sebaceous
          d)   sebaceous and ceruminous

b   6.   Which one of the following is not true of apocrine sweat glands?
          a)   they are located in the axilla, scrotum, and labia majora
          b)   their secretion is very similar to that of other kinds of
                sweat glands
          c)   in females, these glands become hyperactive in conjunction with
                the menstrual cycle
          d)   all of the above are true of apocrine sweat glands

a   7.   Which one of the following is not true of sebaceous glands?
          a)   their mode of secretion is usually apocrine
          b)   they produce a substance which prevents the skin and hair from
                drying out
          c)   they usually empty their secretion into a hair follicle
          d)   they become active during adolescence

d   8.   Which one of the following structures listed would be pierced thirdly by the point of a pin entering a hair follicle at right angles to its long axis?
a)   medulla
b)   internal root sheath
c)   connective tissue sheath
d)   cuticle

d   9.   Erection of body hair is due to:
a)   a structure in the epidermis
b)   a structure that runs from the connective tissue covering a hair follicle to the reticular layer of the dermis
c)   a structure made of skeletal muscle
d)   none of the above is correct

c   10.   Dirt under the finger nails is between the:
a)   lunula and nail
b)   eponychium and nail
c)   nail and hyponychium
d)   matrix and the nail bed

c   11.   Which one of the following is not a function of the skin?
a)   protection
b)   sensation
c)   absorption
d)   maintain homeostasis

a   12.   The most abundant kind of organism on the skin are the:
a)   bacteria
b)   fungi
c)   protozoa
d)   viruses

d   13.   Which one of the following is not true of warts?
a)   the causative organism is a virus
b)   they may spread
c)   they are most common in adolescents and young adults
d)   plantar warts occur on the palm of the hand

d   14.   Chicken pox and _____ are caused by the same organism.
a)   eczema
b)   impetigo
c)   cold sores
d)   shingles

e   15.   If a person had burned the upper right limb. then according to the rule of nines, _____ percent of his or her body would have been burned.
- a)   18%
- b)   9%
- c)   1%
- d)   27%

a   16.   Which one of the following does not contain any blood vessels?
- a)   epithelial portion of hair follicle
- b)   hypodermis
- c)   reticular layer of the dermis
- d)   papillary layer of the dermis

b   17.   The secretion of sweat glands contains all but which one of the following substances?
- a)   urea
- b)   lipids
- c)   sodium chloride
- d)   phosphates
- e)   all of the above are secretions of sweat glands

c   18.   If a layer of the skin is made up of spindle shaped cells containing keratohyalin, then the layer is the:
- a)   stratum spinosum
- b)   stratum lucidum
- c)   stratum granulosum
- d)   none of the above is correct

d   19.   Which one of the following is not correct?
- a)   the pH of the skin gives it an ability to retard the growth of microbes
- b)   the pH of the skin is between 4.0 and 6.8
- c)   when subjected to repeated trauma the skin becomes thickened
- d)   the epidermis contains several kinds of sensory receptors

d   20.   Which one of the following is not a consequence of long exposure to the sun?
- a)   cancers
- b)   marked wrinkling
- c)   development of abnormal kinds of collagen
- d)   all of the above are effects of long term exposure to the sun's rays

NAME_____

SECTION _____

c 1.   Which one of the following is not a function of the skeleton?
        a)   support
        b)   act as a mineral reservoir
        c)   hemostasis
        d)   movement

a 2.   Which one of the following is not a flat bone?
        a)   tarsal
        b)   sternum
        c)   rib
        d)   parietal bone

a 3.   Which one of the following is not a facial bone?
        a)   ethmoid
        b)   zygomatic
        c)   maxillary
        d)   nasal

b 4.   Which one of the following is not part of the axial skeleton?
        a)   rib
        b)   clavicle
        c)   skull
        d)   vertebra

b 5.   Which one of the following listed would be pierced thirdly by the
        point of a pin entering the diaphysis of a long bone?
        a)   periosteum
        b)   endosteum
        c)   yellow bone marrow
        d)   compact bone

d 6.   The spaces in compact bone that contain bone cells are called:
        a)   diploe
        b)   lamellae
        c)   canaliculi
        d)   lacunae

b 7.   Which one of the following does not contain a blood vessel?
        a)   Volkmann's canals
        b)   canaliculi
        c)   Haversian canals
        d)   nutrient foramen

a   8.   The tensile strength of bone is due to:
    a)   collagen fibers
    b)   calcium
    c)   phosphate
    d)   none of the above is correct

b   9.   Which one of the following bones is not formed by intramembranous ossification?
    a)   parietal
    b)   rib
    c)   sternum
    d)   all of the above are formed by intramembranous ossification

d   10.   A fracture where the broken ends of the bone do not penetrate the skin is known as a _____ fracture.
    a)   impacted
    b)   depressed
    c)   comminuted
    d)   simple

c   11.   Deposition of calcium in tissues not usually calcified is known as:
    a)   osteoporosis
    b)   osteomyelitis
    c)   metastatic calcification
    d)   spina bifida

b   12.   If excessive growth hormone is secreted after the epiphyseal cartilages have been replaced by bone, a condition known as _____ may result.
    a)   pituitary giant
    b)   acromegaly
    c)   achrondroplasia
    d)   pituitary dwarf

b   13.   In women the amount of calcium in the bones steadily decreases after _____ years of age.
    a)   35
    b)   40
    c)   45
    d)   55

a   14.   Which one of the following listed occurs thirdly in endochondral ossification?
    a)   primary ossification center
    b)   perichondrium
    c)   periosteum
    d)   epiphyseal plate formation

b   15.   Cells which reabsorb previously deposited bone are known as _____
           and are stimulated by _____.
           a)   osteoprogenitor cells, calcitonin
           b)   osteoclasts, parathormone
           c)   osteocytes, growth hormone
           d)   none of the above is correct

b   16.   Increase in bone length is due to changes in the:
           a)   epiphyseal line
           b)   epiphyseal plate
           c)   perichondrium
           d)   periosteum

a   17.   Which one of the following listed occurs thirdly in the
           intramembranous ossification of bone?
           a)   trabeculae formed
           b)   osteoblast development
           c)   formation of collagen fibers
           d)   development of outer layer of compact bone

b   18.   Which one of the following listed occurs fourthly in the repair of a
           simple fracture?
           a)   replacement of procallus
           b)   production of spongy bone
           c)   fibroblast invasion
           d)   formation of compact bone
           e)   reabsorption of external bony callus

d   19.   Which one of the following is not a part of the inorganic part of the
           bone?
           a)   calcium
           b)   phosphate
           c)   hydroxyapatite
           d)   chondroitin

a   20.   Which one of the following substances is not necessary for production
           of new bone tissue?
           a)   vitamin C
           b)   calcium
           c)   phosphorus
           d)   calcitonin
           e)   all of the above are necessary for the production of new bone
                 tissue

b   21.   A rounded prominence that articulates with another is a:
           a)   fossa
           b)   condyle
           c)   tubercle
           d)   tuberosity

a   22. Which one of the following bones does not make up part of the
         calvarium?
         a)   temporal
         b)   frontal
         c)   occipital
         d)   parietal

c   23. Which one of the following bones does not contribute to the floor of
         the cranial cavity?
         a)   frontal
         b)   ethmoid
         c)   inferior nasal concha
         d)   sphenoid
         e)   occipital

e   24. If a bone has features such as the sella turcica, optic foramina, and
         pterygoid processes, then the bone is known as _____
         bone.
         a)   maxillary
         b)   ethmoid
         c)   temporal
         d)   frontal
         e)   none of the above is correct

c   25. Which one of the following bones does not contribute to part of the
         orbit of the eye?
         a)   lacrimal
         b)   sphenoid
         c)   nasal
         d)   palantine
         e)   frontal

e   26. The bone that forms the posterior inferior portion of the nasal septum
         is the:
         a)   ethmoid
         b)   nasal
         c)   sphenoid
         d)   palatine
         e)   vomer

e   27. Which one of the following bones does not articulate with the palatine
         bones?
         a)   maxillary
         b)   inferior nasal conchae
         c)   ethmoid
         d)   sphenoid
         e)   all of the above articulate with the palatine bone

b   28.   If a vertebra has transverse foramen and a bifurcated spinous process,
          then it is a _____ vertebra.
          a)   thoracic
          b)   cervical
          c)   lumbar
          d)   sacral
          e)   coccygeal

d   29.   A typical vertebra has an anterior portion known as the:
          a)   lamina
          b)   pedicle
          c)   spinous process
          d)   body

d   30.   The median end of the clavicle articulates with the:
          a)   xiphoid process
          b)   scapula
          c)   first rib
          d)   manubrium

d   31.   The head of a rib articulates with the _____ of a thoracic
          vertebra.
          a)   costotransverse facet
          b)   superior articular facet
          c)   inferior articular facet
          d)   demifacet
          e)   none of the above is correct

b   32.   Which one of the following features is not on the same bone as the
          others?
          a)   glenoid fossa
          b)   deltoid tuberosity
          c)   acromion process
          d)   infraspinous fossa

e   33.   The capitulum articulates with the:
          a)   trochlea
          b)   ulna
          c)   semilunar notch
          d)   olecranon fossa
          e)   none of the above is correct

a   34.   The proximal row of wrist bones has the _____ as the one that
          is third from the lateral side.
          a)   triquetrum
          b)   scaphoid
          c)   lunate
          d)   pisiform

d   35.   After sitting for a long time on a hard chair, the bone feature that starts to hurt is the:
- a)   inferior ramus of the pubis
- b)   ischial spine
- c)   posterior inferior iliac spine
- d)   ischial tuberosity

b   36.   Which one of the following is not a characteristic of the male pelvis?
- a)   more massive
- b)   acute obtuse pubic angle
- c)   oval obturator foramen
- d)   pelvic inlet heart shaped
- e)   acetabulum faces laterally

a   37.   Which one of the following features is not found on the posterior surface of the femur?
- a)   patellar surface
- b)   linea aspera
- c)   intercondylar fossa
- d)   gluteal tuberosity

b   38.   The distal end of the tibia has a downward projection on its inward side known as the:
- a)   tibial tuberosity
- b)   medial malleolus
- c)   intercondylar eminence
- d)   medial condyle
- e)   anterior crest

e   39.   The heel bone is the:
- a)   cuboid
- b)   talus
- c)   navicular
- d)   cuneiform
- e)   none of the above is correct

e   40.   The space bounded anteriorly by the symphysis pubis and posteriorly by the coccyx, ischial spines, and tuberosities is known as the:
- a)   pelvic inlet
- b)   false pelvis
- c)   true pelvis
- d)   pelvic brim
- e)   pelvic outlet

NAME_____

SECTION _____

e   1.   Which one of the following is not a function of the skeleton?
          a)   movement
          b)   hemopoiesis
          c)   support
          d)   mineral reservoir
          e)   all of the above are functions of the skeleton

a   2.   Which one of the following is not a long bone?
          a)   tarsal
          b)   phalange
          c)   ulna
          d)   fibula

d   3.   Which one of the following is not a cranial bone?
          a)   parietal
          b)   occipital
          c)   sphenoid
          d)   inferior nasal concha

b   4.   Which one of the following is not part of the appendicular skeleton?
          a)   scapula
          b)   ribs
          c)   ilium
          d)   tibia

a   5.   Which one of the following is not an irregular bone?
          a)   carpal
          b)   vertebra
          c)   scapula
          d)   ischium

a   6.   Which one of the following listed would be pierced thirdly by the
          point of a pin entering an epiphysis?
          a)   spongy bone
          b)   articular cartilage
          c)   compact bone
          d)   yellow bone marrow

c   7.   Which structure contains all the others?
          a)   canaliculi
          b)   lacuna
          c)   osteon
          d)   lamella

d   8.   Which one of the following is not found in both spongy and compact bone?
- a) osteocytes
- b) lacunae
- c) canaliculi
- d) concentric lamellae

c   9.   The compressional strength of bone is due to:
- a) elastic fibers
- b) collagen fibers
- c) salts of calcium and phosphate
- d) none of the above is correct

c   10.   Which one of the following bones is not formed by endochondral ossification?
- a) tibia
- b) ulna
- c) frontal
- d) femur

b   11.   A fracture where the bone is splintered is known as a _____ fracture.
- a) compound
- b) comminuted
- c) simple
- d) impacted

a   12.   The condition in adults where the rate of bone absorption remains normal but there is a gradual reduction in the rate of bone formation is known as _____.
- a) osteoporosis
- b) rickets
- c) osteomalacia
- d) osteomyelitis

d   13.   If under the influence of hereditary factors as well as hormonal levels the epiphyseal cartilages function for only a short time, the result may be a person with short arms and legs but a normal trunk and head. This condition is known as:
- a) pituitary dwarf
- b) acromegaly
- c) osteomalacia
- d) achondroplasia

d 14. In men the amount of calcium in the bones steadily decreases after
        _____ years of age.
        a) 40
        b) 50
        c) 55
        d) 60

a 15. Which one of the following listed occurs thirdly in endochondral
        ossification?
        a) secondary ossification center
        b) hyaline cartilage template
        c) formation of periosteum
        d) epiphyseal line

a 16. Cells which form new deposits of bone are known as _____
        and are stimulated by _____.
        a) osteoblasts, calcitonin
        b) osteoclasts, parathormone
        c) osteoblasts, parathormone
        d) osteoclasts, calcitonin

c 17. Increase in the thickness of a long bone is due to changes in the:
        a) epiphyseal plate
        b) epiphyseal line
        c) periosteum
        d) perichondrium

b 18. Which one of the following listed occurs thirdly in intramembranous
        ossification of bone?
        a) osteoprogenitor cells
        b) trabeculae
        c) increased activity by peripheral osteoblasts
        d) fibroblast activity

d 19. Which one of the following listed is not part of the organic material
        in bone?
        a) collagen fibers
        b) chondroitin
        c) glycoproteins
        d) hydroxyapatite

b 20. Which one of the following listed occurs thirdly in the repair of a
        simple fracture?
        a) reabsorbing the external bony callus
        b) formation of fibrocartilaginous callus
        c) procallus formation
        d) fibroblast invasion

e   21.   A sharp prominent bony ridge is known as a:
         a)   epicondyle
         b)   line
         c)   process
         d)   condyle
         e)   none of the above is correct

b   22.   Which one of the following bones does not contribute to the roof of the skull?
         a)   frontal
         b)   temporal
         c)   occipital
         d)   parietal

c   23.   The suture that separates the two parietal bones is known as the:
         a)   lambdoidal
         b)   squamosal
         c)   sagittal
         d)   coronal
         e)   frontal

d   24.   If a bone has features such as the cribiform plate and crista galli, then the bone is known as the:
         a)   sphenoid
         b)   zygomatic
         c)   pterygoid
         d)   ethmoid
         e)   none of the above is correct

e   25.   Which one of the following bones does not contribute to part of the eye orbit?
         a)   zygomatic
         b)   frontal
         c)   palatine
         d)   maxilla
         e)   all of the above contribute to the eye orbit

e   26.   Which one of the following does not contribute to the nasal cavity?
         a)   ethmoid
         b)   vomer
         c)   inferior nasal conchae
         d)   maxillary
         e)   all of the above contribute to the nasal cavity

d   27.   If a bone has features such as the mastoid process, external auditory meatus, and a zygomatic process, then the bone is the:
  a) zygomatic
  b) maxillary
  c) sphenoid
  d) temporal
  e) ethmoid

b   28.   If a vertebra has demifacets and long, slender spinous processes, then the vertebra would be a:
  a) cervical
  b) thoracic
  c) sacral
  d) lumbar
  e) coccygeal

d   29.   The portion of the neural arch between the centrum and the transverse process is known as the:
  a) lamina
  b) spinous process
  c) body
  d) pedicle

b   30.   The lateral end of the clavicle articulates with the:
  a) manubrium
  b) acromion process
  c) spine of the scapula
  d) coracoid process

b   31.   The tubercle of a rib articulates with the:
  a) demifacet
  b) costotransverse facet
  c) superior articular facet
  d) inferior articular facet
  e) none of the above is correct

e   32.   Which one of the following features is not on the same bone as the others?
  a) glenoid fossa
  b) supraspinous fossa
  c) axillary border
  d) inferior angle
  e) all of the above are features of the same bone

a   33. The trochlea articulates with the:
          a)   semilunar notch
          b)   head of the radius
          c)   radial notch
          d)   capitulum
          e)   olecranon process

b   34. The distal row of wrist bones has the _____ as the one that is
          third from the lateral side.
          a)   trapezium
          b)   capitate
          c)   trapezoid
          d)   hamate

e   35. The ischial ramus articulates with the:
          a)   pubic symphysis
          b)   acetabulum
          c)   superior ramus of the pubis
          d)   iliac fossa
          e)   inferior ramus of the pubis

b   36. Which one of the following is not a characteristic of the female
          pelvis?
          a)   pelvic inlet circular
          b)   obturator foramen oval
          c)   anterior iliac spines more widely separated
          d)   acetabulum faces more anteriorly

d   37. Which one of the following features is not found on the posterior
          surface of the femur?
          a)   intertrochanteric crest
          b)   gluteal tuberosity
          c)   linea aspera
          d)   intercondylar eminence

c   38. The distal end of the fibula is flattened to form the:
          a)   head
          b)   styloid process
          c)   lateral malleolus
          d)   medial malleolus

d   39. The tarsal bone that receives the entire weight of the body is known
          as the:
          a)   cuboid
          b)   navicular
          c)   cuneiform
          d)   talus
          e)   calcaneus

b   40. The space bounded by the sacrum and the two os coxe is known as
          the:
          a)   false pelvic cavity
          b)   true pelvic cavity
          c)   pelvic inlet
          d)   pelvic outlet

d   1.   Which one of the following is not true of synarthroses?
         a)   they permit no true movement even though they may give a little
         b)   the bones in such a joint are held together by fibrous
              connective tissue
         c)   a suture is a specific example of such a joint
         d)   a synchondrosis is an example of such a joint

a   2.   Which one of the following is not a distinguishing feature of all
         diarthroses?
         a)   articular discs
         b)   synovial membrane
         c)   articular capsule
         d)   synovial fluid

c   3.   Which one of the following is an example of a syndesmosis?
         a)   knee joint
         b)   joint between bones of the skull
         c)   joint between the distal ends of the tibia and fibula
         d)   the joint between the two pubic bones

a   4.   Ligaments join:
         a)   bone to bone
         b)   muscle to bone
         c)   periosteum to bone
         d)   muscle to periosteum

b   5.   Cylindrical synovial sacs are known as:
         a)   articular capsule
         b)   tendon sheaths
         c)   synovial joints
         d)   none of the above is correct

d   6.   Which one of the following is not an angular movement?
         a)   flexion
         b)   abduction
         c)   extension
         d)   pronation

d   7.   When a bone is moved in an anterior-posterior plan in a way that
         decreases the angle between it and the adjoining bone, the movement
         is known as:
         a)   elevation
         b)   abduction
         c)   pronation
         d)   none of the above is correct

b   8. In the anatomical position the fingers are said to be:
        a) flexed
        b) adducted
        c) circumducted
        d) pronated

c   9. Circumduction is said to be a sequential combination of several
        movements. Which one of the following is not one of those
        movements?
        a) abduction
        b) extension
        c) elevation
        d) flexion

a   10. Lowering the mandible is an example of:
        a) depression
        b) eversion
        c) protraction
        d) pronation

a   11. Sticking out your tongue is an example of:
        a) eversion
        b) inversion
        c) elevation
        d) protraction

d   12. An example of a gliding joint would be:
        a) between the phalanges
        b) the elbow joint
        c) the metacarpophalangeal joint
        d) between the carpals

c   13. Which one of the following is not an example of a hinge joint?
        a) the elbow joint
        b) between the phalanges
        c) the mandible with the skull
        d) the knee joint

a   14. Which one of the following motions is not possible around a biaxial
        joint?
        a) axial rotation
        b) flexion
        c) adduction
        d) circumduction

c 15. Twisting or overstretching a joint causing ligaments to tear will result in a disorder known as:
- a) dislocation
- b) bursitis
- c) sprain
- d) gout

d 16. Which one of the following is not an example of a biaxial joint?
- a) radius and carpals
- b) occipital condyles on the atlas
- c) carpometacarpal joint of the thumb
- d) all of the above are examples of biaxial joints

e 17. Which one of the following is not an example of a uniaxial joint?
- a) elbow joint
- b) atlas around the odontoid process
- c) proximal articulation between the radius and the ulna
- d) between the phalanges
- e) all of the above are examples of uniaxial joints

b 18. The joint between the tibia and the femur is an example of a _____ joint.
- a) diarthrosis (gliding)
- b) diarthrosis (hinge)
- c) diarthrosis (condyloid)
- d) diarthrosis (spheroid)
- e) none of the above is correct

c 19. The joint between the ribs and the sternum is an example of a ____ joint.
- a) diarthrosis (gliding)
- b) amphiarthrosis (symphysis)
- c) amphiarthrosis (synchondrosis)
- d) diarthrosis (hinge)
- e) none of the above is correct

a 20. The joint between the articular surfaces of the vertebrae is an example of a _____ joint.
- a) diarthrosis (gliding)
- b) amphiarthrosis (symphysis)
- c) diarthrosis (hinge)
- d) amphiarthrosis (synchondrosis)
- e) none of the above is correct

a 21. Which of the following is associated with the shoulder joint?
- a) coracohumeral ligament
- b) ligamentum teres
- c) popliteal ligament
- d) both a and b

d   22. Which of the following is associated with the knee joint?
a)   puboferoral ligament
b)   medial collateral ligament
c)   oblique popliteal ligament
d)   both b and c

b   1.   Which one of the following is not true of amphiarthroses?
           a)   slight movement is possible
           b)   a syndesmosis is an example of such a joint
           c)   a symphysis is an example of such a joint
           d)   in such a joint cartilage unites the bones

d   2.   Which one of the following is not a distinguishing feature of all
           synovial joints?
           a)   articular
           b)   synovial membrane
           c)   synovial fluid
           d)   articular disks

a   3.   Which one of the following is an example of a synchondrosis?
           a)   between the ribs and the sternum
           b)   between the two halves of the mandible
           c)   the pad between the vertebrae
           d)   the joint between the two pubic bones

b   4.   Tendons join:
           a)   bone to bone
           b)   muscle to bone
           c)   periosteum to bone
           d)   none of the above is correct

c   5.   Bursae:
           a)   wrap around tendons
           b)   usually are located between the bone and the skin
           c)   usually are located between the tendon and the bone
           d)   do not contain a synovial membrane

a   6.   Which one of the following is not a circular movement?
           a)   adduction
           b)   pronation
           c)   rotation
           d)   supination

c   7.   Bringing the limbs posteriorly beyond the plane of the body is known as:
           a)   depression
           b)   adduction
           c)   hyperextension
           d)   none of the above is correct

c   8.   In the anatomical position the hands are said to be:
          a)   flexed
          b)   abducted
          c)   supinated
          d)   pronated

d   9.   The joint motion known as _____ delineates a cone-shaped movement.
          a)   pronation
          b)   depression
          c)   rotation
          d)   circumduction

c   10.   Shrugging the shoulders is an example of:
          a)   inversion
          b)   retraction
          c)   elevation
          d)   supination

b   11.   Twisting the foot so that the sole of the foot faces outward with its outer margin raised is an example of:
          a)   retraction
          b)   eversion
          c)   elevation
          d)   rotation

b   12.   An example of a gliding joint would be:
          a)   the elbow
          b)   between the articular processes of the vertebrae
          c)   between the phalanges
          d)   the metacarpophalangeal joint

b   13.   Which one of the following is not an example of a uniaxial joint?
          a)   between the radius and the carpals
          b)   the carpometacarpal joint of the thumb
          c)   the occipital condyles on the atlas
          d)   the metatarsophalangeal joint

b   14.   Which one of the following is an example of a saddle joint?
          a)   between the radius and the carpals
          b)   the carpometacarpal joint of the thumb
          c)   the occipital condyles on the atlas
          d)   the metatarsophalangeal joint

c   15.   Two conditions that affect women more than men are:
          a)   gout and degeneration of the intervertebral disks
          b)   rheumatoid arthritis and osteoarthritis
          c)   Heberden's nodes and rheumatoid arthritis
          d)   osteoarthritis and gout

b 16. Which one of the following is an example of a saddle joint?
a) articulation between radius and the carpals
b) the carpometacarpal joint of the thumb
c) the occipital condyles on the atlas
d) the metatarsophalangeal joints
e) none of the above is an example of such a joint

a 17. Which one of the following is an example of a pivot joint?
a) the proximal articulation between the radius and the ulna
b) the interphalangeal joints
c) the mandible with the temporal bone
d) the tibia with the fibula distally
e) none of the above is correct

a 18. The joint between the mid-radius and the ulna (interosseous membrane) is an example of a:
a) synarthrosis (syndesmosis)
b) synarthrosis (suture)
c) synarthrosis (synchondrosis)
d) none of the above is correct

d 19. The joint between the tarsals is an example of a:
a) diarthrosis (hinge)
b) diarthrosis (condyloid)
c) diarthrosis (spheroid)
d) diarthrosis (gliding)
e) none of the above is correct

c 20. The joint between the proximal tibia and fibula is an example of a _____ joint.
a) hinge
b) condyloid
c) gliding
d) spheroid
e) none of the above is correct

d 21. Which of the following is associated with the elbow joint?
a) articular capsule
b) ulnar collateral ligament
c) radial collateral ligament
d) all of the above

d 22. Which of the following is associated with the hip joint?
a) ischiofemoral ligament
b) menisci
c) cruciate ligaments
d) glenoid labrum

a   1.   Which one of the following is not a characteristic of skeletal muscle?
           a)   not striated
           b)   voluntary
           c)   controlled by the somatic nervous system
           d)   all of the above are characteristics of skeletal muscle

a   2.   The connective tissue sheath around an entire muscle is known as:
           a)   epimysium
           b)   perimysium
           c)   endomysium
           d)   fascia

b   3.   The type of attachment where a muscle has its outer covering
           attached to the periosteum of the bone is known as:
           a)   attachment via a tendon
           b)   direct attachment
           c)   attachment via an aponeurosis
           d)   attachment by a ligament

b   4.   Which one of the following is least related to the others?
           a)   origin
           b)   belly
           c)   proximal
           d)   less movable end

d   5.   A muscle that has a radiate pattern of fibers is the:
           a)   sartorius
           b)   deltoid
           c)   tibialis anterior
           d)   pectoralis

b   6.   Starting with the largest listed below, which one of the following is
           the third largest in size?
           a)   fiber
           b)   sarcomere
           c)   myofibril
           d)   I band

b   7.   The I band is made up entirely of:
           a)   thick filaments
           b)   thin filaments
           c)   fibers
           d)   myofibrils

d   8.   Which one of the following makes up the thick filaments?
          a)   tropomyosin
          b)   actin
          c)   troponin
          d)   myosin

c   9.   Lateral sacs are part of the:
          a)   plasma membrane
          b)   sarcomere
          c)   triad
          d)   T tubules

c   10.   Which one of the following events listed below occurs thirdly in the contraction of a muscle cell?
          a)   spread of a stimulatory impulse via the T tubule
          b)   excitation of a myoneural junction
          c)   myosin links with actin
          d)   ATP breaks down

a   11.   The substance that serves essentially as a quickly available energy reserve when muscles are actively contracting is:
          a)   creatine phosphate
          b)   glucose
          c)   ADP
          d)   lactic acid

b   12.   The muscle whose contraction opposes a particular movement is known as a/an:
          a)   fixator
          b)   antagonist
          c)   prime mover
          d)   synergist

c   13.   Flexion of the elbow is an example of a class _____ lever system.
          a)   I
          b)   II
          c)   III
          d)   IV

d   14.   A rare chronic condition characterized by muscle weakness and believed to involve an immunological block of the myoneural junction is known as:
          a)   muscular dystrophy
          b)   muscular atrophy
          c)   muscular hypertrophy
          d)   myasthenia gravis

e   15.   A cross section of a myofibril reveals that each thick filament is
          surrounded by _____ thin filaments.
          a)   2
          b)   3
          c)   4
          d)   5
          e)   6

a   16.   The type of muscle contraction in which tension develops but no
          appreciable shortening occurs is _____.
          a)   isometric
          b)   isotonic
          c)   isosmotic
          d)   hypotonic

c   17.   Calcium ions released from lateral sacs bond to:
          a)   G actin
          b)   tropomyosin
          c)   troponin
          d)   myosin

c   18.   The molecule released by the end of the motor neuron at the
          myoneural junction is known as:
          a)   ATP
          b)   calcium ions
          c)   acetylcholine
          d)   glucose

b   19.   Myosin can bond with:
          a)   troponin
          b)   G actin
          c)   ADP
          d)   two of the above are correct

e   20.   Two substances are needed for the contraction of a muscle cell. They
          are ATP and:
          a)   potassium ion
          b)   sodium ion
          c)   glucose
          d)   oxygen
          e)   none of the above is correct

c   21.   Elevated respiration after intense physical activity is necessary to
          provide oxygen for all but which one of the following?
          a)   production of ATP
          b)   conversion of lactic acid to glucose
          c)   resynthesis of creatine
          d)   the oxidative breakdown of glucose

c 22. The processes associated with excitation-contraction coupling take
        place during the _____.
        a) period of relaxation
        b) period of contraction
        c) latent period
        d) refractory period

e 23. An inability of a muscle to contract when stimulated by a threshold or
        greater stimulus is known as:
        a) oxygen debt
        b) psychological fatigue
        c) contracture
        d) asynchronous motor unit summation
        e) none of the above is correct

c 24. Which one of the following is not true of multi-unit smooth muscle?
        a) found in large arteries
        b) each cell responds independently to stimulation
        c) self-excitable
        d) not responsive to stretch

d 25. Which one of the following means the same as spatial summation?
        a) wave summation
        b) tetanus
        c) asynchronous motor unit summation
        d) multiple motor unit summation

   26. A sustained contraction of a muscle is due to:
        a) multiple motor unit summation
        b) a muscle twitch
        c) incomplete tetanus
        d) asynchronous motor unit summation

b 27. The greater the starting length of the muscle, the ____ the force of
        contraction.
        a) weaker
        b) stronger
        c) there is no relationship between muscle length and force of
            contraction

a 28. Which one of the following is not a characteristic of fast
        twitch fatigable fibers?
        a) well supplied with blood vessels
        b) contain much glycogen
        c) geared to anaerobic metabolic processes
        d) do not contain large amounts of myoglobin

b   29.   Which one of the following is not true of skeletal muscle as compared to smooth muscle? Skeletal muscle fibers can:
        a)   generate less tension under stretching than smooth muscle
        b)   shorten far more than smooth muscle
        c)   contract more rapidly
        d)   stretch less before exhibiting marked changes in tension

d   30.   Which one of the following is not true of both smooth and skeletal muscle physiology?
        a)   have thick filaments of myosin
        b)   thin filaments have actin and tropomyosin
        c)   use an active transport pump to put the calcium ions back in their storage site so that the fibers can relax
        d)   have the calcium bond to troponin molecules

d   1.   Which one of the following is not a characteristic of visceral muscle?
          a)   involuntary
          b)   under control of the autonomic nervous system
          c)   associated with internal organs
          d)   striated

c   2.   The connective tissue sheath around a fasciculus is known as:
          a)   epimysium
          b)   endomysium
          c)   perimysium
          d)   fascia

c   3.   When a muscle is joined to a bone via a cable-like extension of its
          connective tissue sheath, one calls that attachment _____.
          a)   direct
          b)   by an aponeurosis
          c)   by a tendon
          d)   by a ligament

d   4.   Which one of the following is least related to the others?
          a)   insertion
          b)   distal
          c)   more movable end
          d)   the place where the muscle arises

c   5.   A muscle that has a longitudinal pattern of fibers is the:
          a)   vastus medialis
          b)   deltoid
          c)   sartorius
          d)   gluteus maximus

b   6.   Which one of the following listed is the smallest in size?
          a)   fasciculus
          b)   filament
          c)   fiber
          d)   myofibril

a   7.   The length of the A band is determined by the length of the:
          a)   thick filaments
          b)   thin filaments
          c)   H zones
          d)   Z lines

c   8.   Which one of the following is not part of a thin filament?
a)   actin
b)   tropomyosin
c)   myosin
d)   troponin

a   9.   T tubules are part of the:
a)   plasma membrane
b)   lateral sacs
c)   sarcoplasmic reticulum
d)   sarcomere

d   10.   Which one of the following events listed occurs thirdly in the contraction of a muscle cell?
a)   release of calcium ions
b)   myosin head swivels
c)   acetylcholine released
d)   tropomyosin moves from its blocking position

d   11.   The energy source that is directly responsible for muscle cell contraction is:
a)   protein
b)   glucose
c)   glycogen
d)   ATP

c   12.   The muscle that is mostly responsible for a particular movement is known as the:
a)   synergist
b)   antagonist
c)   prime mover
d)   fixator

b   13.   Raising the body on its toes is an example of a class ____ lever system.
a)   I
b)   II
c)   III
d)   IV

b   14.   A group of genetically transmitted diseases characterized by progressive muscular weakness is known as:
a)   myasthenia gravis
b)   muscular dystrophy
c)   muscular atrophy
d)   muscular hypertrophy

b   15.   Each thin filament is surrounded by _____ thick filaments.
- a) 2
- b) 3
- c) 5
- d) 6
- e) none of the above is correct

b   16.   The type of muscle contraction in which there is an appreciable shortening is known as _____.
- a) isometric
- b) isotonic
- c) isomotic
- d) hypertonic

e   17.   Myosin heads bond to _____.
- a) calcium ions
- b) ATP
- c) troponin
- d) G actin
- e) two of the above are correct

d   18.   The substance which binds to receptor sites on the sarcolemma to cause an action potential is:
- a) sodium ion
- b) calcium ion
- c) potassium ion
- d) none of the above is correct

b   19.   In a relaxed muscle, calcium ions are stored in the ___.
- a) T tubules
- b) lateral sacs
- c) sarcomere
- d) troponins

e   20.   In order for muscle contraction to occur, two substances are necessary: calcium ions and _____.
- a) DNA
- b) creatine phosphate
- c) glucose
- d) ADP
- e) none of the above is correct

b   21.   After periods of intense physical exercise, the body converts lactic acid back to:
- a) creatine phosphate
- b) glucose
- c) ATP
- d) creatine

a   22. A brief response of a muscle to a single stimulus of at least threshold intensity is called a _____.
         a)   muscle twitch
         b)   multiple motor unit summation
         c)   wave summation
         d)   tetanus

a   23. Which one of the following does not contribute to muscle fatigue?
         a)   depletion of ADP
         b)   buildup of lactic acid
         c)   changes in pH
         d)   lack of oxygen

e   24. Which one of the following is not true of single-unit smooth muscle?
         a)   connected by gap junctions
         b)   sensitive to stretch
         c)   self-excitable
         d)   found in the uterus
         e)   all of the above are true

d   25. Which one of the following is not true of asynchronous motor unit summation?
         a)   produces a smooth contraction of the muscle as a whole
         b)   not all of the motor units contract at once
         c)   motor units are alternately stimulated
         d)   maximum muscle tension is produced

b   26. The inclusion of more muscle fibers into the contractile state in order to lift a heavy weight is physiologically termed _____.
         a)   tetanus
         b)   spatial summation
         c)   tetanus
         d)   isometric contraction

b   27. The greater the load, the _____ the speed of shortening.
         a)   faster
         b)   slower
         c)   there is no relationship between the load on a muscle and the speed of shortening

c   28. Which one of the following is not a characteristic of slow twitch, fatigue-resistant fibers?
         a)   high capacity for oxidative metabolism
         b)   contain much myoglobin
         c)   not well supplied by blood vessels
         d)   can meet their needs for energy by metabolizing glucose

d   29.   Which one of the following is not true of smooth muscle fibers as compared to skeletal muscle fibers? Smooth muscle fibers can:
- a) generate tension under greater stretching than skeletal muscle
- b) shorten far more than skeletal muscle
- c) contract more slowly
- d) stretch less before exhibiting and marked changes in tension

b   30.   Which one of the following is not true of smooth muscle as compared to skeletal muscle? In smooth muscle:
- a) calcium ions are stored in the endoplasmic reticulum
- b) calcium ions when released bond to troponin molecules
- c) calcium ions can get inside the fiber from the extra-cellular fluid
- d) the thin fibers contain actin and tropomyosin

b   1.   Which one of the following muscles is involved in chewing?
         a)   buccinator
         b)   masseter
         c)   corrugator
         d)   obicularis oculi

a   2.   Which one of the following muscles does not have its insertion on the
         humerus?
         a)   biceps brachii
         b)   deltoid
         c)   latissimus dorsi
         d)   pectoralis major

c   3.   Which one of the following muscles does not help to extend the
         vertebral column?
         a)   multifidus
         b)   rotatores
         c)   scalenes
         d)   interspinales

a   4.   Which one of the following muscles does not have its insertion on the
         linea alba?
         a)   rectus abdominis
         b)   external oblique
         c)   internal oblique
         d)   transversus abdominis

c   5.   Which one of the following muscles has its insertion on the lateral
         third of the clavicle, the acromion process, and the spine of the
         scapula?
         a)   rhomboideus major
         b)   pectoralis major
         c)   trapezius
         d)   serratus anterior

a   6.   The muscle that has its origin on the medial half of the clavicle, the
         sternum, the costal cartilages of the upper six ribs, and the
         aponeurosis of the external oblique is the:
         a)   pectoralis major
         b)   pectoralis minor
         c)   trapezius
         d)   internal oblique

b   7.   Which one of the, following muscles does not have at least part of its insertion on the greater tubercle of the humerus?
- a) pectoralis major
- b) deltoid
- c) supraspinatus
- d) infraspinatus

d   8.   Which one of the following muscles does not help in the flexion of the arm?
- a) deltoid
- b) pectoralis major
- c) coracobrachialis
- d) teres major

c   9.   Which one of the following does not aid in flexing the forearm?
- a) biceps brachii
- b) brachialis
- c) anconeus
- d) brachioradialis

b   10.   Which one of the following muscles does not help to abduct the wrist?
- a) extensor pollicis brevis
- b) extensor carpi ulnaris
- c) flexor carpi radialis
- d) extensor carpi radialis brevis

a   11.   Which one of the following muscles is not part of the hamstrings?
- a) rectus femoris
- b) semitendinosus
- c) semimembranosus
- d) biceps femoris

b   12.   Which one of the following muscles does not aid in flexion of the thigh?
- a) sartorius
- b) gracilis
- c) rectus femoris
- d) tensor fasciae lata

d   13.   Which one of the following muscles does not assist in lateral rotation of the thigh?
- a) gluteus maximus
- b) adductor magnus
- c) obturator internus
- d) gluteus medius

a   14. Which one of the following does not aid in dorsiflexion of the foot?
        a)   peroneus longus
        b)   tibialis anterior
        c)   extensor hallucis longus
        d)   extensor digitorum longus

c   15. Which one of the following muscles has its origin on the lateral
        condyle and the proximal two-thirds of the shaft of the tibia, and the
        interosseous membrane?
        a)   gastrocnemius
        b)   soleus
        c)   tibialis anterior
        d)   peroneus brevis

b   16. Which one of the following muscles has its origin on the posterior
        gluteal line of the ilium and the posterior surface of the sacrum and
        the coccyx?
        a)   adductor longus
        b)   gluteus maximus
        c)   gluteus medius
        d)   gluteus minimus

d   17. The muscle involved in raising the eyebrows is the:
        a)   occipitalis
        b)   orbicularis oculi
        c)   corrugator
        d)   frontalis

d   18. The muscle that closes and protrudes the lips is the:
        a)   mentalis
        b)   platysma
        c)   zygomaticus
        d)   none of the above is correct

d   19. If a muscle has its origin on the manubrium of the sternum and the
        medial portion of the clavicle, then the muscle could be the _____.
        a)   trapezius
        b)   transversus thoracis
        c)   diaphragm
        d)   none of the above is correct

b   20. Which one of the following muscles does not belong with the others in
        functioning as a group?
        a)   digastric
        b)   sternohyoid
        c)   stylohyoid
        d)   mylohyoid
        e)   geniohyoid

c   21.   Which one of the following muscles is most responsible for quiet
          inspiration?
          a)   external intercostals
          b)   internal intercostals
          c)   diaphragm
          d)   transversus thoracis

d   22.   The muscle whose fibers run horizontally, encircling the abdominal
          cavity, is the _____.
          a)   rectus abdominis
          b)   external oblique
          c)   internal oblique
          d)   transversus abdominus
          e)   none of the above is correct

d   23.   The urethra, rectum, and vagina are surrounded by all but which one
          of the following muscles?
          a)   iliococcygeus
          b)   puococcygeus
          c)   levator ani
          d)   coccygeus

d   24.   Which one of the following muscles does not have the same action as
          the others?
          a)   trapezius
          b)   rhomboidei
          c)   levator scapulae
          d)   pectoralis minor

b   25.   _____ muscles insert on the humerus, but only ____ of those
          have no attachments to the scapula.
          a)   7,2
          b)   9,2
          c)   6,3
          d)   5,2
          e)   none of the above is correct

d   26.   Which one of the following muscles is not part of the anterior group
          that act on the wrist, hands, and fingers?
          a)   pronator teres
          b)   palmaris longus
          c)   flexor pollicis longus
          d)   supinator

c 27. If a muscle has its origin on the outer surface of the ilium, between the posterior and anterior gluteal lines, and its insertion on the lateral surface of the greater trochanter of the femur, then the muscle is the _____.
a) gluteus minimus
b) gluteus maximus
c) gluteus medius
d) none of the above is correct

a 28. If a muscle has its origin on the inferior ramus of the pubis, and the ischium and ischial tuberosity, then the muscle is known as the ___.
a) adductor magnus
b) adductor longus
c) adductor brevis
d) inferior gemellus

c 29. Which one of the following muscles does not have at least part of its insertion on the greater trochanter of the femur?
a) piriformis
b) inferior gemellus
c) gluteus maximus
d) obturator internus

d 30. Which one of the following is not a muscle in the medial compartment of the thigh?
a) adductor magnus
b) gracilis
c) pectineus
d) sartorius

d 31. The urogenital diaphragm is composed primarily of which of the following muscles?
a) deep transverse perinei
b) sphincter urethrae
c) superficial transverse perinei
d) both a and b

c 1. Which one of the following muscles does not have its origin on the humerus?
   a) triceps brachii
   b) brachioradialis
   c) deltoid
   d) brachialis

d 2. Which one of the following muscles is involved in drinking liquid through a straw?
   a) lateral pterygoid
   b) masseter
   c) temporalis
   d) buccinator

d 3. Which one of the following muscles does not help to extend the vertebral column?
   a) spinalis thoracis cervicis
   b) longissimus thoracis
   c) cervicis capitis
   d) intertransversarii

b 4. Which one of the following muscles does not have at least part of its origin on the iliac crest?
   a) transversus abdominis
   b) rectus abdominis
   c) internal oblique
   d) quadratus lumborum

d 5. Which one of the following muscles has its origin on the outer surface of the first nine ribs?
   a) pectoralis major
   b) pectoralis minor
   c) subclavius
   d) serratus anterior

c 6. The muscle that has its origin on the spinous processes of the lower six thoracic and lumbar vertebrae, the sacrum, and the posterior iliac crest is the:
   a) pectoralis major
   b) trapezius
   c) latissimus dorsi
   d) external oblique

a   7. Which one of the following muscles does not have at least part of its origin on a fossa of the scapula?
    a) coracobrachialis
    b) subscapularis
    c) infraspinatus
    d) supraspinatus

c   8. Which one of the following muscles does not help in flexion of the arm?
    a) deltoid
    b) pectoralis major
    c) coracobrachialis
    d) teres major

c   9. Which one of the following does not aid in extending the forearm?
    a) triceps brachii
    b) anconeus
    c) brachioradialis
    d) two of the above do not aid in extending the forearm

c   10. Which one of the following muscles does not help to abduct the wrist?
    a) extensor carpi radialis longus
    b) extensor pollicis longus
    c) extensor carpi ulnaris
    d) abductor pollicis longus

a   11. Which one of the following muscles is not part of the hamstrings?
    a) rectus femoris
    b) semitendinosus
    c) semimembranosus
    d) biceps femoris

c   12. Which one of the following muscles does not aid in extension of the thigh?
    a) gluteus maximus
    b) biceps femoris
    c) gluteus medius
    d) adductor magnus

d   13. Which one of the following does not assist in abduction of the thigh?
    a) tensor fasciae latae
    b) gluteus medius
    c) gluteus minimus
    d) gluteus maximus

b   14. Which one of the following does not aid in plantar flexion of the
          foot?
          a)   soleus
          b)   tibialis anterior
          c)   gastrocnemius
          d)   peroneus longus

d   15. Which one of the following muscles has its insertion on the medial
          surface of the first cuneiform and first metatarsal?
          a)   gastrocnemius
          b)   peroneus longus
          c)   extensor digitorum longus
          d)   tibialis anterior

c   16. Which muscle has its insertion on the middle third of the linea aspera
          of the femur?
          a)   adductor magnus
          b)   adductor brevis
          c)   adductor longus
          d)   gluteus maximus

b   17. Which one of the following muscles wrinkles the skin between the
          eyebrows but is not involved in frowning?
          a)   corrugator
          b)   procerus
          c)   frontalis
          d)   orbicularis oculi

c   18. Which one of the following muscles raises and protrudes the lower lip?
          a)   orbicularis oris
          b)   depressor labii inferioris
          c)   mentalis
          d)   risorius

c   19. If the action of a muscle is to flex the cervical vertebral column, then
          the muscle could be the _____.
          a)   multifidus.
          b)   rotatores
          c)   sternocleidomastoid
          d)   semispinalis cervicis

e   20. Which one of the muscles listed does not function with the others as
          a group?
          a)   sternohyoid
          b)   thyrohyoid
          c)   omohyoid
          d)   sternothyroid
          e)   all of the above function as a group

b 21. Which one of the following muscles does not aid in expiration?
   a) transversus thoracis
   b) external intercostals
   c) subcostales
   d) internal intercostals

e 22. The muscle whose fibers run upward and medially to become
       continuous with the internal intercostal muscles is the _____.
   a) transversus abdominis
   b) external abdominal oblique
   c) rectus abdominis
   d) quadratus lumborum
   e) none of the above is correct

a 23. Which one of the following does not contribute to the pelvic
       diaphragm?
   a) coccygeus
   b) levator ani
   c) iliococcygeus
   d) pubococcygeus

a 24. The anterior muscle that anchors the scapula to the thorax has three
       antagonists. Which one of the muscles listed below is not an
       antagonist of that anterior muscle?
   a) subclavius
   b) trapezius
   c) rhomboidei
   d) levator scapulae

c 25. _____ muscles arise from the scapula and insert on the
       humerus.
   a) 2
   b) 5
   c) 7
   d) 9

b 26. Which one of the following muscles is not part of the posterior group
       that acts on the wrist, hand, and fingers?
   a) extensor digitorum communis
   b) palmaris longus
   c) supinator
   d) abductor pollicis brevis

a   27.   If a muscle has its origin on the posterior gluteal line of the ilium, and the posterior surface of the sacrum and coccyx, then the muscle is the _____.
- a) gluteus maximus
- b) gluteus medius
- c) gluteus minimus
- d) psoas major
- e) none of the above is correct

e   28.   Which one of the following muscles do not have at least a part of its origin on the ischial tuberosity?
- a) semitendinosus
- b) semimembranosus
- c) quadratus femoris
- d) inferior gemellus
- e) all of the above are correct

c   29.   Which one of the following muscles has its insertion on the lesser trochanter of the femur?
- a) piriformis
- b) inferior gemellus
- c) psoas major
- d) obturator internus

b   30.   Which one of the following muscles is not found in the anterior compartment of the thigh?
- a) sartorius
- b) gracilis
- c) rectus femoris
- d) vastus lateralis

d   31.   The pelvic diaphragm is composed primarily of which of the following muscles?
- a) coccygeus
- b) pubococcygeus
- c) iliococcygeus
- d) all of the above

NAME_____

SECTION _____

c   1.   The structure that responds to a change in the external environment
         is a/an:
         a)   effector
         b)   sensory nerve cell
         c)   receptor
         d)   motor nerve cell

a   2.   Which one of the following is not correct of the CNS? The CNS:
         a)   receives input from motor nerve cells
         b)   is the integrative and control center of the nervous system
         c)   lies within the dorsal cavity
         d)   is completely surrounded by bony structure

b   3.   Which one of the following terms does not belong with the others?
         a)   parasympathetic division
         b)   afferent division
         c)   sympathetic division
         d)   involuntary nervous system

d   4.   The embryonic tissue that gives rise to the nervous system is called:
         a)   mesoderm
         b)   endoderm
         c)   pachyderm
         d)   ectoderm

c   5.   The cells which form a segmented covering around the processes of
         many of the neurons in the PNS are called:
         a)   nerve cells
         b)   glial cells
         c)   Schwann cells
         d)   ependymal cells

b   6.   The connective tissue sheath that surrounds several fasciculi is called
         the:
         a)   perineurium
         b)   epineurium
         c)   endoneurium
         d)   exoneurium

c   7.   Nissl bodies are:
         a)   mitochondria
         b)   Golgi apparatus
         c)   rough endoplasmic reticulum
         d)   microtubules

a 8. A nerve fiber is a/an:
   a) axon together with certain sheaths
   b) nerve cell
   c) dendrite
   d) neuron

d 9. Which one of the following is not found in the CNS?
   a) myelinated fibers
   b) unmyelinated fibers
   c) oligodendrocytes
   d) neurilemma

c 10. Which one of the following is not true of the neurons carrying impulses from the eye?
   a) sensory
   b) afferent
   c) unipolar
   d) bipolar

d 11. Meissner's corpuscles probably function as sensory receptors for:
   a) pain
   b) heat
   c) cold
   d) light touch

a 12. Which one of the following is believed to be a mechanoreceptor?
   a) Golgi tendon organs
   b) taste buds
   c) Ruffini's corpuscles
   d) end bulbs of Krause

e 13. The PNS is made up of all but which one of the following?
   a) the afferent division
   b) the somatic nervous system
   c) the autonomic nervous system
   d) ganglia
   e) nuclei

c 14. Which one of the following is not under the control of the ANS?
   a) smooth muscle
   b) glands
   c) skeletal muscle
   d) cardiac muscle

c 15. A bundle of axons in the peripheral nervous system is called a:
   a) neuron
   b) ganglion
   c) nerve
   d) tract

d   16.   The neuroglia refer to the:
          a)   afferent neurons
          b)   efferent neurons
          c)   internuncial neurons
          d)   none of the above is correct

c   17.   Each nerve fiber is individually wrapped in a thin connective-tissue
          sheath called the:
          a)   perineurium
          b)   epineurium
          c)   endoneurium
          d)   fasciculus

d   18.   The portion of the neuron that is the "receptive" part is the:
          a)   axon
          b)   nerve cell body
          c)   nerve
          d)   dendrite

c   19.   The myelin sheath of peripheral axons develops from:
          a)   endoneurium
          b)   oligodendrocytes
          c)   neural crest cells
          d)   Schwann cells

c   20.   A mixed nerve is:
          a)   one that goes to both the skin surface and to the viscera
          b)   one that has its pathway mixed with other nerves
          c)   one that carries both sensory and motor fibers
          d)   one that carries large and small motor fibers

d   21.   The cells which form the blood brain barrier are known as:
          a)   oligodendrocytes
          b)   microglia
          c)   ependymal cells
          d)   astrocytes

d   1.   The structure that responds to impulses from the CNS by either
         secreting or contracting is a:
         a)   receptor
         b)   motor neuron
         c)   sensory neuron
         d)   effector

c   2.   Which one of the following is not correct about the PNS?
         a)   the autonomic nervous system is a part of the efferent
              division of the PNS
         b)   the PNS includes 12 pairs of cranial nerves
         c)   the somatic nervous system of the PNS is efferent only
         d)   the PNS contains ganglia

c   3.   Which one of the following terms does not belong with the others?
         a)   visceral sensory nerve cell
         b)   somatic sensory nerve cell
         c)   efferent division
         d)   receptors

b   4.   Which one of the following is not present by the fourth week of
         gestation?
         a)   mesencephalon
         b)   telencephalon
         c)   rhombencephalon
         d)   prosencephalon

a   5.   Most of the supporting cells that provide for the neurons in one way
         or another are derived from:
         a)   mesoderm
         b)   ectoderm
         c)   endoderm
         d)   echinoderm

a   6.   The connective tissue sheath that surrounds one fasciculus is called
         the:
         a)   perineurium
         b)   epineurium
         c)   exoneurium
         d)   endoneurium

d   7.   Which one of the following is not found in an axon?
         a)   neurofibrils
         b)   mitochondria
         c)   microtubules
         d)   Nissl bodies

b   8.   Which one of the following is not true of dendrites?
     a)   they are the receptive portion of a nerve cell
     b)   they tend to be unbranched
     c)   they are the part of the neuron where the nerve impulse originates
     d)   they often attach directly to the cell body

d   9.   Which one of the following is not found in the PNS?
     a)   white matter
     b)   unmyelinated fibers
     c)   neurilemma
     d)   oligodendrocytes

c   10.   Which one of the following is not true of neurons carrying impulses away from the CNS?
     a)   efferent
     b)   multipolar
     c)   sensory
     d)   motor

b   11.   Ruffini's corpuscles are believed to function as _____ receptors.
     a)   deep pressure
     b)   heat
     c)   cold
     d)   pain

c   12.   Which one of the following is least related to the others listed?
     a)   interoceptor
     b)   proprioceptor
     c)   exteroceptor
     d)   visceroceptor

d   13.   The PNS is made up of all but which one of the following?
     a)   cranial nerves
     b)   spinal nerves
     c)   the efferent division
     d)   centers
     e)   the autonomic nervous system

d   14.   Which one of the following is not true of the autonomic nervous system?
     a)   it is efferent
     b)   it is motor
     c)   some of its effectors are glands
     d)   it is voluntary
     e)   all of the above are true

c 15. A cluster of cell bodies in the peripheral nervous system is known as
a:
a) nerve
b) center
c) ganglion
d) nucleus

d 16. Which one of the following is not true of glial cells:
a) transfer nutrients from the blood to neurons
b) may be phagocytic
c) participate in the blood brain barrier
d) are unable to undergo cell division

a 17. Neuroglia cells that line the ventricles of the brain are called:
a) ependymal cells
b) oligodendrocytes
c) microglia
d) satelite cells

e 18. The neuron process called the axon _____.
a) varies in length
b) may have several collateral branches
c) may arise from a cone-shaped process on the cell body
d) all of the above are correct.

a 19. Which one of the following does not arise from neural crest cells?
a) glial cells
b) sensory neurons
c) Schwann cells
d) sympathetic motor nerves

a 20. Which one of the following is not true of sensory nerves?
a) efferent
b) conduct impulses toward the CNS
c) may contain processes of bipolar neurons
d) may contain processes of unipolar neurons

b 21. Free nerve endings function primarily as:
a) motor neuron endings
b) pain receptors
c) heat receptors
d) chemoreceptors

b   1.   The positive charge on the surface of an unstimulated neuron is
           primarily due to the accumulation of which one of the following ions?
           a)   potassium
           b)   sodium
           c)   calcium
           d)   magnesium
           e)   chlorine

e   2.   Which one of the following is the correct statement of Ohm's law?
           a)   I = R/E
           b)   R = E/I
           c)   E = R/I
           d)   E = I/R
           e)   none of the above is correct

d   3.   In an unstimulated neuron, a/an _____ opposes the inward movement
           of positively charged ions.
           a)   diffusion gradient
           b)   active transport mechanism
           c)   electric force
           d)   two of the above are correct

d   4.   Which one of the following is true of an unstimulated neuron?
           a)   chloride ions move along a diffusion gradient from inside
                 to outside
           b)   sodium ions move readily across the membrane from outside to
                 inside
           c)   electrical forces facilitate the movement of potassium outside
           d)   movements in response to concentration gradients and electrical
                 forces are passive movements

d   5.   The differences in the ionic compositions of the extracellular and
           intracellular fluids are maintained by:
           a)   movements in response to electric forces
           b)   passive movements
           c)   diffusion gradients
           d)   active transport mechanisms

a   6.   Which one of these is most related to the concept of "local excitatory
           state"?
           a)   partial depolarization
           b)   refractory period
           c)   maximal stimulus
           d)   all or none law

c   7.   If a stimulus of sufficient intensity is applied to a neuron, the cell
        may reach "threshold" and trigger which one of the following
        potentials?
        a)   generator
        b)   local
        c)   action
        d)   IPSP

b   8.   The local potential of an axon:
        a)   can be as great as the resting potential
        b)   is proportional to the magnitude of the stimulus
        c)   is an all or none phenomenon
        d)   is the propagated event; i.e., it travels along the length of the
             axon from the site of origination

c   9.   When a neuron reaches threshold and an action potential is generated:
        a)   potassium ions rapidly enter the cell
        b)   permeability to both potassium and sodium decreases dramatically
        c)   sodium ions rapidly enter the cell
        d)   sodium ions rapidly leave the cell

d   10. A subthreshold stimulus delivered to a neuron produces:
        a)   no electrical or permeability changes in the cell membrane
        b)   a propagated action potential
        c)   a proportional increase in the membrane potential; i.e., a slight
             hyperpolarization
        d)   a proportional decrease in the membrane potential; i.e., a slight
             hypopolarization

b   11. Which one of the events listed below occurs thirdly in the formation
        of a nerve impulse?
        a)   the inside of the cell becomes positively charged
        b)   the potassium channels open
        c)   sodium ions leave the cell
        d)   the sodium channels open
        e)   the sodium-potassium pump becomes activated

b   12. Which one of the following is true of the absolute refractory period?
        a)   another action potential can be evoked if a stronger stimulus is
             used
        b)   the duration of this period corresponds to the period of sodium
             permeability changes
        c)   the duration of this period corresponds to the period of
             potassium permeability changes
        d)   refractory periods have no effect on the number of action
             potentials that can be generated in a given time period

d   13.   Speaking of a particular axon, which one of the following is true of
           the action potentials developed?
           a)   they are of different magnitudes, depending on the stimulus
                 strength
           b)   their speed of conduction is faster the greater the stimulus
           c)   their speed of conduction is slower the smaller the stimulus
           d)   none of the above is correct
           e)   two of the above are correct

a   14.   Which one of the following is true?
           a)   action potentials are an all or none phenomenon
           b)   large diameter axons conduct action potentials at a slower rate
                 than smaller diameter axons
           c)   the action potentials in unmyelinated axons show saltatory
                 conduction
           d)   a nerve impulse is a graded response

c   15.   Which one of these statements about nerve fibers is not correct?
           a)   nerve impulses can be conducted in either direction along a nerve
                 fiber
           b)   assuming a uniform condition of the nerve fiber throughout its
                 length, a nerve impulse is as strong at one end as at the other
           c)   nerve fibers can fatigue rather readily if given a rapid series of
                 stimuli
           d)   nerve impulse frequency is limited mostly by the duration of the
                 refractory period

e   16.   Which one of the following is true of an excitatory synapse?
           a)   when its receptors are triggered, channels open that increase the
                 permeability of the membrane to potassium ions
           b)   its EPSP is an all or none phenomenon
           c)   its receptors must be stimulated by a generator potential
           d)   when its receptors are triggered, channels close and keep sodium
                 ions out
           e)   none of the above is correct

d   17.   The neurotransmitter in the brain involved in the maintenance of
           arousal and in the regulation of mood is called:
           a)   GABA
           b)   serotonin
           c)   acetylcholine
           d)   norepinephrine

c 18. Nerve impulses in many different stimulatory presynaptic cells
traveling to one postsynaptic cell may arrive there about the same
time. This is known as _____.
  a) divergence
  b) temporal summation
  c) spatial summation
  d) saltatory conduction
  e) none of the above is correct

b 19. A stronger than normal excitatory stimulus is required to elicit a
nerve impulse in a postsynaptic neuron after which one of the
following has occurred in the postsynaptic neuron?
  a) EPSP
  b) IPSP
  c) generator potential
  d) facilitation

a 20. Generator potentials:
  a) can summate
  b) have a long absolute refractory period
  c) have a short absolute refractory period
  d) generally are of shorter duration than an action potential
  e) are an all or none phenomenon

a 21. Which one of the following is true of adaptation?
  a) the responsiveness of the receptor decreases with time
  b) the same number of action potentials are being formed in the
sensory neuron
  c) it causes sensations to diminish in intensity when the stimulus is
removed
  d) pain is an example of a sensation that shows adaptation

e 22. Which one of the following is not true of receptors?
  a) they may be specialized endings of neurons
  b) they may be individual specialized cells
  c) they act as transductors
  d) they tend to be more sensitive to one kind of stimulus and less
sensitive to others
  e) they are the places in the body where sensations are "felt"

a 23. Which one of the following is not a graded potential?
  a) action potential
  b) generator potential
  c) receptor potential
  d) EPSP
  e) IPSP

a   24.   Which one of the following is not a way of discriminating different stimulus intensities?
a)   varying the size of the action potential in the sensory neuron
b)   varying the number of sensory neurons stimulated
c)   varying the number of action potentials generated by a stimulus depending on its strengths
d)   all of the above are ways of discriminating different stimulus intensities

c   25.   The release of acetylcholine at a neuromuscular junction can cause a local depolarization on the muscle cell membrane called a:
a)   generator potential
b)   local potential
c)   end plate potential
d)   IPSP

c   1. Which one of the following is true of the concentrations of ions inside
         and outside of an unstimulated neuron?
         a)   more potassium outside than inside
         b)   more sodium inside than outside
         c)   more chloride outside than inside
         d)   none of the above is correct

d   2. When positive charges and negative charges are separated from one
         another, they have the potential to perform work. The measurement
         of this potential is known as:
         a)   resistance
         b)   current
         c)   Ohm's law
         d)   voltage

c   3. In an unstimulated neuron a/an _____ opposes the movement of
         chloride ions.
         a)   diffusion gradient
         b)   active transport mechanism
         c)   electric force
         d)   two of the above are correct

d   4. Which one of the following is true of an unstimulated neuron?
         a)   potassium ions move along a diffusion gradient from outside to
              inside
         b)   sodium ions move readily by diffusion across the membrane from
              inside to outside
         c)   electrical forces facilitate the movement of potassium outside
         d)   none of the above is true

c   5. The membrane potential of an unstimulated neuron is maintained by:
         a)   concentration gradients
         b)   electrical forces
         c)   active transport
         d)   passive forces

c   6. The local potential of an axon:
         a)   can be as great as the membrane potential
         b)   is an all or none phenomenon
         c)   is proportional to the magnitude of the stimulus
         d)   is basically a hyperpolarization of the membrane potential
         e)   is a propagated event; i.e., it travels along the length of the
              axon

c   7.   If a below threshold stimulus is applied to an axon, which one of the following potentials will occur?
- a)   resting
- b)   action
- c)   local
- d)   generator
- e)   IPSP

c   8.   The action potential of an axon:
- a)   can be as great as the resting potential
- b)   is proportional to the magnitude of the stimulus
- c)   is an all or none phenomenon
- d)   is basically the same as a generator potential

c   9.   When action potential reaches threshold:
- a)   potassium ions rapidly enter the cell
- b)   permeability to both potassium and sodium ions decreases dramatically
- c)   initially sodium ions rapidly enter the cell
- d)   initially chloride ions leave the cell

c   10.   A greater than threshold stimulus delivered to a neuron produces an action potential of _____ magnitude than a threshold stimulus.
- a)   greater
- b)   lesser
- c)   the same

e   11.   Which one of the events listed below occurs fourthly in the formation of a nerve impulse?
- a)   the inside of the cell becomes positively charged
- b)   the potassium channels open
- c)   sodium ions leave the cell
- d)   the sodium channels open
- e)   the sodium-potassium pump becomes activated

d   12.   During a period that roughly corresponds to the period of increased potassium permeability, a stimulus stronger than that normally required to reach threshold may be able to initiate another nerve impulse. This period is known as the:
- a)   all or none period
- b)   absolute refractory period
- c)   latent period
- d)   relative refractory period

b 13. Which one of the following nerve fibers would conduct nerve impulses most rapidly?
    a) large diameter, unmyelinated
    b) large diameter, myelinated
    c) small diameter, unmyelinated
    d) small diameter, myelinated

c 14. Which one of the following is not true?
    a) nerve impulses are an all or none phenomenon
    b) if an axon is stimulated halfway along its length, the nerve impulse will travel in both directions
    c) nerve impulses can go from the dendrite of a postsynaptic neuron to the axon of a presynaptic neuron
    d) all of the above are true

b 15. Which one of the following statements about an axon is correct?
    a) some nerve impulses traveling along a given axon travel faster than others
    b) nerve fibers do not fatigue very easily
    c) the strength of a nerve impulse varies from one end of an axon to the other
    d) nerve impulses pass from one internodal area to the next

d 16. Which one of the following is true of an inhibitory synapse?
    a) the binding of a neurotransmitter molecule to receptors of a postsynaptic neuron closes sodium gates tighter
    b) the postsynaptic neuron shows an increased permeability to calcium ions
    c) the postsynaptic neuron has its sodium gates open
    d) the postsynaptic neuron has its potassium gates open

e 17. A common inhibitory neurotransmitter in the brain is:
    a) serotonin
    b) substance P
    c) glutamic acid
    d) enkephalins
    e) GABA

e 18. When many nerve impulses arrive in a short time at a synapse between a single stimulatory presynaptic cell and a single postsynaptic cell, this is called:
    a) convergence
    b) divergence
    c) facilitation
    d) spatial summation
    e) none of the above is correct

a   19.   Inhibition of a postsynaptic neuron occurs because its cell membrane has been:
- a)   hyperpolarized
- b)   depolarized
- c)   hypopolarized
- d)   put into a refractory period

c   20.   Which one of the following is not true of generator potentials?
- a)   it is a graded potential
- b)   the rate of application of a stimulus can affect the size of the generator potential
- c)   a stimulus that builds in intensity slowly produces a greater generator potential
- d)   generator potentials last longer than nerve impulses

a   21.   During adaptation nerve impulse frequency:
- a)   diminishes while stimulus intensity remains constant
- b)   diminishes while stimulus intensity diminishes
- c)   stays constant while the stimulus intensity diminishes
- d)   none of the above is true

a   22.   The nature or kind of stimulus as "felt" by the body is determined by:
- a)   the kind of receptor stimulated
- b)   the kind of stimulus acting on the receptor
- c)   the quality of the nerve impulse generated
- d)   the speed of the nerve impulse generated

e   23.   Which one of the following is not a graded potential?
- a)   end plate potential
- b)   resting potential
- c)   generator potential
- d)   receptor potential
- e)   all of the above are graded potentials

e   24.   Transmission at neuromuscular junctions and transmission at synapses are alike in that both:
- a)   are characterized by two-way conduction
- b)   rely on the release of sodium ions to stimulate the cell across the "gap"
- c)   are not similar to any great extent
- d)   use electrical impulses rather than a chemical to "bridge" the gap
- e)   result in a change in the potential of the membrane across the "gap"

e   25.   Arrange the following list in sequence and pick as your answer the
          third one.
          a)   response
          b)   receptor
          c)   effector
          d)   stimulus
          e)   CNS

d　　1.　Which one of the following is not a part of the telencephalon?
　　　　　a)　temporal lobe
　　　　　b)　amygdaloid nucleus
　　　　　c)　pre-central gyrus
　　　　　d)　optic chiasma

c　　2.　Tracts that connect various areas of the cerebral cortex within the
　　　　　same hemisphere are known as:
　　　　　a)　projection tracts
　　　　　b)　commissural tracts
　　　　　c)　association tracts
　　　　　d)　the corpus collosum

c　　3.　The groove that separates the two cerebral hemispheres is called the:
　　　　　a)　lateral fissure
　　　　　b)　central fissure
　　　　　c)　longitudinal fissure
　　　　　d)　transverse fissure

a　　4.　The primary motor area is located in the:
　　　　　a)　pre-central gyrus
　　　　　b)　temporal lobe
　　　　　c)　parietal lobe
　　　　　d)　occipital lobe

d　　5.　The visual area is located in the:
　　　　　a)　temporal lobe
　　　　　b)　parietal lobe
　　　　　c)　frontal lobe
　　　　　d)　occipital lobe

c　　6.　The area considered to be the site of origin of the higher intellectual
　　　　　activities that are characteristic of humans is known as the:
　　　　　a)　somatic association area
　　　　　b)　visual association area
　　　　　c)　frontal association area
　　　　　d)　auditory association area

a　　7.　Which one of the following is not true of the basal ganglia?
　　　　　a)　they are part of the pyramidal system
　　　　　b)　they are involved in somatic motor function
　　　　　c)　they have inhibitory neurons
　　　　　d)　disorders of the basal ganglia cause symptoms associated with
　　　　　　　Parkinson's disease

d   8.   The structure that serves as a relay station for olfactory neurons is the:
  a) tuber cinereum
  b) thalamus
  c) pineal body
  d) mammillary body

b   9.   The structure that contains neurons that transport hormones from the hypothalamus to the pituitary gland is the:
  a) pineal body
  b) tuber cinereum
  c) mammillary body
  d) optic chiasma

b   10.   Which one of the following is not a function of the left brain?
  a) verbal processes
  b) recognition of visual patterns
  c) performing rational tasks
  d) performing sequential processes

b   11.   Which one of the following structures is not part of or originates from the mesencephalon?
  a) cerebral aqueduct
  b) cranial nerve II
  c) the cerebral peduncles
  d) the superior colliculi

c   12.   The superior colliculi serve as relay stations for nerve fibers from cranial nerve:
  a) I
  b) VIII
  c) II
  d) IX

d   13.   Which one of the following statements is not correct?
  a) the superior cerebellar peduncles carry mostly efferent fibers
  b) the inferior cerebellar peduncles carry mostly afferent fibers
  c) the cerebellum coordinates activity of the skeletal muscles
  d) the functions with which the cerebellum are concerned are above the level of consciousness

a   14.   The pons contains nuclei for several cranial nerves. Which one of the following cranial nerves would not have its nucleus in the pons?
  a) IV
  b) VIII
  c) VII
  d) VI

c 15. Which one of the following structures listed is passed thirdly by a drop of cerebrospinal fluid formed in a lateral ventricle?
a) third ventricle
b) sub-arachnoid space
c) foramina of Luschka
d) cerebral aqueduct

d 16. Which one of the following is not a function of centers in the medulla?
a) heart rate
b) vomiting
c) respiration
d) thirst

a 17. Which one of the following structures listed would be pierced thirdly by the point of a pin entering the brain from the outside?
a) pia matter
b) grey matter
c) dura matter
d) arachnoid

c 18. Lumbar punctures are usually made between the _____ and _____lumbar vertebrae.
a) 1st and 2nd
b) 2nd and 3rd
c) 3rd and 4th
d) 4th and 5th

c 19. The group of descending nerve roots at the inferior end of the spinal cord is called the:
a) filum terminale
b) conus medullaris
c) cauda equina
d) denticulate ligaments

d 20. Inflammation of the spinal cord is known as:
a) meningitis
b) encephalitis
c) Parkinsonism
d) myelitis

c 21. Degeneration of the posterior funiculi and the dorsal roots due to the microbe that causes syphilis is known as:
a) poliomyelitis
b) multiple sclerosis
c) tabes dorsalis
d) Parkinsonism

c 22. The posterior grey columns contain:
  a) the grey commissure
  b) the central canal
  c) axons of sensory neurons
  d) axons of motor neurons

d 23. Which one of the following is not part of the grey matter of the brain?
  a) nerve cell bodies
  b) neuroglia
  c) unmyelinated neurons
  d) myelinated processes of neurons

d 24. Which one of the following is not found in dorsal roots?
  a) sensory neurons
  b) axons
  c) parts of afferent neurons
  d) dendrites

b 25. Which one of the following is not true of sensory pathways?
  a) the first order neuron is afferent
  b) the second order neuron has its cell body in the thalamus
  c) the third order neuron has its axon ending in the cerebral cortex
  d) the first order neuron has its cell body in the dorsal root ganglion

a 26. Which one of the following is not a sensory tract?
  a) lateral corticospinal tract
  b) fasciculus gracilis
  c) spinothalamic tracts
  d) spinocerebellar tracts

b 27. The fasciculus gracilis carries sensory information about:
  a) pain
  b) muscle-position sense
  c) temperature
  d) pressure

a 28. Which one of the following is not true of the corticospinal tracts?
  a) they are called lower motor neurons
  b) most of their neurons cross to the opposite side of the cord in the medulla
  c) they carry motor nerve impulses
  d) their nerve cell bodies lie in the pre-central gyrus

c   29.   Which one of the following items listed would be passed thirdly in a polysynaptic reflex arc?
    a)   effector
    b)   dorsal root ganglion
    c)   ventral root
    d)   internuncial neuron

b   30.   The perception of pain "felt" in toes even after the foot has been amputated is known as:
    a)   referred pain
    b)   phantom pain
    c)   somatic pain
    d)   visceral pain

c   31.   Which kind of sleep shows the same kind of activity as the awake brain?
    a)   slow-wave sleep
    b)   coma sleep
    c)   paradoxical sleep
    d)   fast-wave sleep

a   32.   Which one of the following is not true of the tendon reflex?
    a)   the receptors are sensitive to muscle length
    b)   the frequency of nerve impulses in afferent neurons associated with the receptors increases as a result of muscle contraction
    c)   the afferent neurons synapse with inhibitor neurons within the spinal cord
    d)   alpha motor neurons supply the muscle associated with the tendon

c   33.   The ridges of the cerebellum are called:
    a)   sulci
    b)   gyri
    c)   folia
    d)   fissures

e   34.   Broca's area of the brain is associated with the ability to:
    a)   play a musical instrument
    b)   maintain a state of wakefulness
    c)   remember
    d)   hear sounds
    e)   none of the above is correct

e   35.   Which one of the following is not one of the basal nuclei?
    a)   caudate
    b)   putamen
    c)   lentiform
    d)   amygdaloid
    e)   thalamus

c 36. The massa intermedia interconnects the:
    a)   cerebellar hemispheres
    b)   epithalamus with the pineal body
    c)   left and right thalami
    d)   corpus callosum with the basal ganglia

e 37. Which one of the following would not be one of the structures that play an important role in regulating emotional behavior?
    a)   fornix
    b)   cingulate gyrus
    c)   amygdaloid nucleus
    d)   hippocampus
    e)   frontal lobe

a 38. The structure concerned with maintaining the state of wakefulness is the:
    a)   reticular formation
    b)   red nucleus
    c)   limbic system
    d)   cerebellum

e 39. The _____ produce cerebrospinal fluid.
    a)   dural sinuses
    b)   fornix
    c)   arachnoid villi
    d)   meninges
    e)   none of the above is correct

b 40. Which one of the following is not correct?
    a)   short term memory requires that information be rehearsed
    b)   every 5 to 20 minutes during sleep one's brain goes into periods of REM sleep
    c)   consolidation of information within long term memory takes some time
    d)   a person who is in a permanent coma as a result of destruction of a portion of their medulla shows an EEG characteristic of the sleep state

NAME_____

SECTION _____

a   1.   Which one of the following is not part of the telencephalon?
         a)   pituitary gland
         b)   caudate nucleus
         c)   post-central gyrus
         d)   frontal lobe

b   2.   Tracts that carry ascending or descending impulses in the CNS are
         known as:
         a)   association tracts
         b)   projection tracts
         c)   commissural tracts
         d)   none of the above is correct

d   3.   The groove that separates each temporal lobe from the lower portion
         of the frontal and parietal lobe is known as the:
         a)   central fissure
         b)   longitudinal fissure
         c)   transverse fissure
         d)   lateral fissure

c   4.   The primary sensory area is located in the:
         a)   pre-central gyrus
         b)   temporal lobe
         c)   parietal lobe
         d)   occipital lobe

b   5.   The auditory area is located in the:
         a)   frontal lobe
         b)   temporal lobe
         c)   occipital lobe
         d)   parietal lobe

d   6.   The area that makes it possible to determine the shape, texture, and
         orientation of an object is known as the:
         a)   visual association area
         b)   frontal association area
         c)   auditory association area
         d)   somatic association area

c   7.   Which one of the following is not true of the basal ganglia?
         a)   they are part of the extrapyramidal system
         b)   they are involved in somatic motor function
         c)   they have neurons that are excitatory
         d)   disorders of the basal ganglia cause symptoms associated with
              Parkinson's disease

c   8.   The structure that acts as a major sensory relay and integration center of the brain for all tracts except olfaction is the:
  a)   pineal body
  b)   mammillary body
  c)   thalamus
  d)   hypothalamus

c   9.   The structure that contains nuclei that regulate body temperature, water balance, appetite, and gastrointestinal activity is the:
  a)   epithalamus
  b)   thalamus
  c)   hypothalamus
  d)   pituitary gland

a   10.   Which one of the following structures is not part of or does not originate from the mesencephalon?
  a)   cerebellar peduncles
  b)   inferior colliculi
  c)   red nucleus
  d)   cranial nerve III

c   11.   The inferior colliculi serve as relay stations for nerve fibers from cranial nerve:
  a)   I
  b)   II
  c)   VIII
  d)   IX

a   12.   Which one of the following statements is not correct?
  a)   the superior cerebellar peduncles carry mostly afferent fibers
  b)   the cerebellum coordinates activities of skeletal muscle
  c)   the functions with which the cerebellum are concerned are below the level of consciousness
  d)   the cerebellum receives input from proprioceptors

c   13.   The metencephalon contains nuclei for several cranial nerves. Which one of the following cranial nerves would not have its nucleus in the pons?
  a)   XII
  b)   XIII
  c)   XI
  d)   X

a   14.   Which one of the following is not a function of centers in the medulla?
  a)   appetite
  b)   coughing
  c)   dilation of blood vessels
  d)   heart rate

d 15. Which one of the following structures listed is passed thirdly by a drop of cerebrospinal fluid formed in a lateral ventricle?
    a)   aqueduct of Sylvius
    b)   sub-arachnoid space
    c)   foramen of Monroe
    d)   fourth ventricle

b 16. Which one of the following structures listed would be pierced thirdly by a point of a pin entering the brain from the outside?
    a)   dura matter
    b)   arachnoid
    c)   pia matter
    d)   bone

a 17. The spinal cord ends at the level of the _____ and _____ lumbar vertebrae.
    a)   1st, 2nd
    b)   12th thoracic, 1st
    c)   2nd, 3rd
    d)   3rd, 4th

b 18. The inferior tapered end of the spinal cord is called the:
    a)   filum terminale
    b)   conus medullaris
    c)   cauda equina
    d)   denticulate ligament

d 19. The chronic condition resulting in destruction of myelin sheaths and their replacement by hard plaques that interfere with normal transmission is known as:
    a)   syringomyelia
    b)   myelitis
    c)   tabes dorsalis
    d)   multiple sclerosis

b 20. Degeneration of cells in the basal ganglia resulting in useless contractions of skeletal muscles is known as:
    a)   tabes dorsalis
    b)   Parkinsonism
    c)   multiple sclerosis
    d)   syringomyelia

a 21. The anterior grey columns contain:
    a)   cell bodies of motor neurons
    b)   cell bodies of sensory neurons
    c)   cell bodies of internuncial neurons
    d)   axons of sensory neurons

b   22.   Which one of the following is not part of the grey matter of the spinal cord?
- a)   neuroglia
- b)   myelinated processes of neurons
- c)   nerve cell bodies
- d)   unmyelinated neurons

a   23.   Which one of the following is not found in ventral roots?
- a)   dendrites
- b)   motor neurons
- c)   axons
- d)   efferent neurons

c   24.   Which one of the following is not true of sensory spinal tracts?
- a)   they are ascending tracts
- b)   their second order neuron has its cell body in the spinal cord of the medulla
- c)   their third order neuron has its cell body in the cerebral cortex
- d)   their first order neuron has its dendrite as a part of a spinal nerve

d   25.   Which one of the following is not a motor tract?
- a)   lateral corticospinal tract
- b)   ventral corticospinal tract
- c)   extrapyramidal tract
- d)   spinocerebellar tract

c   26.   The ventral spinothalamic tracts carry impulses from receptors sensitive to:
- a)   muscle position sense
- b)   pain
- c)   pressure
- d)   temperature

b   27.   Which one of the following is not true of the extrapyramidal tracts?
- a)   some are inhibitory
- b)   they influence muscles involved in fine movements of the body
- c)   they influence coordination and balance
- d)   they include the rubrospinal and vestibulospinal tracts

d   28.   Which one of the following items listed below would be passed thirdly in a polysynaptic reflex arc?
- a)   dorsal root
- b)   effector
- c)   receptor
- d)   ventral root

e   29. Which one of the following is not a function of the right brain?
a)   recognition of complex visual patterns
b)   expression of emotion
c)   artistic abilities
d)   processing of information in a holistic manner
e)   all of the above are functions of the right brain

d   30. The incorrect interpretation of pain as having come from regions far from the actual site of pain is known as:
a)   visceral pain
b)   somatic pain
c)   phantom pain
d)   referred pain

e   31. Which one of the following is not correct?
a)   sleep results when the RAS activity declines past a certain level
b)   there is positive feedback between other parts of the brain and the RAS
c)   the RAS receives signals from various receptors
d)   the RAS sends signals to muscles
e)   all of the above are correct

a   32. Which one of the following is not true of the stretch reflex?
a)   muscle spindles provide information about tension
b)   muscle spindles respond both to the rate and the magnitude of a length change by a muscle
c)   afferent neurons stimulate alpha motor neurons
d)   afferent neurons synapse with inhibitory neurons
e)   it is characterized by reciprocal inhibition

a   33. Which one of the following is not part of the metencephalon?
a)   cerebral peduncles
b)   vermis
c)   pons
d)   4th ventricle
e)   all of the above are part of the metencephalon

a   34. The perception of taste is located:
a)   in the parietal lobe deep in the lateral fissure near the insula
b)   along the upper margin of the temporal lobe
c)   on the medial surface of the temporal lobe
d)   in the post-central gyrus

e 35. Which one of the following includes all the others?
       a) internal capsule
       b) caudate nucleus
       c) putamen
       d) globus pallidus
       e) corpus striatum

c 36. The corpus callosum interconnects the:
       a) cerebellar hemispheres
       b) thalami
       c) cerebral hemispheres
       d) corpus striatum

c 37. Which one of the following is not part of the limbic system?
       a) fornix
       b) mammillary bodies
       c) septum pellucidum
       d) amygdaloid nucleus
       e) all of the above are part of the limbic system

c 38. Which one of the following is not true of the reticular formation?
       a) it extends throughout the brain stem
       b) it receives impulses from the cerebellum
       c) it receives impulses from some selected ascending sensory tracts
           as they pass through the medulla
       d) injury to this system often produces coma
       e) it sends impulses to the cerebral cortex

d 39. Which one of the following besides the ependymal cells produce
       cerebrospinal fluid?
       a) dural sinuses
       b) arachnoid villi
       c) meninges
       d) choroid plexi

c 40. Which one of the following is correct?
       a) short term memory does not require that information be
           rehearsed
       b) during REM sleep muscle tone is greatly increased
       c) long term memory involves changes that are both structural as
           well as biochemical
       d) a state of altered consciousness from which a person can be
           aroused by appropriate stimuli is known as a coma

c   1.   Which one of the following is not true of cranial nerve I?
         a)   sensory only
         b)   its receptors lie ventral to the cribriform plate
         c)   impulses go to the parietal lobe
         d)   it arises from the telencephalon

d   2.   Which one of the following listed is passed thirdly by a nerve impulse?
         a)   optic tract
         b)   optic chiasma
         c)   occipital lobe
         d)   lateral geniculate bodies

a   3.   Which one of the following is not true of cranial nerve III?
         a)   it innervates the lateral rectus muscle
         b)   it supplies the levator palpebrae superioris
         c)   it supplies the sphincter of pupil
         d)   it supplies the inferior oblique

a   4.   Which one of the following is not true of cranial nerve IV?
         a)   it is called the trigeminal nerve
         b)   it supplies the superior oblique muscle
         c)   its site of exit from the skull is the superior orbital fissure
         d)   its site of exit from the brain is the midbrain

c   5.   Which one of the following is not true of cranial nerve V?
         a)   it is motor for the muscles of mastication
         b)   it is sensory for the scalp and the skin of the nose
         c)   it is sensory for taste for the anterior two-thirds of the tongue
         d)   it is sensory for cutaneous sense for the cheek and tongue

c   6.   Which one of the following is not true of cranial nerve VI?
         a)   its site of exit from the brain is the medulla
         b)   its site of exit from the skull is the superior orbital fissure
         c)   it supplies the medial rectus muscle of the eye
         d)   it carries proprioceptive fibers from the lateral rectus muscle

b   7.   Which one of the following is not true of cranial nerve VII?
         a)   it is sensory for taste for the anterior two-thirds of the tongue
         b)   it is sensory for cutaneous sensation on the tongue
         c)   it has parasympathetic fibers going to salivary glands
         d)   it is motor to the muscles of facial expression

a   8.   Which one of the following is not true of the eighth cranial nerve?
     a)   it exits the brain at the medulla
     b)   its name is the vestibulocochlear nerve
     c)   it is sensory only
     d)   it is sensory for hearing and balance

b   9.   Which one of the following is not true of the glossopharyngeal nerve?
     a)   it is sensory for taste for the posterior third of the tongue?
     b)   its preganglionic parasympathetic fibers synapse in the optic ganglia
     c)   it is proprioceptive from innervated muscles
     d)   it is sensory for the pharynx and middle ear cavity

b   10.   Which one of the following is not true of the tenth cranial nerve?
     a)   it leaves the skull via the jugular foramen
     b)   it has extensive parasympathetic fibers going to the muscles of the pharynx and larynx
     c)   it is sensory for the skin of the external ear
     d)   it is sensory for taste from the posterior part of the tongue

c   11.   Which one of the following is not true of the eleventh cranial nerve?
     a)   it supplies the sternocleidomastoid muscle
     b)   it supplies the trapezius muscle
     c)   parts of its fibers exit through the foramen magnum
     d)   it supplies the muscles of the pharynx

c   12.   Which one of the following is not true of cranial nerve XII?
     a)   it exits the skull via the hypoglossal canal
     b)   some of its fibers supply some of the muscles of the neck
     c)   it exits from the posterior surface of the medulla
     d)   it is proprioceptive from innervated muscles

b   13.   Which one of the following is incorrect?
     a)   the second through seventh pairs of cervical spinal nerves emerge above the vertebrae for which they are named
     b)   the first pair of cervical spinal nerves exits between the atlas and axis
     c)   the 8th pair of cervical spinal nerves emerges between the 7th cervical and 1st thoracic vertebrae
     d)   all lumbar spinal nerves exit below the vertebrae for which they are named

d   14.   Which one of the following is not mixed, that is, contains only sensory or only motor fibers?
     a)   dorsal rami
     b)   ventral rami
     c)   spinal nerve
     d)   ventral root

a   15.   The largest nerve in the body is a branch of the _____ plexus.
        a)   sacral
        b)   lumbar
        c)   cervical
        d)   thoracic

d   16.   The plexuses are formed from:
        a)   dorsal rami
        b)   dorsal roots
        c)   ventral roots
        d)   ventral rami

c   17.   Which one of the following is not true of the brachial plexuses?
        a)   they are formed from the last 4 cervical nerves and the first thoracic nerve
        b)   each one can be divided into 3 cords
        c)   the musculocutaneous nerve is a branch of the posterior cord
        d)   they innervate the skin and muscles of the upper arms

a   18.   Which one of the following is not true of a lumbar plexus?
        a)   it is formed from the first five lumbar nerves and some fibers from the 12th thoracic nerve
        b)   one of its branches is the saphenous nerve
        c)   nerves from this plexus supply the lower abdomen and anterior and medial portions of the lower limb
        d)   all of the above are true

b   19.   Which one of the following is not true of shingles?
        a)   it is caused by a virus
        b)   it involves the ventral root ganglia
        c)   it causes pain
        d)   it causes fluid-filled vessicles on the skin along the path of the peripheral sensory neurons that are infected

d   20.   Which of the following is most essential for nerve regeneration?
        a)   intact myelin sheath
        b)   viability of axons distal to the injury
        c)   presence of cell bodies in the periphery
        d)   Schwann cells

d   21.   Which of the following is a branch of the posterior cord of the brachial plexus?
- a) axillary nerve
- b) ulnar nerve
- c) radial nerve
- d) both a and c

a   22.   Which of the following is a branch of the lumbar plexus?
- a) femoral nerve
- b) sciatic nerve
- c) tibial nerve
- d) both a and c

d   1.   Which one of the following listed is passed thirdly by a nerve impulse?
         a)   temporal lobe
         b)   cribriform plate
         c)   olfactory bulb
         d)   olfactory tract

b   2.   Which one of the following is not true of cranial nerve II?
         a)   its function is sensory only for vision
         b)   it enters from the skull via the superior orbital fissure
         c)   it enters the brain at the diencephalon
         d)   impulses from it may go to the superior colliculi

d   3.   Which one of the following is not true of cranial nerve III?
         a)   it exits from the mesencephalon
         b)   it supplies the ciliary muscle of the lens
         c)   it supplies the medial rectus
         d)   it innervates the superior oblique

b   4.   Which one of the following is not true of cranial nerve IV?
         a)   it is called the trochlear nerve
         b)   it supplies the inferior oblique muscle
         c)   it carries proprioceptive fibers for the muscle it innervates
         d)   it joins the midbrain

b   5.   Which one of the following is not true of the fifth cranial nerve?
         a)   it is sensory for the skin of the chin and lower jaw
         b)   it has parasympathetic fibers
         c)   its site of exit from the brain is the pons
         d)   it is both sensory and motor in function

c   6.   Which one of the following is not true of cranial nerve VI?
         a)   it is known as the abducens nerve
         b)   it originates just below the pons
         c)   it exits the skull through the foramen rotundum
         d)   except for a proprioceptive function for the muscle it supplies, it
              is motor only

b   7.   Which one of the following is not true of the seventh cranial nerve?
         a)   its fibers supply the muscles of facial expression
         b)   it is sensory for taste from the posterior one-third of the tongue
         c)   it is proprioceptive for the muscles of facial expression
         d)   it joins the brain at the lower border of the pons

b 8. Which one of the following is not true of the eighth cranial nerve?
   a) it joins the brain at the metencephalon
   b) it passes through the external auditory meatus
   c) it is sensory for hearing
   d) it is sensory for equilibrium

d 9. Which one of the following is not true of cranial nerve IX?
   a) it is motor for pharyngeal muscles
   b) it has parasympathetic fibers to the parotid salivary gland
   c) it exits the skull via the jugular foramen
   d) it is sensory for taste on the anterior third of the tongue

c 10. Which one of the following is not true of the tenth cranial nerve?
   a) it is motor for the pharyngeal and laryngeal muscles
   b) it is visceral sensory from the thoracic and abdominal organs
   c) it joins the brain at the pons
   d) it is sensory for taste from the rear of the tongue

c 11. Which one of the following is not true of the eleventh cranial nerve?
   a) it supplies the muscles of the larynx
   b) it supplies the trapezius muscle
   c) its fibers arise only from the medulla
   d) it supplies the sternocleidomastoid muscles

a 12. Which one of the following is not true of cranial nerve XII?
   a) its fibers pass under the muscles of the cheek
   b) its fibers supply the extrinsic muscles of the tongue
   c) its fibers supply the intrinsic muscles of the tongue
   d) its fibers join the brain at the medulla

a 13. Which one of the following is incorrect?
   a) there are 7 pairs of cervical spinal nerves
   b) there are 12 pairs of thoracic spinal nerves
   c) there are 5 pairs of lumbar spinal nerves
   d) there are 5 pairs of sacral spinal nerves

c 14. Which one of the following is mixed; i.e., contains sensory and motor fibers?
   a) dorsal roots
   b) ventral roots
   c) dorsal rami
   d) ventral root ganglia

d 15. The diaphragm is supplied by a nerve arising from the:
   a) cervical plexus
   b) lumbosacral plexus
   c) brachial plexus
   d) thoracic plexus

d   16.   The distribution of sensory fibers to the skin forms uniform regions
          called:
          a)   plexuses
          b)   dorsal rami
          c)   ventral rami
          d)   none of the above is correct

e   17.   Which one of the plexuses contributes to the phrenic nerve?
          a)   lumbar
          b)   sacral
          c)   brachial
          d)   thoracic
          e)   cervical

a   18.   Which one of the following is not true of the sacral plexus?
          a)   it is formed from the last lumbar and first four sacral spinal
               nerves
          b)   some of its nerve fibers supply the lower back
          c)   the sciatic nerve is actually two nerves wrapped in a common
               sheath, the common peroneal and the tibial nerves
          d)   sitting on a hard surface with a thick wallet for long periods of
               time can cause inflammation of the sciatic nerve

b   19.   Short attacks of excruciating pain along the course of the trigeminal
          nerve are called:
          a)   neuritis
          b)   tic douloureux
          c)   Bell's palsy
          d)   multiple sclerosis
          e)   none of the above is correct

d   20.   Which one of the following listed occurs thirdly in the regeneration of
          a peripheral nerve?
          a)   new myelin sheath formed
          b)   Schwann cells form cords
          c)   nerve dies back to next intact proximal node of Ranvier
          d)   sprouts grown into endoneurial tubes
          e)   return of sensory function to denervated structure

b   21.   Which of the following is a branch of the medial cord of the brachial
          plexus?
          a)   musculocutaneous nerve
          b)   ulnar nerve
          c)   radial nerve
          d)   both a and b

d   22.   Which of the following is a branch of the sacral plexus?
       a)   femoral nerve
       b)   sciatic nerve
       c)   tibial nerve
       d)   both c and d

d　1.　Sympathetic preganglionic neurons arise from the _____ parts
　　　of the CNS.
　　　a)　cranial and sacral
　　　b)　thoracic and cervical
　　　c)　sacral and lumbar
　　　d)　thoracic and lumbar

a　2.　Which one of the following is not true of the ANS?
　　　a)　it is both sensory and motor in function
　　　b)　it is part of the PNS
　　　c)　its effectors are smooth muscle, cardiac muscle and glands
　　　d)　it is normally an involuntary system

c　3.　Which one of the following structures listed is passed thirdly by an
　　　autonomic nerve impulse?
　　　a)　white ramus communicantes
　　　b)　lateral grey column
　　　c)　paravertebral ganglion
　　　d)　grey ramus communicantes

c　4.　Which one of the following is not true of the sympathetic division of
　　　the ANS?
　　　a)　it shuts down digestive tract function
　　　b)　there is usually a two-neuron pathway between the CNS and the
　　　　　effector
　　　c)　it has terminal ganglia
　　　d)　sympathetic postganglionic fibers secrete norepinephrine

b　5.　Which one of the following cranial nerves does not have any
　　　parasympathetic function?
　　　a)　III
　　　b)　VI
　　　c)　IX
　　　d)　X

b　6.　The pelvic nerve is made up of:
　　　a)　sympathetic preganglionic fibers
　　　b)　parasympathetic preganglionic fibers
　　　c)　sympathetic postganglionic fibers
　　　d)　parasympathetic postganglionic fibers

d　7.　Which one of the following structures is not cholinergic?
　　　a)　parasympathetic preganglionic fibers
　　　b)　parasympathetic postganglionic fibers
　　　c)　sympathetic preganglionic fibers
　　　d)　sympathetic postganglionic fibers (most)

c 8. The highest center of ANS coordination is the:
 a) thalamus
 b) pons
 c) hypothalamus
 d) cerebrum

a 9. A disease characterized by episodes of pallor or cyanosis of the extremities as the result of exaggerated vasomotor responses is known as:
 a) Raynaud's Disease
 b) Achalasia
 c) Achrondroplasia
 d) Hirschsprung's Disease

a 10. Which one of the following listed is an effect of sympathetic stimulation?
 a) contraction of the arrector pili muscles
 b) contraction of the urinary bladder
 c) dilation of the blood vessels supplying salivary glands
 d) constriction of the blood vessels supplying skeletal muscle

a 11. Upon entering chain ganglia, preganglionic sympathetic neurons may take various paths. Which one of the following is not one of those paths?
 a) they may travel up or down within the chain ganglia before synapsing with parasympathetic postganglionic neurons at a higher or lower level
 b) they may synapse with the cell bodies of postganglionic neurons in the chain ganglion located at the same level at which the preganglionic neurons entered the chain.
 c) they may pass through the chain ganglia without synapsing
 d) all of the above are correct

b 12. Which one of the following is not true?
 a) some preganglionic neurons form pathways called splanchnic nerves
 b) splanchnic nerves lead to terminal ganglia
 c) one of the splanchnic nerves leads to the celiac ganglion
 d) there are no postganglionic sympathetic neurons innervating the adrenal medulla

c 13. What sort of nerve fiber forms the white ramus communicantes?
 a) parasympathetic postganglionic fibers
 b) parasympathetic preganglionic fibers
 c) sympathetic preganglionic fibers
 d) sympathetic postganglionic fibers

a   14.   Terminal ganglia:
- a) innervate smooth muscle
- b) are found in the sympathetic division
- c) are found in the dorsal root
- d) are found on either side of the vertebral bodies
- e) are found just anterior to the aorta

c   15.   Which one of these is not a function of the sympathetic division of the ANS?
- a) dilate the pupil
- b) reduce salivary gland secretion
- c) reduce sweat gland secretion
- d) accelerate heart rate
- e) all of the above are correct

c   16.   Which one of the following is NOT true of BOTH ordinary somatic efferent activity and autonomic function?
- a) the impulses pass through ventral roots of spinal nerves
- b) the fiber emerging from the CNS releases acetylcholine at its endings
- c) two nerve fibers exist between the cord and the effector
- d) the cell body of the fiber emerging from the CNS lies in the lateral grey column of the spinal cord

d   17.   Receptors that are sensitive to acetylcholine are called:
- a) muscarinic
- b) $alpha_1$
- c) nicotine
- d) both a and c

c   1.   Parasympathetic preganglionic neurons arise from the _____ parts
         of the CNS.
         a)   thoracic and cranial
         b)   thoracic and lumbar
         c)   cranial and sacral
         d)   cervical and cranial

b   2.   Which one of the following is not true about the ANS?
         a)   it normally functions below the conscious level
         b)   it innervates skeletal muscle
         c)   many neurons of the ANS travel in spinal and cranial nerves
         d)   it is part of the efferent division of the PNS

d   3.   Which one of the following structures listed is passed thirdly by an
         autonomic nerve impulse?
         a)   prevertebral ganglion
         b)   ventral root
         c)   white ramus communicantes
         d)   splanchnic nerve

a   4.   Which one of the following is not true of the parasympathetic division
         of the ANS?
         a)   its postganglionic fibers are adrenergic
         b)   it is known as the "feed and breed" division because of the
              effectors it turns on
         c)   it has long preganglionic fibers
         d)   its postganglionic fibers are short

c   5.   Which one of the following cranial nerves does not have any
         parasympathetic function?
         a)   III
         b)   VII
         c)   XI
         d)   X

c   6.   The splanchnic nerve is made up of:
         a)   parasympathetic preganglionic fibers
         b)   parasympathetic postganglionic fibers
         c)   sympathetic preganglionic fibers
         d)   sympathetic postganglionic fibers

d   7.   Which one of the following is adrenergic?
         a)   parasympathetic preganglionic fibers
         b)   parasympathetic postganglionic fibers
         c)   sympathetic preganglionic fibers
         d)   most sympathetic postganglionic fibers

a   8.   Which one of the following structures is not supplied by both divisions
         of the ANS?
         a)   sweat glands
         b)   digestive tract
         c)   urinary bladder
         d)   eye

c   9.   A disease characterized by difficulty in swallowing accompanied by a
         feeling that food is sticking in the esophagus is known as:
         a)   Raynaud's Disease
         b)   Hirschsprung's Disease
         c)   Achalasia
         d)   Achrondroplasia

d   10.   Which one of the following is the result of parasympathetic
         stimulation?
         a)   constriction of blood vessels of the viscera
         b)   relaxation of the bronchioles
         c)   relaxation of the urinary bladder
         d)   dilation of the blood vessels of the salivary glands

a   11.   Upon entering the chain ganglia, the sympathetic preganglionic axons
         may take various paths. Which one of the following is not one of
         those paths?
         a)   they may pass through the chain ganglia without synapsing until
              they reach the terminal ganglia
         b)   they may travel up or down within the chain ganglia before
              synapsing with sympathetic postganglionic neurons at a higher or
              lower level
         c)   they may synapse with the cell bodies of postganglionic neurons
              if preganglionic neurons enter the chain
         d)   all of the above are correct

c   12.   Which one of the following is not true?
         a)   the adrenal medulla acts as a sympathetic postganglionic neuron
         b)   the ventral ramus of a spinal nerve generally has a chain
              ganglion associated with it
         c)   the terms prevertebral, collateral, and chain ganglia all mean the
              same thing
         d)   one of the splanchnic nerves leads to the inferior mesenteric
              ganglion
         e)   the solar plexus is another term for the celiac ganglion

d   13.   What sort of fiber forms the grey ramus communicantes?
         a)   parasympathetic preganglionic neuron
         b)   parasympathetic postganglionic neuron
         c)   sympathetic preganglionic neuron
         d)   sympathetic postganglionic neuron
         e)   none of the above is correct

e   14. The cell bodies of parasympathetic postganglionic neurons lie within:
     a)   prevertebral ganglia
     b)   dorsal root ganglia
     c)   chain ganglia
     d)   the celiac ganglion
     e)   none of the above is correct

c   15. What is meant by "double innervation" with respect to the ANS?
     a)   there are two neurons between the CNS and the effector
     b)   two different kinds of neurotransmitter molecules are used
     c)   effectors are generally controlled by both divisions of the ANS
     d)   impulses go through two ganglia between the CNS and the effector

d   16. Which one of the following is not true?
     a)   there are synapses in autonomic ganglia but none in the dorsal root ganglia
     b)   the autonomic ganglia are motor but the dorsal root ganglia are sensory
     c)   autonomic ganglia contain cell bodies of neurons that are efferent
     d)   autonomic ganglia contain cell bodies of multipolar neurons, and so do dorsal root ganglia

d   17. Receptors that are sensitive to norepinephrine or epinephrine are called:
     a)   muscarinic
     b) $alpha_1$
     c) $beta_2$
     d)   both b and c

NAME_____

SECTION _____

a   1.   The optic cup develops into the:
          a)   retina
          b)   cornea
          c)   fibrous tunic
          d)   vascular tunic

c   2.   The person who said "don't shoot till you see the whites of their eyes" was speaking of the:
          a)   cornea
          b)   conjunctiva
          c)   sclera
          d)   iris

b   3.   Which one of the following is not part of the vascular tunic of the eye?
          a)   choroid
          b)   ora serrata
          c)   ciliary body
          d)   iris

d   4.   Which one of the following listed is passed thirdly by a ray of light entering the eye from the front?
          a)   retina
          b)   pupil
          c)   anterior chamber
          d)   vitreous humor

b   5.   Cone cells are found primarily in the:
          a)   fovea centralis
          b)   macula
          c)   optic disk
          d)   ora serrata

c   6.   Which one of the following listed would be pierced thirdly by the point of a pin entering the back of the eyeball from the outside?
          a)   rod and cone cells
          b)   sclera
          c)   bipolar cells
          d)   ganglion cell layer

c   7.   Which one of the following is not true?
          a)   cones function at high light levels
          b)   rods function at low light levels
          c)   many cones will converge on one bipolar neuron cell
          d)   vision with cones is more precise than with rods

a   8.   The muscle that closes the eyelids is the:
       a)   orbicularis oculi
       b)   levator palpebrae superioris
       c)   superior oblique
       d)   superior rectus

c   9.   A sty is an infection of the:
       a)   cornea
       b)   ciliary glands
       c)   meibomian glands
       d)   lacrimal glands

a   10.   Which one of the following listed is passed thirdly by a lacrimal secretion?
       a)   lacrimal canaliculus
       b)   nasolacrimal duct
       c)   lacrimal gland
       d)   punctum

d   11.   The muscle whose contraction turns the eye medially is the:
       a)   superior rectus
       b)   inferior oblique
       c)   superior oblique
       d)   none of the above is correct

d   12.   The greatest bending of light rays occurs as the light enters the:
       a)   aqueous humor
       b)   vitreous humor
       c)   lens
       d)   cornea

c   13.   The contraction of the ciliary muscles enables the:
       a)   tension of the suspensory ligament to increase
       b)   the ciliary body to move backward and inward
       c)   the lens to get rounder
       d)   the pupil to constrict

a   14.   Binocular vision permits:
       a)   depth perception
       b)   convergence
       c)   accommodation
       d)   refraction

c   15.   Which one of the following listed is passed thirdly by a nerve impulse?
       a)   optic chiasma
       b)   occipital lobe
       c)   optic tract
       d)   optic nerve

b   16.   The condition of nearsightedness is known as:
          a)   astigmatism
          b)   myopia
          c)   presbyopia
          d)   hyperpia

a   17.   The passageway between the nasopharynx and the middle ear is known
          as the:
          a)   auditory tube
          b)   oval window
          c)   round window
          d)   mastoid sinus

c   18.   Which of the following structures listed is passed thirdly by a sound
          wave?
          a)   oval window
          b)   tympanic membrane
          c)   incus
          d)   malleus

c   19.   The receptors for hearing are located in the:
          a)   utricle
          b)   saccule
          c)   organ of Corti
          d)   ampulla

d   20.   The scala tympani joins the _____ at the helicotrema.
          a)   scala media
          b)   cochlear duct
          c)   semicircular duct
          d)   scala vestibuli

c   21.   The greater the amplitude of a sound wave the:
          a)   higher the pitch
          b)   lower the pitch
          c)   lower the sound
          d)   quieter the sound

d   22.   Which one of the following events listed occurs thirdly?
          a)   the round window bulges outward
          b)   the vestibular membrane bulges into the scala media
          c)   the oval window bulges inward
          d)   the basilar membrane bulges into the scala tympani

b   23. Which one of the following is correct?
     a)   downward movement of the basilar membrane causes the hair
          cells to depolarize
     b)   high frequency sounds cause maximum movement of the basilar
          membrane near the base of the cochlea
     c)   the hairs of the sensory cells of the organ of Corti are embedded
          in the basilar membrane
     d)   the amplitude of a sound wave decreases as it travels along the
          basilar membrane

b   24. Which one of the following is incorrect?
     a)   the utricle and saccule contain receptors for static balance
     b)   the static balance receptors are called cristae
     c)   the static balance receptors contain otoliths
     d)   the receptors for static balance contain hair cells

c   25. The receptors for dynamic equilibrium provide information about all
         but which one of the following conditions?
     a)   acceleration
     b)   deceleration
     c)   rotation at a constant velocity
     d)   direction of rotation

c   26. A feeling that the external environment is revolving is known as:
     a)   nystagmus
     b)   dizziness
     c)   vertigo
     d)   otitis media

d   27. Which one of the following would not be a cause of conductive
         deafness?
     a)   ear wax
     b)   inflammation of the ear drum
     c)   adhesion between the ossicles
     d)   destruction of the organ of Corti

c   28. Which one of the following is not correct?
     a)   receptors for smell are located just below the cribriform plate
     b)   processes of the receptor neurons pass through holes in the
          cribriform plate
     c)   the sensory neurons for smell are unipolar in shape
     d)   air moving through the nose usually does not pass through the
          portion of the nasal cavity where the receptors are located

a   29. Which one of the following is incorrect?
     a)   the sense of taste is perceived in the pre-central gyrus
     b)   taste buds are found in the larynx and pharynx
     c)   cranial nerves seven and nine carry sensory neurons for taste
     d)   there are four specific taste modalities

e   30.   The structure that drains away excess aqueous humor from the eye is the:
        a)   lacrimal canaliculus
        b)   anterior chamber
        c)   foramen of Monroe
        d)   optic foramen
        e)   none of the above is correct

c   31.   Which one of the following listed colors of light has the shortest wavelength?
        a)   white
        b)   orange
        c)   blue
        d)   red
        e)   green

b   32.   Which one of the following is not correct concerning the image that falls on the light-sensitive part of the eye?
The image is:
        a)   reduced in size
        b)   right side up
        c)   reversed
        d)   smaller the farther away the object is from the eye

e   33.   The condition where the movements of the two eyes are not properly coordinated with one another when a person looks at an object is known as:
        a)   emmetropia
        b)   binocular vision
        c)   hyperopia
        d)   astigmatism
        e)   none of the above is correct

e   34.   The ability of the eye to distinguish detail is known as:
        a)   light adaptation
        b)   dark adaptation
        c)   emmetropia
        d)   neural adaptation
        e)   none of the above is correct

a   35.   Which one of the following is not filled with perilymph?
        a)   scala media
        b)   scala tympani
        c)   scala vestibuli
        d)   vestibule

d   36.   The human ear is most sensitive to sound waves with frequencies between:
a)   100-2000cps
b)   1000-2000cps
c)   1000-3000cps
d)   1000-4000cps
e)   1000-20,000cps

c   37.   Which one of the following is not true of basilar fibers?
a)   they project from the spiral lamina toward the outer wall of the cochlea
b)   they increase in length from the base of the cochlea to the helicotrema
c)   they decrease in length from the base of the cochlea to the helicotrema
d)   they decrease in thickness and rigidity from the base of the cochlea to the helicotrema

a   38.   High frequency sounds are detected best at the _____ of the cochlea.
a)   base
b)   middle
c)   helicotrema end

d   39.   Which one of the following includes all the others?
a)   saccule
b)   utricle
c)   semicircular canals
d)   vestibular apparatus

b   40.   When starting to rotate to the LEFT, which way do the hairs in the ampulla of the horizontal semicircular canals bend? (All directions are relevant to the subject; assume that both sets of flagella are directed straight ahead before rotation.)
a)   both bend to the left
b)   both bend to the right
c)   the hairs in the ampulla of the right ear bend left, while those in the left ear bend right
d)   the hairs in the right ear bend right, while those in the left ear bend left

NAME_____

SECTION _____

d   1.   The cornea develops from:
         a)   lens vesicles
         b)   the optic cup
         c)   the sides of the diencephalon
         d)   cells superficial to the lens vesicles

c   2.   The sclera makes up the other _____ of the eye.
         a)   posterior 3/4
         b)   anterior 1/6
         c)   posterior 5/6
         d)   posterior 4/5

c   3.   Which one of the following is not a part of the vascular tunic of the eye?
         a)   ciliary muscle
         b)   ciliary processes
         c)   macula
         d)   iris

d   4.   Which one of the following listed is passed thirdly by a ray of light entering the eye from the front?
         a)   fovea centralis
         b)   aqueous humor
         c)   conjunctiva
         d)   lens

c   5.   A part of the retina lacking both rod and cone cells is the:
         a)   fovea centralis
         b)   ora serrata
         c)   optic disk
         d)   macula

a   6.   Which one of the following listed would be pierced thirdly by the point of a pin entering the back of the eyeball from the outside?
         a)   ganglion cell layer
         b)   rod and cone cells
         c)   choroid
         d)   vitreous humor

b   7.   Which one of the following is not true?
         a)   rods are more peripherally distributed than cones
         b)   vision with rods is interpreted as color vision
         c)   many rods will converge on one bipolar neuron cell
         d)   vision with rods is less precise than with cones

c   8.   The muscle that opens the eye is the:
- a) orbicularis oculi
- b) orbicularis oris
- c) levator palpebrae superioris
- d) superior rectus

c   9.   A cyst is an infection of the:
- a) sebaceous glands
- b) ciliary glands
- c) meibomian glands
- d) lacrimal glands

c   10.   Which one of the following listed is passed thirdly by lacrimal secretions?
- a) conjunctival sac
- b) nasal cavity
- c) lacrimal sac
- d) lacrimal canaliculus

b   11.   The muscle whose contraction turns the eye upward and laterally is the:
- a) medial rectus
- b) inferior oblique
- c) superior oblique
- d) lateral rectus

c   12.   Which one of the following listed does not refract light?
- a) lens
- b) cornea
- c) pupil
- d) medial rectus

d   13.   Relaxation of the ciliary muscles enables the:
- a) tension of the suspensory ligament to lessen
- b) ciliary body to move backwards and inward
- c) pupil to constrict
- d) lens to get flatter

d   14.   The adjustment that eyes make to focus on objects closer than 20 feet involves several factors. Which one of the following listed below is NOT one of those factors?
- a) convergence
- b) pupillary constriction
- c) accommodation
- d) all of the above are involved

c   15. Which one of the following listed is passed thirdly by a nerve impulse?
a)   retina
b)   thalamus
c)   optic chiasma
d)   optic nerve

b   16. The condition resulting in an abnormal increase in intraocular pressure due to deficient drainage of aqueous humor is known as:
a)   cataract
b)   glaucoma
c)   astigmatism
d)   myopia

c   17. The structure which provides a means of equilizing the air pressure in the middle ear chamber is known as the:
a)   oval window
b)   mastoid sinus
c)   auditory tube
d)   round window

d   18. Which one of the following structures listed is passed thirdly by a sound wave?
a)   stapes
b)   external auditory meatus
c)   scala vestibuli
d)   oval window

d   19. Which one of the following structures does not contain receptors for balance?
a)   saccule
b)   utricle
c)   ampulla
d)   scala media

b   20. The basilar membrane separates the scala media from the:
a)   scala vestibuli
b)   scala tympani
c)   helicotrema
d)   modiolus

d   21. The higher the frequency of a sound wave, the:
a)   louder the sound
b)   quieter the sound
c)   lower the tone
d)   higher the tone

a 22. Which one of the following listed occurs thirdly?
   a)   pressure waves form in the scala tympani
   b)   pressure waves form in the scala vestibuli
   c)   pressure waves develop in the scala media
   d)   the round window bulges outward

c 23. Which one of the following is incorrect?
   a)   upward movement of the basilar membrane causes the hair cells
        to depolarize
   b)   high frequency sounds cause maximum movement of the basilar
        membrane near the base of the cochlea
   c)   the amplitude of a sound wave decreases as it travels along the
        basilar membrane
   d)   the hairs of the sensory cells of the organ of Corti are embedded
        in the tectorial membrane

c 24. Which one of the following is incorrect?
   a)   the ampullae contain receptors for dynamic balance
   b)   the receptors for dynamic balance contain hair cells
   c)   the dynamic balance receptors are called maculae
   d)   movement of endolymph is important in stimulating the sensory
        receptors for dynamic balance

a 25. Which one of the following is correct?
   a)   when the head begins to rotate, there is a backwards flow of
        endolymph
   b)   rotation at a constant velocity results in an increased sensation
        of movement
   c)   deceleration causes the endolymph to flow forward
   d)   the motion of the endolymph ceases at the same time as that of
        the semicircular duct itself

d 26. A jerking movement of the eyes following rotation where the eyes
      move rapidly in the direction opposite to that of the rotation is
      known as:
   a)   motion sickness
   b)   vertigo
   c)   dizziness
   d)   nystagmus

d 27. Which one of the following would not be a cause of perceptive
      deafness?
   a)   destruction of organ of Corti cells
   b)   damage to the temporal lobe
   c)   damage to the inferior colliculi
   d)   thickening of the oval window

b   28.   Which one of the following is incorrect?
- a) the odorous substance must dissolve in the mucous layer that covers the receptors
- b) processes of the receptor neurons pass through the olfactory tracts
- c) the sense of smell is perceived on the medial sides of the temporal lobes
- d) the receptor neurons for smell are bipolar in shape

c   29.   Which one of the following is incorrect?
- a) the sense of taste is perceived in the post-central gyrus
- b) tastebuds are found on the roof of the mouth and in the larynx
- c) cranial nerves five and nine carry sensory neurons for taste
- d) the sense of taste is closely associated with the sense of smell

b   30.   The structure that produces aqueous humor is known as the:
- a) canal of Schlemm
- b) ciliary processes
- c) anterior chamber
- d) anterior cavity
- e) none of the above is correct

d   31.   Which one of the following listed colors of light has the longest wavelength?
- a) white
- b) orange
- c) blue
- d) red
- e) green

e   32.   Which one of the following is not correct of the image that falls on the light-sensitive part of the eye?
- a) inverted
- b) reversed
- c) smaller the farther away the object is from the eye
- d) reduced in size
- e) all of the above are correct

a   33.   The condition where the normal eye focuses on parallel light rays from distances greater than 20 feet is known as:
- a) emmetropia
- b) hyperopia
- c) diplopia
- d) presbyopia
- e) myopia

e 34. As one gets older, the near point gets _____ while the far
        point gets _____.
        a) closer; farther
        b) farther; nearer
        c) closer; nearer
        d) farther; farther
        e) none of the above is correct

d 35. Which one of the following is not filled with endolymph?
        a) scala media
        b) membranous labyrinth
        c) cochlear duct
        d) the vestibule

d 36. The range of human hearing is _____ cps.
        a) 20-400
        b) 20-2000
        c) 1000-4000
        d) 20-20,000
        e) none of the above is correct

a 37. Which one of the following is not true of the basilar membrane?
        a) it contains basilar fibers which extend completely across the
            cochlea from the spiral lamina
        b) basilar fibers increase in length from the base of the cochlea to
            the helicotrema
        c) basilar fibers decrease in thickness from the base of the cochlea
            to the helicotrema
        d) when a portion of the basilar membrane vibrates, the fluid
            between the vibrating portion of the membrane and the oval and
            round windows must move, too

b 38. Low frequency sounds are detected best at the _____ of
        the cochlea.
        a) base
        b) apex
        c) middle

e 39. Which one of the following includes all the others?
        a) spiral ganglion cells
        b) modiolus
        c) round window
        d) scala tympani
        e) cochlea

a   40.   Which one of the following is not a place to which impulses would be routed from the organ of Corti?
       a)   superior colliculi
       b)   medulla
       c)   thalamus
       d)   temporal lobe

d   1.   Which one of the following is not true of endocrine glands?
          a)   they secrete their product into the blood stream
          b)   their product is a hormone
          c)   they are, generally speaking, ductless glands
          d)   all of the above are true of endocrine glands

d   2.   Which one of the following listed occurs thirdly?
          a)   adenylate cyclase is activated
          b)   receptor sites are stimulated on the target cell membrane
          c)   increase in cyclic AMP concentrations
          d)   ATP breakdown

a   3.   The neurohypophysis of the pituitary is functionally connected to the
          brain by the:
          a)   infundibulum
          b)   adenohypophysis
          c)   hypophyseal portal veins
          d)   primary capillary plexus

c   4.   Which one of the following is incorrect?
          a)   ADH is produced by the hypothalamus
          b)   ADH promotes reabsorption of water from urine-forming
               structures of the kidney
          c)   oversecretion of ADH leads to diabetes insipidus
          d)   neurophysin is used to transport ADH to the pars nervosa

a   5.   Which one of the following is incorrect?
          a)   FSH is known as ICSH in males
          b)   FSH is a glycoprotein
          c)   in males LH causes its target organ to produce androgens
          d)   surging levels of LH lead to ovulation in women

b   6.   Which one of the following is incorrect?
          a)   TSH is produced by the adenohypophysis
          b)   TSH is a steroid
          c)   TSH undersecretion can lead to abnormal skeletal ossification in
               the young as well as to mental retardation
          d)   TSH undersecretion can lead to the formation of a goiter

c   7.   Which one of the following is correct?
          a)   ACTH stimulates the adrenal medulla to secrete cortisol
          b)   ACTH is a protein
          c)   ACTH oversecretion leads to symptoms of Cushing's Syndrome
          d)   ACTH stimulates the production of epinephrine

b   8.   Which one of the following is incorrect?
    a)   GH causes the retention of amino acids and favors their incorporation into protein
    b)   GH is an antagonist of ACTH and TSH
    c)   GH promotes glucose formation from liver glycogen

b   9.   Which one of the following is correct?
    a)   prolactin is produced by the pars intermedia
    b)   prolactin is under the control of both PRF and PIF
    c)   prolactin has an important function in males
    d)   prolactin is a glycoprotein

d   10.   Which one of the following is not correct?
    a)   the pars distalis is not directly connected to the brain
    b)   the neurosecretory cells in the brain that make releasing or inhibiting substances put their products in the vicinity of the primary capillary plexus
    c)   if the chemical identity of a releasing or inhibiting substance has not been established, that substance is called a factor
    d)   the releasing or inhibiting substances get into the circulatory system via the tuberohypophyseal tract

b   11.   Which one of the following is incorrect about the thyroid hormones?
    a)   large doses of thyroxin inhibit protein synthesis
    b)   they potentiate the glycogen synthesis and glucose-utilizing effects of insulin
    c)   their influences are modified by the catecholamines
    d)   they have a significant calorigenic effect on the tissues of the brain, uterus, and testes

a   12.   Which one of the following is incorrect?
    a)   thyroid hormones travel free in the blood stream
    b)   some of the thyroid hormones are made up of iodine molecules bound to an amino acid
    c)   thyrotropin controls some of the thyroid hormones
    d)   calcitonin is a polypeptide

c   13.   Which one of the following is incorrect?
    a)   calcitonin lowers blood calcium levels
    b)   release of calcitonin is triggered by an increase in the level of calcium in the blood
    c)   calcitonin is made in the colloid of the thyroid gland
    d)   calcitonin lowers blood phosphate levels

a   14.   Which one of the following is incorrect? PTH:
    a)   is released when plasma phosphate levels are high
    b)   increases plasma calcium concentration
    c)   increases phosphate excretion in the urine
    d)   is a polypeptide

c   15. Epinephrine is produced by the:
    a)   zona fasciculata
    b)   zona glomerulosa
    c)   adrenal medulla
    d)   zona reticularis

c   16. Which one of the following is not a function of epinephrine?
    a)   it stimulates the release of ACTH
    b)   it increases the systolic pressure and cardiac output
    c)   it dilates vessels in the skin and mucous membranes
    d)   it dilates vessels in skeletal muscle

a   17. Which one of the following statements about mineralocorticoids is incorrect?
    a)   they promote reabsorption of potassium from urine-forming structures of the kidneys
    b)   they promote the reabsorption of sodium
    c)   undersecretion of mineralocorticoids can lead to decreased fluid volume
    d)   oversecretion of mineralocorticoids may lead to hypertension

d   18. Which one of the following statements about glucocorticoids is incorrect?
    a)   they lead to hyperglycemia
    b)   they promote gluconeogenesis
    c)   they accelerate the breakdown of proteins
    d)   they are released by angiotensinogen

a   19. Which one of the following is not true of the adrenal gland?
    a)   it produces androgenic substances and larger amounts of estrogenic substances
    b)   these sex hormones are steroids
    c)   the sex hormones produced by the adrenal gland are derived from cholesterol
    d)   the sex hormones produced by the adrenal gland are made in its cortex

a   20. Glucagon is produced by the _____ cells in the pancreatic islets.
    a)   alpha
    b)   beta
    c)   delta
    d)   chromaffin

c   21. Which one of the following statements about insulin is incorrect?
    a)   it is a protein
    b)   it is a hypoglycemic agent
    c)   it favors lipid breakdown
    d)   it facilitates the movement of amino acids into cells

a 22. Which one of the following statements about glucagon is incorrect?
    a)   it promotes hypoglycemia
    b)   it is a polypeptide
    c)   it stimulates glycogenolysis
    d)   it stimulates lipolysis

a 23. Which one of the following statements about insulin release is incorrect?
    a)   release is increased as a result of parasympathetic neuron action
    b)   it is inhibited by adrenal medullary hormones
    c)   the more blood glucose, the greater the amount of insulin release
    d)   secretin and gastrin promote insulin release

c 24. Which one of the following is not a symptom or sign of a person suffering from diabetes mellitus?
    a)   increased levels of blood glucose
    b)   sugar in urine
    c)   decrease in the amount of urine production
    d)   protein breakdown is increased

b 25. Which one of the following is not true of hormones?
    a)   they often are bound to specific carrier molecules
    b)   the binding of a hormone to a carrier molecule does not prevent excretion of that hormone from the kidneys
    c)   the binding of a hormone to a carrier molecule can provide a storage system for the hormone
    d)   usually only a small fraction of the hormone is present in the free form

d 26. Which one of the following substances is not a mediator of hormonal effects?
    a)   cyclic AMP
    b)   cyclic GMP
    c)   calcium ions
    d)   cyclic ADP

a 27. The precursor substance from which ACTH and other substances can be formed is:
    a)   pro-opiomelanocortin
    b)   B-LPH
    c)   MSH
    d)   enkephalin

d 28. The mechanism by which a particular pituitary hormone may influence the brain's secretion of releasing or inhibiting substances is known as:
    a)   positive feedback
    b)   long feedback loop
    c)   short positive feedback loop
    d)   none of the above is correct

c   29. Delta cells of the pancreas produce:
     a)   somatomedin
     b)   somatotropin
     c)   somatostatin
     d)   somatotrophic hormone

d   30. Which one of the following is not true of prostaglandins?
     a)   they are made up of fatty acids
     b)   they have a wide range of effects
     c)   they occur in very small amounts in the body
     d)   they are made by a limited variety of tissues

c   1.   Which one of the following is incorrect?
          a)   hormones are represented by several categories of chemicals
          b)   hormones are produced by glands that pass their secretion into
                the blood stream
          c)   the nervous and endocrine systems are not very closely related in
                terms of function
          d)   hormones are considered to be chemical messengers

c   2.   Which one of the following listed occurs thirdly?
          a)   hormone molecules bind to receptor sites
          b)   genes become activated
          c)   hormone-receptor molecules migrate to the nucleus
          d)   receptor proteins are structurally changed

b   3.   The adenohypophysis is functionally connected to the brain by the:
          a)   infundibulum
          b)   hypophyseal portal veins
          c)   primary capillary plexus
          d)   hypothalamus

d   4.   Which one of the following is incorrect with regard to oxytocin?
          a)   it is produced by the hypothalamus
          b)   it stimulates contraction of the smooth muscles of the uterus
          c)   it aids in "let down" of milk during lactation
          d)   it has an important function in the male

c   5.   Which one of the following is incorrect?
          a)   undersecretion of the gonadotropins in males may cause
                hypogonadotropic eunuchoidism
          b)   FSH and LH are both present in males
          c)   FSH stimulates the interstitial cells of the testes
          d)   progesterone is produced by the corpus luteum

d   6.   Which one of the following is incorrect about TSH?
          a)   undersecretion in the adult can lead to a condition known as
                myxedema
          b)   it is a glycoprotein
          c)   an excess of this hormone can lead to thyrotoxicosis
          d)   it is produced by the neurohypophysis

a   7.   Which one of the following is correct about ACTH?
          a)   undersecretion leads to symptoms of Addison's disease
          b)   it is a steroid
          c)   it stimulates the release of norepinephrine
          d)   it stimulates the production of mineralocorticoids

c 8. Which one of the following is incorrect about GH?
    a) it increases the release of fatty acids from fat cells
    b) it may act synergistically with ACTH and TSH
    c) undersecretion may cause acromegaly in adults
    d) oversecretion may cause gigantism

b 9. Which one of the following is not true of prolactin?
    a) it is involved in milk secretion in females
    b) undersecretion may cause failure to lactate after giving birth
    c) it has many metabolic effects that resemble those of GH
    d) it is produced by the hypothalamus

c 10. Which one of the following is correct?
    a) the pars distalis is directly connected to the brain
    b) if the chemical identity of a releasing or inhibiting substance is not known, it is called a hormone
    c) FSH-RF and LH-RF are the same compound
    d) there is little control by the brain over the release of adenohypophyseal hormones

b 11. Which one of the following is correct?
    a) the thyroid gland differs from all other endocrine glands in having extracellular storage sites
    b) small doses of thyroxin inhibit protein synthesis
    c) the thyroid produces more T_4 than T_3 hormone
    d) the thyroid hormones stimulate lipid synthesis, mobilization, and degradation

b 12. Which one of the following is a correct statement about the T_3 and T_4 hormones?
    a) they decrease the body's heat production
    b) they are made in the colloid of the follicles
    c) they are directly regulated by TRH
    d) they travel in the blood not bound to plasma proteins

b 13. Which one of the following is not true of calcitonin?
    a) its release is triggered by an increase in blood calcium
    b) it raises blood calcium levels
    c) it is a polypeptide
    d) it lowers blood phosphate levels

b 14. Which one of the following statements is not correct?
    a) PTH and calcitonin are antagonists of each other in terms of their effect on calcium levels in the blood
    b) PTH is produced by oxyphilic cells
    c) PTH increases plasma calcium levels
    d) PTH helps in the absorption of calcium from the gut

d 15. Mineralocorticoids are produced by the:
    a) zona fasciculata
    b) zona reticularis
    c) medulla of the adrenal gland
    d) zona glomerulosa

d 16. Which one of the following is not a function of norepinephrine?
    a) vasoconstrictor
    b) raises systolic blood pressure
    c) raises diastolic blood pressure
    d) exerts metabolic effects

c 17. Which one of the following statements about mineralocorticoids is incorrect?
    a) they promote the reabsorption of sodium
    b) they promote the excretion of potassium
    c) oversecretion can contribute to Addison's disease
    d) they are released by Angiotensin II

b 18. Which one of the following statements about glucocorticoids is incorrect?
    a) they are released by ACTH
    b) they are produced by cells of the zona fasciculata and glomerulosa
    c) they cause fatty acids to be mobilized
    d) they elevate blood glucose levels

b 19. Which one of the following is not due at least in part to adrenal cortical dysfunction?
    a) Cushing's disease
    b) Addison's disease
    c) neuroblastoma
    d) hyperaldosteronism

b 20. Insulin is produced by _____ cells in the pancreatic islets.
    a) alpha
    b) beta
    c) delta
    d) gamma

c 21. Which one of the following statements about insulin is incorrect?
    a) it favors the synthesis of proteins
    b) it facilitates the uptake of glucose by cells
    c) it raises blood sugar levels
    d) it favors lipid formation

d   22. Which one of the following statements about glucagon is not true?
         a)   it may promote breakdown of protein
         b)   it increases the substances available for gluconeogenesis
         c)   it has effects that are generally antagonistic to insulin
         d)   it is released when there are high blood glucose levels

b   23. Which one of the following statements about glucagon release is not true?
         a)   amino acids stimulate its release
         b)   glucagon output is stimulated by activation of the para-sympathetic division of the ANS
         c)   low blood glucose levels stimulate glucagon release
         d)   norepinephrine stimulates glucagon release

a   24. Which one of the following is not a symptom or sign of a person suffering from diabetes mellitus?
         a)   alkalosis
         b)   ketosis
         c)   polyuria
         d)   weight loss

d   25. Which one of the following is true of hormones?
         a)   steroid hormones travel free in the blood stream not bound to a carrier molecule
         b)   more hormone exists in the free rather than the bound state
         c)   the active form of the hormone is when the hormone is bound to its carrier molecule
         d)   Unbound hormones are more easily excreted by the kidneys

a   26. Which one of the following is not true of cyclic AMP?
         a)   it is formed from adenylate cyclase
         b)   it activates protein kinases
         c)   its effect depends on the characteristics of the cell being activated
         d)   it is considered to be an intracellular rather than an extracellular mediator

d   27. Pro-opiomelanocortin is a precursor of all but which one of the following kinds of substances?
         a)   ACTH
         b)   b-lph
         c)   MSH
         d)   FSH

d   28.   The process by which the body maintains optimum internal operating
           conditions for each cell is known as:
           a)   positive feedback
           b)   hemostasis
           c)   homeostasis
           d)   negative feedback

d   29.   Which one of the following hormones listed is produced by both the
           brain and the pancreas?
           a)   somatomedin
           b)   somatotropin
           c)   somatotrophic hormone
           d)   somatostatin

c   30.   Which one of the following is not an effect of prostaglandins?
           a)   stimulate uterine contractility
           b)   inhibit gastric secretion
           c)   contract the smooth muscle in the airways to the lung
           d)   increase urine flow

a　　1.　As the amount of adipose tissue decreases, the volume of blood per unit of body weight:
　　　　a)　increases
　　　　b)　decreases
　　　　c)　remains the same
　　　　d)　there is no relationship between the amount of adipose tissue and the volume of blood per unit of body weight

a　　2.　The most abundant constituent of plasma is:
　　　　a)　water
　　　　b)　electrolytes
　　　　c)　glucose
　　　　d)　plasma proteins

d　　3.　Myeloblasts give rise to
　　　　a)　lymphocytes
　　　　b)　monocytes
　　　　c)　erythrocytes
　　　　d)　granulocytes
　　　　e)　none of the above

c　　4.　The cell type that can differentiate into all the different formed elements of the blood is the:
　　　　a)　myeloblast
　　　　b)　megakaryocyte
　　　　c)　hemocytoblast
　　　　d)　lymphoblast

b　　5.　Which one of the following is not true?
　　　　a)　a lowered hematocrit may be observed following excessive blood loss
　　　　b)　the higher the hematocrit, the lower the viscosity of the blood
　　　　c)　increasing the viscosity of the blood makes it more difficult for the heart to move the blood
　　　　d)　hematocrits are higher in males than in females

a　　6.　Which one of the following cells listed is the development stage just before the fully formed erythrocyte?
　　　　a)　reticulocyte
　　　　b)　hemocytoblast
　　　　c)　normoblast
　　　　d)　basophil erythroblast

b   7.   Which one of the following is not true?
         a)   only a small amount of the oxygen carried per liter of blood is
               dissolved in the plasma
         b)   the heme groups of hemoglobin are proteins
         c)   one red blood cell may contain several million hemoglobin
               molecules
         d)   carbon dioxide will bond with the protein portion of the
               hemoglobin

c   8.   The cells that dispose of old damaged erythrocytes are the:
         a)   platelets
         b)   monocytes
         c)   macrophages
         d)   neutrophils

a   9.   The heme minus its iron may ultimately end up in which one of the
          following substances?
         a)   bilirubin
         b)   protein
         c)   amino acids
         d)   erythropoietin

a   10.   An inability to absorb adequate amounts of vitamin B_{12} from the
          digestive tract because of insufficient quantities of functional intrinsic
          factor leads to:
         a)   anemia
         b)   polycythemia
         c)   leukemia
         d)   hemophilia

c   11.   Platelets function in:
         a)   phagocytosis
         b)   detoxifying foreign proteins
         c)   blood clotting
         d)   none of the above

b   12.   The engulfment of foreign organisms by certain types of blood cells is
          known as:
         a)   diapedesis
         b)   phagocytosis
         c)   pinocytosis
         d)   polycythemia

a   13.   The leukocyte that is converted to macrophages in the area of
          damaged tissue is the:
         a)   monocyte
         b)   lymphocyte
         c)   neutrophil
         d)   basophil

b   14.   Males normally have as much as _____ grams of hemoglobin per 100 ml. of blood.
- a)   25
- b)   18
- c)   12
- d)   9

d   15.   When iron is transported in the blood stream outside of the red blood cells it is in the form of a molecule called:
- a)   hemoglobin
- b)   ferritin
- c)   hemosiderin
- d)   none of the above is correct

b   16.   In iron deficiency anemia:
- a)   the erythrocytes produced are larger than normal
- b)   the erythrocytes produced contain less hemoglobin than normal
- c)   the erythrocytes produced contain no globin protein
- d)   none of the above is correct

c   17.   Which of the following is not a granulocyte?
- a)   neutrophils
- b)   eosinophils
- c)   monocytes
- d)   all of the above are granulocytes

c   18.   Which one of the following is not true of an eosinophil?
- a)   phagocytic
- b)   capable of ameboid movement
- c)   produces antibodies
- d)   numbers increase during parasitic infections and allergic reactions

d   19.   The substance which renders platelets sticky is:
- a)   heparin
- b)   histamine
- c)   fibrinogen
- d)   ADP

b   20.   During blood clotting, which conversions occur?
- a)   fibrin to fibrinogen
- b)   prothrombin to thrombin
- c)   antibodies to antigens
- d)   the intrinsic mechanism to the extrinsic mechanism
- e)   hemostasis to homeostasis

d   21.   Which one of the following is not involved in the activation of factor X?
        a)   factor VII
        b)   calcium ions
        c)   tissue thromboplastin
        d)   factor V

d   22.   In the intrinsic mechanism, which one of the following do not occur?
        a)   activation of factor XII
        b)   activation of factor XI
        c)   activation of factor X
        d)   all of the above occur

d   23.   Which one of the following is not an anti-clotting factor?
        a)   heparin
        b)   plasmin
        c)   antithrombin III
        d)   factor VIII

c   24.   Which substance can decompose fibrin and dissolve clots?
        a)   erythropoietin
        b)   biliverdin
        c)   plasmin
        d)   none of the above

c   25.   The structures that actively function in clot retraction are the:
        a)   neutrophils
        b)   basophils
        c)   platelets
        d)   monocytes
        e)   lymphocytes

b   26.   A differential blood count provides information concerning what?
        a)   the abundance of red blood cells
        b)   the relative abundance of each type of leukocyte
        c)   the viscosity of the blood
        d)   the clotting time

NAME_____

SECTION _____

b 1. As the amount of adipose tissue increases, the volume of blood per unit of body weight:
 a) increases
 b) decreases
 c) remains the same
 d) there is no relationship between the two

d 2. The blood plasma contains:
 a) electrolytes
 b) proteins
 c) water
 d) all of the above

b 3. Neutrophils are formed from:
 a) megakaryocytes
 b) mycloblasts
 c) monoblasts
 d) none of the above

d 4. Which one of the following types of cells may give rise to the others?
 a) myeloblast
 b) megakaryocyte
 c) lymphoblast
 d) none of the above is correct

c 5. Which one of the following is not true?
 a) erythrocytes have the shape of a biconcave disk
 b) a cubic millimeter of normal blood may contain 4.3 - 5.8 million erythrocytes
 c) the percentage of plasma by volume in whole blood is known as the hematocrit
 d) erythrocytes contribute to the total viscosity of the blood

d 6. Which one of the following cells does not contain a nucleus?
 a) normoblast
 b) basophil erythroblast
 c) polychromatophil erythroblast
 d) reticulocyte

c   7.   Which one of the following is not true?
a)   the protein portion of hemoglobin is called globin
b)   oxygen is carried by the heme portion of hemoglobin
c)   carbon dioxide-iron complexes involving hemoglobin are called carbamino compounds
d)   most of the oxygen transported per liter of blood is carried by the hemoglobin molecules.

d   8.   The average life of an erythrocyte is:
a)   one month
b)   six months
c)   one year
d)   none of the above

b   9.   During erythrocyte destruction, the iron from hemoglobin ultimately ends up in which one of the following substances?
a)   bilirubin
b)   globin
c)   erythropoietin
d)   none of the above

a   10.   Anemia would probably have which one of the following effects on the level of erythropoietin in the blood?
a)   make it increase
b)   make it decrease
c)   make it remain the same
d)   have no effect

d   11.   Platelets are formed from cells called:
a)   thrombocytes
b)   erythrocytes
c)   granulocytes
d)   megakaryocytes

c   12.   Which of the following is an agranulocyte?
a)   neutrophil
b)   basophil
c)   monocyte
d)   eosinophil

d   13.   The leukocyte that releases histamine is the:
a)   eosinophil
b)   neutrophil
c)   monocyte
d)   basophil

c   14. Females normally have at least _____ grams of hemoglobin
         per 100 ml. of blood.
         a)   23
         b)   18
         c)   12
         d)   10
         e)   8

b   15. Most of the iron in the body exists in:
         a)   ferritin
         b)   hemoglobin
         c)   hemosiderin
         d)   transferrin

a   16. The disease resulting from an inadequate production of erythrocytes
         due to destruction of red bone marrow is known as:
         a)   aplastic anemia
         b)   sickle cell anemia
         c)   pernicious anemia
         d)   iron deficiency anemia

d   17. Which cells are important in specific immune responses, including
         antibody products?
         a)   eosinophils
         b)   basophils
         c)   neutrophils
         d)   lymphocytes
         e)   monocytes

d   18. Which one of the following is not true of a monocyte?
         a)   capable of ameboid movement
         b)   capable of leaving blood vessels
         c)   develops into macrophages
         d)   capable of antibody production

c   19. Platelets
         a)   are pinched-off portions of macrophages
         b)   are nucleated cells
         c)   adhere to collagen
         d)   secrete erythropoietin

d   20. Which of the following occurs during the formation of a clot?
         a)   fibrinogen is converted to fibrin
         b)   prothrombin is converted to thrombin
         c)   factor X is activated
         d)   all of the above occur

e   21.   In the intrinsic clotting mechanism, fibrinogen is converted to fibrin
          by:
          a)   plasmin
          b)   activated factor VI
          c)   activated factor XI
          d)   activated factor X
          e)   none of the above

c   22.   Factor VII, calcium ions, and tissue thromboplastin are all involved in
          activating factor _____.
          a)   V
          b)   VIII
          c)   X
          d)   XII

a   23.   Heparin
          a)   is secreted by most cells and basophils
          b)   inhibits the activity of antithrombin III
          c)   converts prothrombin to thrombin
          d)   none of the above

c   24.   Plasmin is involved in:
          a)   clot retraction
          b)   the conversion of fibrinogen to fibrin
          c)   the decomposition of fibrin
          d)   the intrinsic reaction
          e)   none of the above

b   25.   The structure that contains actomyosinlike contractile proteins which
          aid clot retraction is the:
          a)   monocyte
          b)   platelet
          c)   lymphocyte
          d)   basophil
          e)   eosinophil

a   26.   A hematocrit provides information concerning what?
          a)   the proportion of erythrocytes in a blood sample
          b)   the clotting time
          c)   the relative abundance of each type of leukocyte
          d)   none of the above

NAME_____

SECTION _____

c 1. By the end of the _____ week, the embryonic heart already has
       developed into the four chambers it will have in the adult.
       a)   3rd
       b)   5th
       c)   7th
       d)   9th
       e)   12th

a 2. The base of the heart lies:
       a)   at the level of the 2nd intercostal space
       b)   at the level of the 5th intercostal space
       c)   at the level of the diaphragm
       d)   dorsal to the mediastinum

b 3. Which one of the following listed is pierced thirdly by the point of a
       pin entering the body from the outside?
       a)   pericardial fluid
       b)   visceral pericardium
       c)   parietal pericardium
       d)   myocardium

a 4. The anterior surface of the heart is mostly formed by the:
       a)   right ventricle
       b)   right atrium
       c)   left ventricle
       d)   left atrium

c 5. Which one of the following does not open into or out of the right
       atrium?
       a)   superior vena cava
       b)   coronary sinus
       c)   pulmonary trunk
       d)   inferior vena cava

d 6. The _____ lines all the chambers of the heart.
       a)   epicardium
       b)   myocardium
       c)   visceral pericardium
       d)   none of the above is correct

d 7. The myocardium is thickest in the:
       a)   right atrium
       b)   right ventricle
       c)   left atrium
       d)   left ventricle

b   8.   Which one of the following structures is not associated with the
         atrioventricular valves?
         a)   chordae tendinae
         b)   trabeculae carneae
         c)   papillary muscles
         d)   endocardium

c   9.   Which one of the following structures listed is passed thirdly by a
         drop of blood?
         a)   left ventricle
         b)   coronary sinus
         c)   bicuspid valve
         d)   right atrium

c   10. The "lub" sound is due to:
         a)   contraction of the atria
         b)   closure of the semilunar valves
         c)   closure of the AV valves
         d)   relaxation of the ventricles

b   11. The abnormal condition where the cusps of valves do not form a tight
         seal when they are closed is known as:
         a)   stenosis
         b)   valvular regurgitation
         c)   myocarditis
         d)   an infarct

d   12. Which one of the following listed occurs thirdly?
         a)   depolarization of the bundle of His
         b)   depolarization of the SA node
         c)   depolarization of Purkinje fibers
         d)   depolarization of bundle branches

a   13. Rapid uncoordinated contraction of the heart is known as:
         a)   fibrillation
         b)   flutter
         c)   bradycardia
         d)   heart block

c   14. The condition where the contraction of one ventricle is delayed is
         known as:
         a)   complete heart block
         b)   partial heart block
         c)   bundle branch block
         d)   none of the above is correct

b   15. Which one of the following is not true of a myocardial infarction?
          a)   the condition is usually due to arteriosclerosis
          b)   muscle cells will regenerate to repair the damaged area
          c)   the disease resulted from constriction of the coronary arteries
          d)   the infarct may affect the conducting system of the heart

b   16. Which one of the following listed would be passed thirdly by a drop of
          blood?
          a)   capillaries in the posterior wall of the left atrium and ventricle
          b)   circumflex artery
          c)   coronary sinus
          d)   left coronary artery

e   17. Which one of the following is not a characteristic of cardiac muscle
          cells?
          a)   cross-striated
          b)   intercalated disks
          c)   sarcomeres
          d)   larger T tubules than skeletal muscle
          e)   all of the above are characteristics of cardiac muscle cells

c   18. Which one of the following is true for cardiac muscle?
          a)   like skeletal muscle, it shows a graded response
          b)   desmosomes allow electrical impulses to pass from cell to cell
          c)   it is self-excitable
          d)   it has a short refractory period

a   19. During periods of intense physical exercise, the myocardium gets most
          of its energy from the metabolism of:
          a)   lactic acid
          b)   fatty acids
          c)   glucose
          d)   amino acids

b   20. During each cardiac cycle the atria are in diastole for _____
          seconds.
          a)   .8
          b)   .7
          c)   .5
          d)   .3

d   21. During each cardiac cycle the ventricles are in systole for _____
          seconds.
          a)   .8
          b)   .7
          c)   .5
          d)   .3
          e)   .1

c 22. The ventricles are _____ % filled before the atria contract.
   a) 30
   b) 55
   c) 70
   d) 85
   e) 90

b 23. The greater the volume of blood in the ventricle before the start of
        ventricular systole, within physiological limits:
   a) the less powerful the ensuing beat
   b) the more powerful the ensuing beat
   c) the more rapidly the next beat will be completed
   d) the more likely an EKG will occur
   e) the more likely a PVC will occur

d 24. Which of the following accounts for the gradual increase in the atrial
        pressure between the opening of the semilunar valve and the opening
        of the AV valve?
   a) flow of blood from ventricle to atrium
   b) flow of blood from atrium to ventricle
   c) atrial systole
   d) atrial filling from veins
   e) bulging of AV valves into the atrial chamber

c 25. Which one of these best states one of the functions of the elasticity
        of the aorta?
   a) ensures a high systolic pressure
   b) ensures that the maximum pressure in the arteries drops rapidly
       before reaching the delicate capillaries
   c) maintains the diastolic pressure at significant levels
   d) ensures a low diastolic pressure

e 26. Rapid repolarization of the ventricles causes which wave of the ECG?
   a) p
   b) q
   c) r
   d) s
   e) none of the above is correct

d 27. Left heart failure leads to edema in the:
   a) arms
   b) legs
   c) abdominal cavity
   d) lungs

a   28. An increase in which one of the following ions would cause the heart
        to become dilated, flaccid, and weak?
        a)   potassium
        b)   sodium
        c)   calcium
        d)   chloride
        e)   phosphate

d   29. Which one of the following does not have the same effect as the
        others listed on heart rate and the force of contraction?
        a)   epinephrine
        b)   norepinephrine
        c)   sympathetic stimulation
        d)   parasympathetic stimulation

c   30. If a patient has a heart rate of 76 beats per minute, her cardiac
        output would be _____ ml. per minute.
        a)   6000
        b)   5850
        c)   5700
        d)   5250

d   31. The faster the heart rate, the higher the cardiac output.
        a)   true
        b)   false
        c)   true up to about 100 beats/minute
        d)   true up to about 190 beats/minute

a   32. Which one of the following is mostly supplied by sympathetic neurons?
        a)   ventricles
        b)   SA node
        c)   AV node
        d)   atria

e   33. Which one of the following is not true of sympathetic stimulation of
        the heart?
        a)   it increases the rate of discharge of the SA node
        b)   it increases the excitability of most parts of the heart
        c)   it alters cardiac muscle cell membrane permeability
        d)   it decreases the conduction time through the AV node
        e)   all of the above are true

a   34. In a resting individual, if all autonomic innervation is blocked, the
        heart rate will:
        a)   increase
        b)   decrease
        c)   remain the same

a 35. Stroke volume equals:
    a)   end-diastolic volume minus end-systolic volume
    b)   end-diastolic volume plus end-systolic volume
    c)   end-diastolic volume times end-systolic volume
    d)   none of the above is correct

c   1.   By the fifth week of development, the heart is a/an:
          a)   simple sac
          b)   straight tube·
          c)   "S" shaped tube
          d)   4 chambered structure

b   2.   The apex of the heart lies:
          a)   at the level of the 2nd intercostal space
          b)   at the level of the 5th intercostal space
          c)   dorsal to the mediastinum
          d)   ventral to the sternum

c   3.   Which one of the following listed is pierced thirdly by the point of a
          pin entering the body from the outside?
          a)   endocardium
          b)   visceral pericardium
          c)   myocardium
          d)   pericardial space

b   4.   The posterior surface of the heart is mostly formed by the:
          a)   right ventricle
          b)   left ventricle
          c)   right atrium
          d)   left atrium

c   5.   Which one of the following structures does not open into or out of
          the left side of the heart?
          a)   aorta
          b)   pulmonary veins
          c)   pulmonary trunk
          d)   all of the above open into or out of the left side of the heart

a   6.   The tissue that lines all the chambers of the heart is made up of:
          a)   simple squamous cells
          b)   simple cuboidal cells
          c)   simple columnar cells
          d)   stratified squamous cells

a   7.   The myocardium is the thinnest in the:
          a)   right atrium
          b)   right ventricle
          c)   left ventricle
          d)   the myocardium does not vary in thickness

c   8.   Which one of the following is not true of the semilunar valves?
    a)   one set is located at the base of the proximal end of the pulmonary trunk
    b)   one set is located at the base of the aorta
    c)   they function like a parachute valve
    d)   when the ventricles contract, the semilunar valves open

a   9.   Which one of the following structures listed is passed thirdly by a drop of blood?
    a)   left ventricle
    b)   aorta
    c)   tricuspid valve
    d)   left atrium

a   10.   The "dup" heart sound is due to:
    a)   closure of the semilunar valves
    b)   contraction of the atria
    c)   closure of the AV valves
    d)   contraction of the ventricles

b   11.   The abnormal condition where valvular growths constrict the passageway between parts of an opened valve is known as a:
    a)   functional murmur
    b)   valvular stenosis
    c)   valvular regurgitation
    d)   flutter

b   12.   Which one of the following listed occurs thirdly?
    a)   contraction of the atria
    b)   depolarization of the bundle of His
    c)   depolarization of the SA node
    d)   depolarization of Purkinje fibers

d   13.   A slower than normal rate of contraction where the heart drops below 60 beats per minute is known as:
    a)   tachycardia
    b)   flutter
    c)   fibrillation
    d)   bradycardia

a   14.   Which one of the following is not true of complete heart block?
    a)   the rate of contraction of the ventricles is faster than the rate of the atria
    b)   the contractions of the ventricles are independent of the contractions of the atria
    c)   the disorder may result from damage to the AV node
    d)   the ventricles contract at their own rate, which is slower than the rate of the atria

c   15. Swelling and inflammation of the cardiac muscle tissue due to bacteria or viruses is known as:
    a) endocarditis
    b) myocardial infarction
    c) myocarditis
    d) arteriosclerosis

a   16. Which one of the following listed would be passed thirdly by a drop of blood?
    a) posterior interventricular artery
    b) coronary sinus
    c) aortic sinus
    d) ascending aorta
    e) right atrium

b   17. Which one of the following is not a characteristic of cardiac muscle cells?
    a) they have actin and myosin proteins
    b) they have a better developed sarcoplasmic reticulum than skeletal muscle
    c) they have one nucleus per cell
    d) they have myofibrils

e   18. Which one of the following listed is shared in common by both cardiac and skeletal muscle cells?
    a) long refractory period
    b) self-excitable
    c) gap junctions between cells
    d) voluntary
    e) none of the above is shared in common

c   19. In a resting individual, the myocardium gets most of its energy from the metabolism of:
    a) lactic acid
    b) glucose
    c) fatty acids
    d) amino acids

e   20. During each cardiac cycle the atria are in systole for _____ seconds.
    a) .8
    b) .7
    c) .5
    d) .3
    e) .1

c 21. During each cardiac cycle the ventricles are in diastole for _____ seconds.
    a) .8
    b) .7
    c) .5
    d) .3
    e) .1

d 22. _____ % of the ventricular capacity comes from contraction of the atria.
    a) 90
    b) 75
    c) 50
    d) 30
    e) 10

a 23. The smaller the volume of blood in the ventricle before the start of ventricular systole, within physiological limits:
    a) the less powerful the ensuing beat
    b) the more powerful the ensuing beat
    c) the more rapidly the next beat will be completed
    d) the more likely an EKG will occur
    e) the more likely a PVC will occur

e 24. Which one of the following accounts for the incisure of the cardiac cycle?
    a) opening of the tricuspid valve
    b) closing of the bicuspid valve
    c) opening of the aortic semilunar valve
    d) closing of the pulmonary trunk semilunar valve
    e) none of the above is correct

d 25. If the aorta were not an elastic artery, there would be:
    a) large increases in the systolic pressure in the systemic circuit
    b) large decreases in the systolic pressure in the systemic circuit
    c) large increases in the diastolic pressure in the systemic circuit
    d) large decreases in the diastolic pressure in the systemic circuit

b 26. Depolarization of the ventricles causes which wave of the EKG?
    a) p
    b) qrs
    c) t
    d) pq
    e) sa

d   27. Right heart failure eventually leads to edema in all but which one of the following locations?
    a)   arms
    b)   legs
    c)   abdominal cavity
    d)   lungs

c   28. An increase in which one of the following ions would cause the heart to contract spastically?
    a)   potassium
    b)   sodium
    c)   calcium
    d)   chloride
    e)   phosphate

d   29. Which one of the following listed does not increase heart rate and force of contraction of the heart?
    a)   epinephrine
    b)   sympathetic stimulation
    c)   norepinephrine
    d)   acetylcholine

e   30. If a patient has a cardiac output of 5700 ml. per minute, her rate of heart beat, assuming a normal stroke volume, would be:
    a)   125
    b)   85
    c)   76
    d)   52
    e)   none of the above is correct

d   31. Decreases in the length of ventricular diastole cause an increase in cardiac output.
    a)   true
    b)   false
    c)   true up to 100 beats per minute
    d)   true up to about 190 beats per minute

a   32. Which one of the following is not supplied by many parasympathetic neurons?
    a)   ventricles
    b)   SA node
    c)   AV node
    d)   atria

d   33.   Which one of the following is not true of parasympathetic stimulation of the heart?
      a)   the right vagus supplies the SA node
      b)   the left vagus supplies the AV node
      c)   it decreases heart rate
      d)   it decreases membrane permeability to potassium ion
      e)   it uses acetylcholine as a chemical transmitter molecule

b   34.   In a _resting_ individual:
      a)   sympathetic stimulation of the heart is dominant
      b)   parasympathetic stimulation of the heart is dominant
      c)   neither is dominant
      d)   the balance between the two determines which is dominant

e   35.   If a patient has an end-diastolic volume of 96 ml. and an end-systolic volume of 17 ml., his stroke volume is _____ ml.
      a)   113
      b)   1632
      c)   56
      d)   5.6
      e)   none of the above is correct

c   1.   With the progressive change from arteries to capillaries, which one of the following is not true?
- a) there is a decrease in the diameter of the vessels
- b) there is a decrease in the thickness of the wall of the vessels
- c) there is an increase in the velocity at which the blood travels through the vessels
- d) there is a decrease in the pressure within the vessels

d   2.   Which kind of vessel has the greatest total cross-sectional area of the circulatory system?
- a) veins
- b) arteries
- c) arterioles
- d) capillaries

d   3.   Which one of the following layers is not part of the tunica intima?
- a) squamous epithelium
- b) basement membrane
- c) connective tissue
- d) smooth muscle

d   4.   The internal elastic lamina separates the:
- a) tunica intima from the tunica externa
- b) the tunica externa from the tunica media
- c) the tunica media from the mesothelium
- d) the tunica intima from the tunica media

b   5.   The aorta is an example of:
- a) a muscular artery
- b) elastic artery
- c) distributing artery
- d) a vessel without an elastic lamina

c   6.   If a blood vessel has a wall made up of an endothelium and a basement membrane, and if the diameter of that vessel is about that of a red blood cell, then that vessel is probably a/an:
- a) venule
- b) muscular artery
- c) capillary
- d) arteriole

c   7.   The thickest part of the wall of a medium-sized vein is made up of the:
- a) tunica media
- b) endothelium
- c) tunica media
- d) tunica externa

c 8. The disease where there are nodular deposits of lipid material in the tunica intima of arteries is known as:
a) arteriosclerosis
b) phlebitis
c) atherosclerosis
d) aneurysms

b 9. The blood in the pulmonary veins is:
a) high in carbon dioxide
b) high in oxygen
c) low in oxygen
d) two of the above are correct

a 10. The vertebral artery arises directly from the:
a) subclavian artery
b) common carotid artery
c) brachiocephalic artery
d) external carotid artery

c 11. Of the vessels listed, which one arises most superiorly?
a) superior mesenteric artery
b) right ovarian artery
c) celiac artery
d) left renal artery

c 12. The external jugular vein dumps blood directly into the:
a) common jugular vein
b) brachiocephalic vein
c) subclavian vein
d) transverse sinus

d 13. The fluid that accumulates in spaces between body cells and bathes them is called:
a) blood
b) lymph
c) plasma
d) interstitial fluid

b 14. Which one of the following structures is not supplied by the internal iliac arteries?
a) urinary bladder
b) colon
c) uterus
d) vagina

c   15.   The muscles of the posterior compartment of the thigh are supplied by
          the:
          a)   popliteal artery
          b)   lateral femoral circumflex arteries
          c)   deep femoral artery
          d)   femoral artery

a   16.   The vein that drains most of the blood from the brain is the:
          a)   internal jugular vein
          b)   external jugular vein
          c)   vertebral vein
          d)   basilar vein

d   17.   The vein that arises from the medial side of the dorsal veins of the
          hand and ascends along the medial-posterior side of the forearm is
          the:
          a)   cephalic
          b)   median cubital
          c)   axillary
          d)   basilic

d   18.   Which one of the following vessels listed is passed thirdly by a drop
          of blood?
          a)   superior vena cava
          b)   median cubital vein
          c)   axillary vein
          d)   brachiocephalic vein

a   19.   The azygos system of veins receives blood from all but which one of
          the following veins?
          a)   splenic vein
          b)   esophageal vein
          c)   intercostal vein
          d)   bronchial vein

b   20.   Which one of these vessels does not dump its blood directly into the
          inferior vena cava?
          a)   lumbar vein
          b)   left testicular vein
          c)   renal vein
          d)   hepatic veins

a   21.   Which one of the following vessels listed is passed thirdly by a drop
          of blood?
          a)   external iliac vein
          b)   peroneal vein
          c)   femoral vein
          d)   common iliac vein

d   22.   Oxygen and nutrients are provided to the lung tissue itself by the:
          a)   pulmonary arteries
          b)   pulmonary veins
          c)   pulmonary trunk
          d)   bronchial vessels

d   23.   Which one of the following structures listed is passed thirdly by a drop of blood?
          a)   inferior vena cava
          b)   hepatic vein
          c)   hepatic portal vein
          d)   liver
          e)   superior mesenteric artery

a   24.   Which one of the following statements is true?
          a)   most of the cutaneous blood flow services to regulate body temperrature
          b)   the nutrient needs of the skin are quite great
          c)   blood flow to the skin is stored in arteriovenous anastomoses
          d)   the nerve impluses that regulate flow of blood to the skin orginate in the thalamas

b   25.   Which one of the following is not true of capillaries?
          a)   preferential channels connect arterioles and venules
          b)   true capillaries branch from thorough-fare channels and connect to venules
          c)   capillaries are .5 to 1.0 mm. long
          d)   the wall of a capillary is made of only an endothelium and a thin basement membrane

a   26.   Which one of the following is not correct?
          a)   the velocity of blood flow in a portion of the circulatory system is directly proportional to the total cross-sectional area of the vessels in that segment
          b)   the velocity of blood flow in the arterioles is lower than that in the arteries
          c)   the flow of blood is the slowest in the capillaries
          d)   pressures are relatively constant in the capillaries and veins

c   27.   Which one of the following is not true of resistance?
          a)   the thicker the blood, the greater the resistance
          b)   the longer the vessel, the greater the resistance
          c)   the larger the diameter, the greater its resistance
          d)   all of the above are true

d   28.   Which one of the following is not correct?
   a)   the sympathetic division of the ANS has vasomotor nerve fibers to blood vessels
   b)   vasoconstrictor fibers release norepinephrine
   c)   vasodilator fibers release acetylcholine
   d)   most vessel dilation occurs as a result of vasodilator activity

b   29.   Which one of the following is not true of the vasomotor center?
   a)   it is located in the pons and medulla
   b)   carbon dioxide stimulates the vasodilator activity of this center
   c)   nerve impulses from baroreceptors inhibit vasoconstrictor activity of the vasomotor center
   d)   its activity is influenced by nerve impulses from higher brain centers

d   30.   Which one of the following is not true of the nerve impulses from baroreceptors?
   a)   they decrease in frequency when there is a drop in arterial pressure
   b)   they cause the activity of the parasympathetic nerves from the cardiac center to increase
   c)   they cause the activity of the sympathetic nerves to the heart decrease
   d)   they stimulate vasoconstrictor activity of the vasomotor center

c   31. The carotid and aortic bodies do not:
   a)   contain chemoreceptors
   b)   send impulses to the vasomotor center
   c)   send impulses to the cardiac center
   d)   cause an increase in arterial pressure when they sense a drop in arterial oxygen

e   32.   Which one of the following would cause a drop in arterial pressure?
   a)   angiotensin
   b)   epinephrine
   c)   vasopressin
   d)   norepinephrine
   e)   none of the above is correct

a   33.   Which one of the following would not be a substance that would cause local vasodilation?
   a)   oxygen
   b)   lactic acid
   c)   carbon dioxide
   d)   hydrogen ions
   e)   ADP

c 34. Which one of the following is not true of capillaries?
    a)   flow of blood through capillaries is slow
    b)   capillaries are not very permeable to protein but are permeable to small particles
    c)   capillary blood flow is continuous and steady
    d)   the fluid that leaves the capillaries is known as interstitial fluid

e 35. The most important means by which substances pass into and out of capillaries is:
    a)   endocytosis and exocytosis
    b)   osmotic pressure
    c)   fluid pressure
    d)   fluid movement
    e)   none of the above is correct

d 36. Which one of the following forces is not an outward moving force that causes material to move out of the capillary?
    a)   capillary hydrostatic pressure
    b)   interstitial hydrostatic pressure
    c)   interstitial osmotic pressure
    d)   capillary osmotic pressure

d 37. Most of the blood volume at any given time is in the:
    a)   heart
    b)   capillaries
    c)   pulmonary vessels
    d)   veins
    e)   arterioles

c 38. Which one of the following would not cause a decrease in venous return?
    a)   drop in blood volume
    b)   leaky right AV valve
    c)   exercise
    d)   circulatory shock due to vasodilation

b 39. Pulmonary systolic pressure is normally about _____ mm. hg.
    a)   8
    b)   22
    c)   58
    d)   80
    e)   120

d   40.   Which one of the following is not true of coronary circulation?
          a)   at rest about 5% of the cardiac output passes through the
               coronary vessels
          b)   during intense exercise coronary blood flows can be almost
               20% of cardiac output
          c)   the heart can only remove about 25% of oxygen in the arterial
               blood that flows to it
          d)   during ventricular systole blood flow in the coronary vessels
               increases

b   41.   Which one of the following is not true of cerebral circulation?
          a)   blood flow is about 15% of cardiac output at rest
          b)   the metabolic rate of the brain varies widely under different
               physiological conditions.
          c)   a drop in cerebral arterial pressure may result in fainting
          d)   an elevation in arterial carbon dioxide will cause dilation of
               cerebral blood vessels

c   42.   Which one of the following is not true of skeletal muscle circulation
          during periods of intense exercise?
          a)   blood flow to skeletal muscle can increase up to 20 times
          b)   the increase in blood flow is partially due to a local
               autoregulatory response
          c)   the increase in blood flow is partially due to activity of
               parasympathetic vasodilator fibers
          d)   vasodilation begins prior to the start of the actual exercise

b   1.   With the progressive change from capillaries to venules to veins, which
         one of the following statements is incorrect?
         a)   the diameters of the individual vessels increase
         b)   the total cross-sectional area increases
         c)   the thickness of the vessel walls increases
         d)   the blood pressure decreases

d   2.   The lowest blood pressure is found in:
         a)   an artery
         b)   venule
         c)   a capillary
         d)   the vena cava

c   3.   Which one of the following listed would not be found in a small
         artery?
         a)   tunica externa
         b)   tunica intima
         c)   vaso vasorum
         d)   tunica media

b   4.   The external elastic lamina separates the:
         a)   tunica media from the tunica intima
         b)   tunica media from the tunica externa
         c)   tunica intima from the tunica externa
         d)   tunica media from the endothelium

d   5.   If a blood vessel has a diameter of less than 0.5 mm, a small lumen,
         and a relatively thick tunica media made up mostly of smooth muscle,
         then that vessel is a/an:
         a)   elastic artery
         b)   distributing artery
         c)   venule
         d)   arteriole

a   6.   If a blood vessel has a wall composed of an endothelium, a basement
         membrane, and a few smooth muscle fibers, then the vessel is probably
         a/an:
         a)   venule
         b)   capillary
         c)   arteriole
         d)   distributing artery

b   7.   Which one of the following is not true of the valves in veins?
          a)   they are most common in the veins of the lower limbs
          b)   they have the same shape as the AV valves
          c)   they are folds of the tunica intima
          d)   they keep the blood in veins from flowing backwards away from the heart

b   8.   A stroke is a condition where:
          a)   there is a local dilation of an artery due to weakness of the artery wall
          b)   there is a region of brain tissue that dies due to lack of blood
          c)   a clot forms on the wall of a vein
          d)   veins become dilated, tortuous, and lengthened

d   9.   The first branch off of the ascending aorta is the:
          a)   brachiocephalic artery
          b)   left common carotid artery
          c)   left subclavian artery
          d)   none of the above is correct

d   10. The basilar artery arises directly from the:
          a)   posterior communicating artery
          b)   internal carotid artery
          c)   posterior cerebral artery
          d)   vertebral artery

b   11. Which one of the following listed is not supplied by the superior mesenteric artery?
          a)   small intestine
          b)   descending colon
          c)   cecum
          d)   ascending colon

a   12. Which one of the following is commonly used to give or receive blood?
          a)   median cubital vein
          b)   basilic vein
          c)   cephalic vein
          d)   radial vein

d   13. Which one of the following structures is not supplied by the internal iliac arteries?
          a)   wall of the pelvis
          b)   external genitalia
          c)   medial region of the thigh
          d)   all of the above are supplied by the internal iliac arteries

c   14.   Which one of the following vessels listed would be passed thirdly by a drop of blood?
- a) anterior tibial artery
- b) common iliac artery
- c) femoral artery
- d) external iliac artery

b   15.   The vertebral vein joins the:
- a) subclavian vein
- b) brachiocephalic vein
- c) internal jugular vein
- d) external jugular vein

c   16   The vein that arises from the lateral side of the dorsal veins of the hand and crosses to the ventral-lateral side of the forearm is the:
- a) basilic
- b) brachial
- c) cephalic
- d) axillary

d   17.   Which one of the following vessels listed is passed thirdly by a drop of blood?
- a) superior vena cava
- b) median cubital vein
- c) brachiocephalic vein
- d) axillary vein

d   18.   The venous blood of the thorax empties into the:
- a) inferior vena cava
- b) brachiocephalic veins
- c) subclavian veins
- d) superior vena cava

c   19.   The inferior vena cava recevies blood directly from which one of the following structures?
- a) pancreas
- b) spleen
- c) liver
- d) small intestine

b   20.   Which one of the following vessels listed is passed thirdly by a drop of blood?
- a) inferior vena cava
- b) femoral vein
- c) dorsal venous arch
- d) great sophenous vein

d   21.   Which one of the following statements is correct?
        a)   pulmonary arteries carry oxygenated blood
        b)   pulmonary veins carry blood high in carbon dioxide
        c)   the pulmonary arteries take blood to the left atrium
        d)   the pulmonary vessels do not supply oxygen to the lung tissue itself

c   22.   Which one of the following structures listed is passed secondly by a drop of blood?
        a)   inferior vena cava
        b)   hepatic vein
        c)   hepatic portal vein
        d)   liver
        e)   superior mesenteric vein

d   23.   Which one of the following statements relative to cutaneous blood flow is correct?
        a)   constriction of arteriovenous anastomoses is under control of the parasympathetic nervous system
        b)   constriction of arteriovenous anastomoses decreases blood flow in the skin
        c)   the walls of the subcutaneous venous plexuses are muscular
        d)   constriction of arteriovenous anastomoses increases blood flow in the skin

c   24.   Which one of the following listed is passed thirdly?
        a)   left subclavian vein
        b)   lacteal
        c)   thoracic duct
        d)   cisterna chyli

d   25.   Which one of the following is true of capillaries?
        a)   they have a tunica media
        b)   they are up to 10 mm. in length
        c)   they have the same structure in most parts of the body
        d)   their total surface area is larger than a tennis court

c   26.   Which one of the following is correct?
        a)   the velocity of blood flow in a portion of the circulatory system is directly proportional to the total cross-sectional area of the vessels in that segment
        b)   the velocity of blood flow in arteries is lower than that of veins
        c)   the flow of blood is the slowest in the capillaries
        d)   pressures vary considerably with capillaries and veins

d 27. Which one of the following has little effect on resistance?
    a) viscosity
    b) vessel length
    c) vessel diameter
    d) blood flow

d 28. Which one of the following is correct?
    a) the parasympathetic division of the ANS has vasomotor nerve fibers to blood vessels
    b) vasoconstrictor fibers release acetylcholine
    c) vasodilator fibers release norepinephrine
    d) most vessel dilation occurs as a result of diminished vasoconstrictor activity

b 29. Which one of the following is true of the vasomotor center?
    a) it is located in the cerebellum
    b) carbon dioxide stimulates vasoconstrictor activity of this center
    c) nerve impulses from baroreceptors stimulate vasoconstrictor activity
    d) its activity is not influenced by input from higher brain centers

c 30. Which one of the following is not true of baroreceptors?
    a) they are located in the carotid sinus and aortic arch
    b) they function as pressure receptors
    c) nerve impulses from them increase in frequency when there is a drop in arterial pressure
    d) they decrease the frequency of sympathetic nerve impulses to the heart

b 31. Which one of the following is true of the carotid and aortic bodies?
    a) they contain baroreceptors
    b) cause an increase in arterial pressure when they sense a drop in arterial oxygen concentration
    c) send impulses to the cardiac center
    d) sense changes in hydrogen ion concentration, which have drastic effects on arterial pressure

a 32. Considering the effects of angiotensin, vasopressin, and epinephrine on arterial pressure, which one of the following is correct?
    a) all of them increase arterial pressure
    b) all of them decrease arterial pressure
    c) two of them increase arterial pressure; the other has no effect
    d) one of them increases arterial pressure; the other two lower it
    e) none of the above is correct

b   33. Which one of the following would not cause a local dilation of
         arterioles?
         a)   lactic acid
         b)   an increase in oxygen
         c)   carbon dioxide
         d)   potassium ions
         e)   histamine

d   34. Which one of the following is correct?
         a)   flow of blood through capillary beds is rapid
         b)   the fluid that leaves the capillary is known as plasma
         c)   capillaries are very permeable even to proteins
         d)   capillary blood flow is intermittent

e   35. The most important means by which substances move into and out of
         capillaries is:
         a)   vasomotion
         b)   hydrostatic pressure
         c)   osmotic pressure
         d)   endocytosis and exocytosis
         e)   none of the above is correct

d   36. Which one of the following forces is an inward moving force that
         causes materials to move back into the capillary?
         a)   capillary hydrostatic pressure
         b)   interstitial hydrostatic pressure
         c)   interstitial osmotic pressure
         d)   capillary osmotic pressure

e   37. _____ % of the total circulating blood volume is in the capillaries.
         a)   64
         b)   15
         c)   9
         d)   7
         e)   5

a   38. Which one of the following would not cause an increase in venous
         return?
         a)   exhalation of air
         b)   exercise
         c)   transfusion
         d)   venous constriction
         e)   all of the above cause an increase

a 39. Pulmonary diastolic pressure is normally about _____mm. of Hg.
   a) 8
   b) 22
   c) 58
   d) 80
   e) 120

e 40. Which one of the following is true of coronary circulation?
   a) flow of blood through the coronary artery during ventricular systole does not fluctuate
   b) atria systole has no effect on the flow of blood through the coronary arteries
   c) under resting conditions, 1250 ml. of blood flows through the coronary arteries
   d) during heavy exercise the heart can increase its ability to remove oxygen from the blood greatly
   e) none of the above is correct

b 41. Which one of the following is true of cerebral circulation?
   a) the brain can store significant amounts of glycogen
   b) irreversible damage occurs if the brain's blood flow is stopped for over 5 minutes
   c) in terms of local autoregulatory mechanisms, the brain is more sensitive to decreases in oxygen than increases in carbon dioxide
   d) total blood flow to the brain can vary greatly over a broad range of mean arterial pressures

a 42. Which one of the following is true of skeletal muscle circulation?
   a) vasodilation and increased blood flow can occur in skeletal muscles in which all sympathetic nerve fibers have been inactivated
   b) parasympathetic vasodilator fibers are important in increasing blood flow to skeletal muscle
   c) vasodilation does not begin until the start of exercise
   d) vasodilation will not start until the muscle has been significantly depleted in oxygen

d   1.   Which one of the following is not a part of the lymphatic system?
          a)   tonsils
          b)   spleen
          c)   thymus
          d)   all of the above are part of the lymphatic system

c   2.   The larger lymphatic vessels are called:
          a)   capillaries
          b)   lymphatic capillaries
          c)   collecting vessels
          d)   none of the above

d   3.   The milky fluid that passes through the lacteals is called:
          a)   interstitial fluid
          b)   chyme
          c)   lymph
          d)   none of the above

c   4.   The main type of fiber found in a lymph node is:
          a)   elastic
          b)   collagen
          c)   reticular
          d)   muscle

b   5.   Germinal centers of lymph nodes contain:
          a)   macrophages
          b)   lymphocytes
          c)   red blood cells
          d)   none of the above

d   6.   Which one of the following would be passed thirdly by a drop of
          lymph?
          a)   efferent lymphatic vessel
          b)   afferent lymphatic vessel
          c)   medullary sinus
          d)   cortical sinus
          e)   subcapsular sinus

b   7.   The cells in a lymph node that produce antibodies are called:
          a)   macrophages
          b)   lymphocytes
          c)   plasma cells
          d)   none of the above

d   8.   The _____ trunk drains the lower limbs and some pelvic organs.
   a)   intestinal
   b)   subclavian
   c)   bronchomediastinal
   d)   none of the above

b   9.   About _____ liters of lymph enter the cardiovascular system every 24 hours.
   a)   1
   b)   3
   c)   5
   d)   7
   e)   9

d   10.   Which one of the following is not a force that moves lymph along a lymphatic vessel?
   a)   contraction of skeletal muscles
   b)   pulsing of nearby arteries
   c)   contraction of the lymphatic vessel
   d)   all of the above are forces that move lymph

b   11.   The spleen lies in the _____ region.
   a)   left inguinal
   b)   left hypochondriac
   c)   left iliac
   d)   left lumbar

a   12.   Which one of the following is not found in red pulp?
   a)   arterioles
   b)   venous sinuses
   c)   splenic cords
   d)   macrophages

a   13.   The spleen holds about _____ ml. of blood.
   a)   200
   b)   500
   c)   800
   d)   none of the above

c   14.   Which one of the following is not true of the thymus?
   a)   it lies in the mediastinum
   b)   it begins to atrophy following puberty
   c)   its cortex contains thymic corpuscles
   d)   it causes lymphocytes to differentiate into cells that carry out cell-mediated immunity
   e)   all of the above are true of the thymus

b   15.   The _____ become enlarged when you suffer a "sore throat."
          a)   palatine tonsils
          b)   pharyngeal tonsils
          c)   lingual tonsils
          d)   adenoids

d   16.   Which one of the following is not a nonspecific defense mechanism?
          a)   skin
          b)   mucous membranes
          c)   inflammatory response
          d)   cell-mediated immunity

b   17.   Which one of the following kinds of cells is not involved in humoral immunity?
          a)   B cells
          b)   T cells
          c)   memory cells
          d)   plasma cells

c   18.   The defense mechanism that depends mainly on macrophages is:
          a)   cell-mediated immunity
          b)   humoral immunity
          c)   the inflammatory response
          d)   none of the above

b   19.   B cells are produced in the:
          a)   spleen
          b)   bone marrow
          c)   thymus
          d)   lymph nodes

d   20.   These cells, produced by the cell-mediated immune system in response to a virus, can attack and destroy virus infected cells.
          a)   B cells
          b)   effector cells
          c)   plasma cells
          d)   memory cells

d   21.   The fluid that accumulates in spaces between body cells and bathes them is called:
          a)   blood
          b)   lymph
          c)   plasma
          d)   interstitial fluid

c   22. Which one of the following is not true of lymph flow?
- a)   valves help prevent backflow
- b)   muscles of the body help move the lymph in its vessels
- c)   it provides a means of rapidly returning fluid back to the bloodstream.
- d)   it helps prevent edema of body tissues

b   23. The lymphatic vessel that drains most of the body is the:
- a)   cisterna chyli
- b)   thoracic duct
- c)   right lymphatic duct
- d)   lacteal

a   24. Which one of the following is not a function of the spleen in the adult?
- a)   make erythrocytes
- b)   act as a blood reservoir
- c)   break down old erythrocytes
- d)   filter the blood

NAME_____

SECTION _____

d 1. Which one of the following is not a function of the lymphatic system?
    a) return plasma proteins to the blood
    b) return excess interstitial fluid to the blood
    c) provide defense against invasion by foreign particles
    d) all of the above are functions of the lymphatic system

d 2. The lymphatic structures that aid in the absorption of material from the digestive tract are called:
    a) collecting vessels
    b) capillaries
    c) lacteals
    d) none of the above

b 3. The main kind of nutrient absorbed by the lymphatic structure in the digestive tract is:
    a) protein
    b) lipid
    c) carbohydrate
    d) vitamins

d 4. Which one of the following areas of the body would not have very many lymph nodes?
    a) groin
    b) axilla
    c) neck
    d) forearm

a 5. Besides lymphocytes, lymphoid tissue contains:
    a) macrophages
    b) basophils
    c) neutrophils
    d) monocytes

e 6. Which one of the following would be passed fourthly by a drop of lymph?
    a) efferent lymphatic vessel
    b) afferent lymphatic vessel
    c) cortical sinus
    d) subcapsular sinus
    e) medullary sinus

c 7. Which one of the following will not be destroyed by the defense mechanisms within a lymph node?
    a) bacteria
    b) viruses
    c) cancer cells
    d) antigens

a 8. Which one of the following lymph trunks is not paired?
   a) intestinal
   b) lumbar
   c) bronchomediastinal
   d) jugular

c 9. About _____ ml. of lymph enter the cardiovascular system each hour.
   a) 3
   b) 8
   c) 12
   d) 18
   e) 24

d 10. Which one of the following does not facilitate the movement of lymph toward the heart?
   a) valves
   b) contraction of skeletal muscles
   c) hydrostatic pressure of lymph
   d) all of the above facilitate the movement of lymph

c 11. Blood vessels enter and leave the spleen at the:
   a) apex
   b) pelvis
   c) hilus
   d) none of the above is correct

c 12. Which one of the following types of cells are not found in BOTH red and white pulp?
   a) lymphocytes
   b) macrophages
   c) erythrocytes
   d) all of the above may be found in both kinds of pulp

d 13. About 200 ml. of blood are contained within the _____ of the spleen.
   a) subcapsular sinuses
   b) capillaries
   c) arterioles
   d) venous sinuses

e 14. Which one of the following is not true of the thymus?
   a) it increases in size during childhood
   b) it is covered by a connective-tissue capsule
   c) it is divided into a cortex and medulla
   d) it is divided into lobules
   e) all of the above are correct

a 15. The _____ are called adenoids when they are enlarged.
   a) pharyngeal tonsils
   b) palatine tonsils
   c) lingual tonsils
   d) none of the above

d 16. Which one of the following is not a specific immune response?
   a) inflammatory response
   b) cell-mediated immunity
   c) humoral immunity
   d) all of the above are specific immune responses

a 17. Cell-mediated immunity requires the presence of:
   a) T cells
   b) B cells
   c) plasma cells
   d) memory cells

e 18. Which one of the following is not true of antibodies?
   a) they are produced by cells found in lymphoid organs
   b) they are produced in response to specific antigens
   c) they may function by enhancing phagocytosis of the specific invading substance
   d) they may function by directly combining with the specific invading substance
   e) all of the above are true of antibodies

b 19. The precursors of T cells are produced in the:
   a) spleen
   b) bone marrow
   c) thymus
   d) liver
   e) none of the above

d 20. The cells that produce antibodies are the:
   a) T cells
   b) memory cells ·
   c) B cells
   d) plasma cells

c 21. Which one of the following is not true of the lymphatic system?
   a) the walls of lymph vessels are composed of endothelium
   b) the lymphatic system is a one-way system only, returning fluid to the bloodstream.
   c) lymphatic capillaries are less permeable than blood capillaries
   d) special lymph capillaries called lacteals are found in the villi of the gut.

c   22. Which one of the following is not found in both a lymph node and the spleen?
- a) fibrous capsule
- b) trabeculae
- c) lymphocytes
- d) red pulp

c   23. Most of the lymph is returned to the bloodstream by passing directly into the:
- a) right subclavian vein
- b) left brachiocephalic vein
- c) left subclavian vein
- d) right brachiocephalic vein

a   24. The lymphatic structure that can release a hormone that influences lymphocytes after they have left the structure where they were produced is the:
- a) thymus
- b) adenoid
- c) palatine tonsils
- d) spleen

e   1. Which of the following are events of the acute, nonspecific
         inflammatory response?
         a) margination and pavementing of leukocytes
         b) vasodilation of small blood vessels
         c) phagocytosis by macrophages
         d) increased permeability of small vessels
         e) all of the above

c   2. The tough, water insoluble protein present in the outer portion of the
         epidermis of the skin is:
         a) epidermim
         b) carotene
         c) keratin
         d) impenetrin

e   3. Which one of the following substances does not contribute to the
         vasodilation of small blood vessels near the site of any injury?
         a) histamine
         b) prostaglandins
         c) kinins
         d) complement components
         e) all of the above contribute

b   4. The process where leukocytes are displaced to the periphery of the
         blood stream where they first come in contact with the endothelial
         lining is known as:
         a) walling-off
         b) margination
         c) pavementing
         d) aggregation
         e) phagocytosis

e   5. During acute, nonspecific inflammation, the first leukocytes to arrive
         at the site of tissue damage are the:
         a) monocytes
         b) eosinophils
         c) macrophages
         d) lymphocytes
         e) neutrophils

b   6. Evidence of large numbers of _____ at the site of tissue
         injury is often indicative of an allergic response.
         a) monocytes
         b) eosinophils
         c) macrophages
         d) lymphocytes
         e) neutrophils

d   7.   Which one of the following substances would be released into a
         digestive vacuole?
         a)   opsonins
         b)   histamine
         c)   carotene
         d)   hydrogen peroxide

b   8.   Which one of the following is not true of kinins?
         a)   they are polypeptides
         b)   they are formed from precursors that are stored in basophils and
              mast cells
         c)   they increase vascular permeability and induce pain
         d)   active factor XII is important in their activation
         e)   the kinin-generating system shows a positive feedback

e   9.   Which one of the following is not true of the complement system?
         a)   they act as chemotactic agents
         b)   they act as opsonins
         c)   they can kill microbes by forming tunnels in their surfaces that
              make the microbe leaky
         d)   active complement components may be generated by a pathway
              initiated by the protein properdin
         e)   all of the above are true of the complement system

d   10. Which one of the following substances are believed to be chemotactic
         agents?
         a)   leukotrienes
         b)   complement components
         c)   kinins
         d)   all of the above are believed to be chemotactic agents

b   11. Which one of the following is not a local symptom of inflammation?
         a)   redness
         b)   swelling
         c)   pain
         d)   heat
         e)   all are local symptoms of inflammation

e   12. Which one of the following is not true of interferons?
         a)   they slow cell division
         b)   they are proteins
         c)   they stimulate the activity of natural killer cells
         d)   there are at least three kinds of interferons
         e)   double-stranded DNA is an effective inducer of interferon
              production

e   13.   Which one of the following would not be an antigenic substance by
           itself?
           a)   protein
           b)   polysaccharide
           c)   nucleic acid
           d)   complex lipid
           e)   hapten

c   14.   Which one of the following is not true of B cells?
           a)   they are lymphocytes
           b)   they can differentiate into antibody producing cells
           c)   they are produced in the thymus, liver, and kidneys
           d)   they are found in large numbers in lymph nodes

a   15.   The immunoglobulins that are particularly involved in anaphylactic-type
           allergic responses are those that belong to class:
           a)   a
           b)   d
           c)   e
           d)   g

c   16.   Which one of the following is true of antibodies?
           a)   their constant regions do not vary from one class of antibody to
                 another
           b)   antibodies function in the same manner as interferons
           c)   they belong to a family of specialized proteins known as globulins
           d)   they do not participate in the basic inflammatory response to any
                 great extent

a   17.   Which one of the following is not true of humoral immunity?
           a)   plasma cells develop into helper T cells
           b)   humoral immune responses are important in providing resistance
                 to bacterial toxins
           c)   the system frequently responds quickly to a second exposure of
                 an antigen
           d)   humoral immunity involves the production of antibodies that
                 travel in the blood and lymph

b   18.   Which one of the following is not true of cell-mediated immunity?
           a)   it involves T cells
           b)   it provides resistance to the extracellular phase of viral
                 infections
           c)   some kinds of T cells can release lymphokines
           d)   cells involved in this kind of immunity are clonal
           e)   it is involved in transplant rejection

a   19. The plasma of a person with the A antigen contains which one of the following antibodies?
         a)   anti-B
         b)   anti-A
         c)   anti-A and anti-B
         d)   neither anti-A nor anti-B

c   20. Clumping of erythrocytes due to antibody reaction is known as:
         a)   hemostasis
         b)   clotting
         c)   agglutination
         d)   hemolysis
         e)   none of the above is correct

b   21. A person of blood type A could donate blood to persons of which one of the following blood types?
         a)   O
         b)   AB
         c)   B
         d)   none of these

d   22. The father is Rh negative and the mother Rh positive. They have had three children without adverse problems due to Rh factor. The mother is pregnant again. In terms of Rh factor, the risk to the fetus now within the uterus:
         a)   is less than before
         b)   is greater than before
         c)   is the same
         d)   never was a problem

d   23. Which one of the following is not an example of active immunity?
         a)   getting the mumps
         b)   getting a tetanus shot
         c)   getting a vaccination
         d)   getting a shot of anti-cobra serum

d   24. Which one of the following is a hypersensitivity due to complement components?
         a)   anaphylactic shock
         b)   exposure of a sensitized individual to poison oak toxin
         c)   exposure of sensitized individual to pollen
         d)   glomerulonephritis

b   25. Which one of the following is not true of anaphylactic shock?
         a)   it is a systemic rather than a localized reaction
         b)   severe hypertension and airway constriction quickly develop
         c)   a countermeasure is the administration of epinephrine
         d)   insect stings and penicillin may cause such a condition in sensitive individuals

e   26. The failure of the body to mount a specific immune response against a
        particular antigen is known as:
        a)   autoimmune response
        b)   cellular immunity
        c)   transplant rejection
        d)   hypersensitivity
        e)   none of the above is correct

d   1.   Which is not one of the events of the acute, nonspecific inflammatory
          response?
          a)   aggregation of leukocytes at site of injury
          b)   slowing of blood flow
          c)   pavementing of leukocytes
          d)   constriction of small blood vessels
          e)   all of the above are events of the acute, nonspecific
                inflammatory response

a   2.   The water-insoluble protein in the epidermis that is resistant to weak
          acids, bases, and protein-digesting enzymes is:
          a)   keratin
          b)   lysozyme
          c)   sebum
          d)   sweat
          e)   ear wax

d   3.   Which one of the following substances contributes to the increased
          permeability of small blood vessels in acute, nonspecific inflammation?
          a)   pyrogens
          b)   interferons
          c)   agglutinins
          d)   kinins
          e)   none of the above is correct

b   4.   The movement of leukocytes through tissues toward the site of injury
          is an example of:
          a)   aggregation
          b)   margination
          c)   positive chemotaxis
          d)   pavementing
          e)   none of the above is correct

b   5.   During the later stages of the acute, nonspecific inflammatory
          response, the kinds of leukocyte phagocytic cells that predominate at
          the site of injury generally are the:
          a)   lymphocytes and neutrophils
          b)   macrophages and monocytes
          c)   basophils and megakaryocytes
          d)   neutrophils and eosinophils
          e)   T cells and B cells

a   6.   The presence of large numbers of eosinophils in a particular area of tissue is often an indication of a/an:
a)   allergic response
b)   acute, nonspecific inflammatory response
c)   specific immune response
d)   autoimmune response
e)   tolerance

d   7.   Which one of the following is a process involved in the formation of a digestive vacuole?
a)   agglutination
b)   margination
c)   pavementing
d)   degranulation
e)   none of the above is correct

d   8.   Which one of the following does not contain release histamine?
a)   mast cells
b)   platelets
c)   basophils
d)   all of the above contain histamine

d   9.   Which one of the following is true of the complement system?
a)   the complement components normally circulate in the blood in an active form
b)   complement components are generated by a pathway initiated by the protein kallikrein
c)   complement components chemotactically attract hepatocytes
d)   complement components can directly attack invading microbes

d   10.   Which of the following is a common systemic symptom of inflammation?
a)   paralysis
b)   shock
c)   nausea
d)   fever
e)   none of the above

e   11.   Bacterial invasions may cause the leukocyte count to rise as high as _____ cells/cubic mm.
a)   700
b)   1200
c)   7000
d)   15,000
e)   25,000

c   12.   Which one of the following is not true of interferons?
a)   they are proteins
b)   double-stranded RNA is a powerful inducer of interferon production
c)   only eosinophils can make interferons
d)   interferons leave the cell where they were produced and bind to receptor sites on other cell membranes

d   13.   Which one of the following is true of interferons?
a)   they speed up cell division of healthy cells
b)   they speed up cell division of abnormal cells
c)   they are produced only by liver cells, kidney cells and lung cells
d)   they stimulate the activity of cells that are believed to be important in immune surveillance

d   14.   Which of the following is not true of antigenic substances?
a)   they generally have molecular weights of 8000-10,000 or more
b)   they react with substances called antibodies to form antigen-antibody complexes
c)   proteins, polysaccharides, and nucleic acids may act as antigenic substances
d)   all of the above are true

a   15.   Which one of the following is not true of T cells?
a)   they are a homogeneous group of monocytes
b)   the thymus gland is important in their production
c)   they continually circulate through the blood, tissues, and lymph
d)   they are present in lymph nodes

e   16.   Which one of the following is not true of antibodies?
a)   the constant regions of their heavy polypeptide chains differ from one class of antibody to another
b)   they combine with antigens to form antigen-antibody complexes
c)   they are produced in response to the presence of an immunogenic antigen
d)   antibodies of the I_gG class can act as opsonins
e)   all of the above are true

d   17.   Which one of the following is true of humoral immunity?
a)   humoral antibodies do not provide resistance to bacterial toxins
b)   when an antigen reaches lymphoid tissue, large numbers of B lymphocytes become helper T cells
c)   the humoral system has no memory component
d)   plasma cells produce antibodies

d   18.   Which one of the following is true of cell-mediated immunity?
          a)   T cells release kinins
          b)   neutrophils serve as a memory component of the system
          c)   it is not involved in the rejection of solid-tissue transplants
          d)   it provides resistance to fungi and intracellular bacteria
          e)   none of the above is true

b   19.   The plasma of a person with the B antigen contains which one of the
          following antibodies?
          a)   anti-B
          b)   anti-A
          c)   anti-A and anti-B
          d)   neither anti-A nor anti-B

d   20.   Which one of the following is not a potential consequence of
          transfusing incompatible blood types?
          a)   hemolysis
          b)   blockage of small blood vessels
          c)   agglutination of erythrocytes
          d)   all of the above are potential consequences of transfusing
                incompatible blood types

c   21.   A person of blood type B could donate blood to persons of which of
          the following blood types?
          a)   O
          b)   A
          c)   AB
          d)   none of the above is correct

c   22.   The father is Rh positive and the mother Rh negative. They have had
          three previous children without adverse problems due to Rh factor.
          One of these children was Rh positive. The mother is pregnant again.
          The risk to the fetus now within the uterus is _____ than in the
          previous pregnancies.
          a)   less
          b)   no more
          c)   greater
          d)   none of the above is correct

d   23.   Which one of the following is not an example of passive immunity?
          a)   antibodies from the mother moving across the placenta to the
                fetus
          b)   getting a shot of anti-cobra serum
          c)   shortly after birth injecting Rh negative mothers with anti-Rh
                antibodies in order to limit Rh sensitization
          d)   vaccination

c   24.   Which one of the following is true of anaphylactic shock?
        a)   it is a localized rather than a systemic reaction
        b)   severe hypertension can result
        c)   in a sensitive person, it may result from a single insect sting
        d)   histamine is injected as a countermeasure

e   25.   Which one of the following is not true of tolerance?
        a)   it is the failure of the body to mount a specific immune response against a particular antigen
        b)   it may result from clonal deletion of T and B cells
        c)   it is important in the body's recognition of itself
        d)   tolerance is more easily established in the fetus than in the adult
        e)   all of the above are true of tolerance

c   26.   Which one of the following is true of the metabolism of foreign chemicals?
        a)   chemicals toxic to the body are always polar substances that are not very soluble in lipids
        b)   toxic substances cannot enter through the skin
        c)   metabolic processing sometimes converts toxic substances to less-toxic ones
        d)   the pathways for processing foreign chemicals are different from those used to process chemicals normally required by the body
        e)   none of the above is true

a   27.   Which one of the following is a hypersensitivity due to effector T cells?
        a)   exposure of a sensitized individual to poison oak toxin
        b)   glomerulonephritis
        c)   anaphylactic shock
        d)   exposure of a sensitized individual to dust

d   1.   Which one of the following listed is passed thirdly by air during
           inspiration?
           a)   nasopharynx
           b)   vestibule
           c)   nasal cavity
           d)   internal nares

a   2.   The roof of the nasal cavity is formed by the:
           a)   ethmoid bone
           b)   superior concha
           c)   middle concha
           d)   nasal bones

b   3.   Which one of the following bones does not contain a paranasal sinus?
           a)   frontal
           b)   inferior concha
           c)   ethmoid
           d)   sphenoid

a   4.   Which one of the following is incorrect regarding the mucous
           membranes of the nasal cavity?
           a)   they have stratified squamous epithelium
           b)   they have an extensive blood supply
           c)   they saturate the air passing over them with water
           d)   they become inflamed due to infections such as colds

c   5.   The following listed are parts of the respiratory system and an
           opening that communicates with that part. Which one is not correctly
           paired with its opening?
           a)   nasopharynx-internal nares
           b)   larynogopharynx-glottis
           c)   laryngopharynx-fauces
           d)   nasopharynx-auditory tube
           e)   nasal cavity-vestibule

d   6.   Adenoids are enlarged:
           a)   lingual tonsils
           b)   palatine tonsils
           c)   fauces
           d)   pharyngeal tonsils

b   7.   Which one of the following listed is an unpaired cartilage of the
           larynx?
           a)   arytenoid
           b)   cricoid
           c)   cuneiform
           d)   corniculate

d   8.   The pitch of the voice is regulated by movement of the:
a)   thyroid cartilage
b)   cricoid cartilage
c)   epiglottis
d)   arytenoid cartilage

d   9.   Which one of the following structures listed is passed thirdly by air during expiration?
a)   trachea
b)   respiratory bronchiole
c)   alveoli
d)   terminal bronchiole

b   10.   Which one of the structures listed would have no cartilage at all in its walls?
a)   primary bronchi
b)   respiratory bronchiole
c)   secondary bronchi
d)   trachea

c   11.   Each bronchopulmonary segment represents the portion of a lung supplied by a specific:
a)   primary bronchus
b)   secondary bronchus
c)   tertiary bronchus
d)   terminal bronchiole

a   12.   Which one of the following structures listed would be pierced thirdly by a point of a pin entering the body from the outside?
a)   pleural fluid
b)   skin
c)   lung
d)   parietal pleura

b   13.   The bronchi are supplied with blood by the:
a)   intercostal arteries
b)   small branches from the descending aorta
c)   branches of the pulmonary artery
d)   they do not need any, because they get all the oxygen they need from the air passing through them

c   14.   Which one of the following parts of a respiratory membrane is passed thirdly by a molecule of oxygen leaving the lung and entering the blood stream?
a)   endothelium
b)   alveolar epithelium
c)   endothelial basement membrane
d)   alveolar basement membrane

a   15. Which one of the following statements is incorrect?
- a) because of the surface tension of the pleural fluid, the lung is held tightly against the visceral pleura
- b) the lung lacks muscles for breathing
- c) air moves into the lungs when the intrapulmonary pressure is less than atmospheric pressure
- d) between breaths the pressure in the lungs is equal to atmospheric pressure

d   16. Which one of the following statements is correct?
- a) contraction of the diaphragm aids in expiration
- b) at rest, expiration is usually an active process
- c) during forced expiration, the external intercostal muscles are contracted
- d) respiration is controlled by neurons whose cell bodies are found in the medulla of the brain

b   17. Vital capacity for young males is about _____ ml.
- a) 5900
- b) 4700
- c) 3000
- d) 1200

b   18. The condition where the lung tissue walls off bacterial infection is known as:
- a) emphysema
- b) tuberculosis
- c) pneumonia
- d) pleurisy

d   19. The condition where the alveoli become over-distended and replaced by fibrous tissue is known as:
- a) tuberculosis
- b) pneumonia
- c) pleurisy
- d) emphysema

c   20. Since atmospheric air is 20.9% oxygen, and since the barometric pressure at 10,000 feet is 523 mm. of Hg, the partial pressure of oxygen at that elevation in the inspired air is _____ mm. of Hg.
- a) 25
- b) 159
- c) 109
- d) 523
- e) 2502

c 21. Which one of the following is something that opposes the collapse of the lungs?
   a) intrapleural pressure
   b) the elastic components of the lung wall
   c) surfactant
   d) the surface tension of the fluid that coats the alveoli
   e) all of the above favor the collapse of the lungs

d 22. Which one of the following is true of quiet exhalation?
   a) intrapulmonary pressures are lowered
   b) the external intercostal muscles contract
   c) the diaphragm muscles contract, making the diaphragm rise
   d) none of the above is true

e 23. Which one of the following does not reduce resistance to air flow?
   a) epinephrine
   b) antihistamine
   c) sympathetic stimulation of the muscles of the respiratory passageways
   d) an increase in the diameter of the respiratory passageways
   e) none of the above is correct

b 24. Children who are born without enough pulmonary surfactant have difficulty breathing. This condition is known as:
   a) emphysema
   b) hyaline membrane disease
   c) acute coryza
   d) pleurisy
   e) none of the above is correct

e 25. A normal figure for minute respiratory volume is ____ ml. per minute.
   a) 500
   b) 800
   c) 3400
   d) 4400
   e) 6000

a 26. The total volume occupied by inhaled air that is not in contact with the alveolar exchange surfaces is known as the:
   a) anatomical dead space
   b) pneumothorax
   c) residual volume
   d) inspiratory reserve volume
   e) none of the above is correct

d   27.   Which one of the following is correct?
     a)   the volume of air that enters the respiratory passageways is more important than the volume of air that enters the alveoli
     b)   carbon dioxide causes pulmonary blood vessels to dilate
     c)   in an area of alveoli whose airflow is excessive in relation to their blood flow, the pulmonary blood vessels in this area will constrict
     d)   none of the above is correct

a   28.   In the alveoli, the partial pressure of oxygen is about ___ mm. of Hg.
     a)   100
     b)   159
     c)   40
     d)   0.3
     e)   none of the above is correct

c   29.   Within the tissues of the body where metabolic processes are taking place, the partial pressure of carbon dioxide is about _____ mm. of Hg.
     a)   159
     b)   100
     c)   45
     d)   40
     e)   0.3

d   30.   Which of the following is correct?
     a)   the higher the temperature, the more oxygen is bound to hemoglobin
     b)   the higher the hydrogen ion concentration, the more oxygen is bound to hemoglobin
     c)   the higher the partial pressure of carbon dioxide, the more oxygen is bound to hemoglobin
     d)   none of the above is correct

c   31.   The smallest amount of carbon dioxide in the blood travels in the form of:
     a)   bicarbonate ions
     b)   carbamino compounds
     c)   dissolved carbon dioxide gas
     d)   carboxyhemoglobin

c   32.   Which one of the following does not occur in the deep tissues of the
          body?
          a)   carbon dioxide diffuses from the tissues and moves into red blood
               cells
          b)   within the erythrocytes, carbonic anhydrase facilitates the
               production of carbonic acid
          c)   bicarbonate ion binds with hemoglobin
          d)   chloride ion enters the erythrocytes

d   33.   Inspiratory neurons arising in the medulla can be inhibited by:
          a)   impulses from stretch receptors in the lungs
          b)   impulses from the apneustic center
          d)   two of the above are correct
          e)   none of the above is correct

d   34.   In normal quiet breathing, inspiration is brought about by:
          a)   inhibition of expiration
          b)   contraction of the internal intercostal muscles
          c)   stimulation of parasympathetic neurons
          d)   contraction of the diaphragm
          e)   none of the above is correct

d   35.   Which one of the following is not a cause of stagnant hypoxia?
          a)   heart failure
          b)   emboli
          c)   arteriosclerosis
          d)   sickle cell anemia
          e)   all of the above are causes of stagnant hypoxia

d   36.   Cutting the nerves that go from the lung stretch receptors to the
          inspiratory center will:
          a)   increase the breathing rate markedly
          b)   decrease the breathing rate markedly
          c)   completely stop breathing
          d)   have no marked effect on breathing rate

b   37.   Which one of the following will lead to a decrease in alveolar
          ventilation?
          a)   increasing the partial pressure of oxygen in inspired air by 25%
          b)   decreasing the partial pressure of carbon dioxide in arterial blood
               by 25%
          c)   increasing the hydrogen ion concentration of arterial blood by
               25%
          d)   two of the above are correct

c 38. Which one of the following is paired correctly?
    a) apneustic center-inhibits inspiration
    b) pneumotaxic center-stimulates inspiration
    c) lung stretch receptors-inhibit respiration
    d) all of the above are correct
    e) two of the above are correct

d 39. Which one of the following is true of the chemoreceptors in the aortic arch and carotid artery?
    a) they are more sensitive to carbon dioxide than to hydrogen ions
    b) an increase in the partial pressure of oxygen increases the discharge rates of these receptors
    c) they are responding to the presence of oxygen in arterial blood rather than to the amount of oxygen within their local tissue environment
    d) none of the above is correct

c 40. Which one of the following is correct?
    a) the carotid and aortic bodies are more important than the central chemoreceptors in terms of their effect on respiratory rates
    b) the chemical to which the central chemoreceptors respond directly is carbon dioxide
    c) the expulsion of carbon dioxide from the lungs can increase the pH of the blood
    d) arterial pH is more important than arterial partial pressure of carbon dioxide in normal respiratory regulation

c   1.   Which one of the following listed is passed thirdly by air during expiration?
     a)   external nares
     b)   nasopharynx
     c)   nasal cavity
     d)   vestibule

c   2.   The floor of the nasal cavity is formed partly by the:
     a)   inferior conchae
     b)   vomer
     c)   maxillary
     d)   ethmoid

d   3.   Which one of the following does not open into or out of the nasal cavities?
     a)   nasolacrimal ducts
     b)   paranasal sinuses
     c)   vestibule
     d)   middle meatus

b   4.   Which one of the following is incorrect regarding mucous membranes of the nasal cavity?
     a)   they have a ciliated pseudostratified columnar epithelium
     b)   infections may pass to the eye via the opening of the frontal sinus
     c)   infection can pass to the mucous membrane of the pharynx
     d)   infection can pass to the middle ear via the auditory tube

d   5.   The following are parts of the pharynx and an opening into or out of that part. Which part is not correctly paired with the right opening?
     a)   oropharynx-fauces
     b)   nasopharynx-auditory tube
     c)   nasopharynx-oropharynx
     d)   oropharynx-esophagus

b   6.   Which one of the following listed does not have a mucous membrane made of the same epithelium as the others?
     a)   mouth
     b)   nasopharynx
     c)   oropharynx
     d)   laryngopharynx

a   7.   Which one of the following listed is a paired cartilage of the larynx?
     a)   arytenoid
     b)   thyroid
     c)   epiglottis
     d)   cricoid

c   8.   The vocal cords are attached to the:
           a)   thyroid and cricoid cartilages
           b)   hyoid bone and cricoid cartilages
           c)   thyroid and arytenoid cartilages
           d)   thyroid and cuneiform cartilages

c   9.   Which one of the following structures listed is passed thirdly by air during inspiration?
           a)   secondary bronchi
           b)   alveolar sac
           c)   respiratory bronchiole
           d)   terminal bronchiole

d   10.   Which one of the following structures is not lined with ciliated pseudostratified squamous epithelium?
           a)   trachea
           b)   terminal bronchiole
           c)   secondary bronchi
           d)   alveoli

a   11.   The region where the bronchi and blood vessels enter and leave the lung is known as the:
           a)   hilus
           b)   base
           c)   apex
           d)   cardiac notch

b   12.   Which one of the following structures listed would be pierced thirdly by a point of a pin entering the body from the outside?
           a)   pleural cavity
           b)   visceral pleura
           c)   parietal pleura
           d)   lungs

c   13.   The alveoli are supplied with blood by the:
           a)   intercostal arteries
           b)   branches of the descending aorta
           c)   branches of the pulmonary artery
           d)   they do not need any, because they are in the lungs

d   14.   Which one of the following parts of the respiratory membrane is passed thirdly by a molecule of carbon dioxide leaving the blood?
           a)   endothelium
           b)   alveolar epithelium
           c)   endothelial basement membrane
           d)   alveolar epithelium basement membrane

c   15.   Which one of the following statements is incorrect?
    a)   at rest about 500 ml. of air enters the lungs with each breath
    b)   only 5% of alveolar air is exchanged with each breath
    c)   the lungs actively expand and contract
    d)   the lungs are suspended in an air-tight cavity

b   16.   Which one of the following statements is correct?
    a)   the greatest increase in the volume of the thoracic cavity is brought about by the elevation of the ribs
    b)   the external intercostal muscles act to increase the size of the thoracic cavity
    c)   when the diaphragm contracts, the size of the thoracic cavity decreases
    d)   diaphragmatic breathing is the major method of lowering intrapulmonary pressures during forced inspiration

d   17.   For young males the inspiratory reserve volume is about ____ ml.
    a)   500
    b)   1200
    c)   4700
    d)   none of the above is correct

c   18.   An acute or chronic inflammation of the bronchial tree caused by bacterial infection or irritants is known as:
    a)   emphysema
    b)   hay fever
    c)   bronchitis
    d)   bronchial asthma

a   19.   Collapse of a lung is known as:
    a)   atelectasis
    b)   pleurisy
    c)   pneumothorax
    d)   pulmonary failure

b   20.   Assuming an atmospheric pressure equivalent to that at sea level, and knowing that the amount of oxygen in air is about one-fifth of the total pressure, then the partial pressure of the oxygen base in air at sea level is ____ Hg.
    a)   175
    b)   150
    c)   200
    d)   225
    e)   none of the above is correct

c   21. Which one of the following is not a force that favors the collapse of the lungs?
- a) intrapleural pressure
- b) the surface tension of the fluid that coats the alveoli
- c) the surface tension between the visceral and the parietal pleura
- d) the elastic components of the wall of the lungs
- e) all of the above are forces that favor the collapse of the lungs

b   22. Which one of the following is not true of quiet exhalation?
- a) intrapulmonary pressure rise
- b) the volume of the thoracic cavity is decreased by processes that involve muscle contraction
- c) the diaphragm rises
- d) all of the above are true

a   23. Which one of the following is not a factor in increasing the resistance to air flow?
- a) epinephrine
- b) histamine
- c) accumulation of mucus within passageways
- d) parasympathetic stimulation of the muscles of the respiratory passageways
- e) all of the above increase resistance to air flow

d   24. Pulmonary surfactant acts to:
- a) lower the surface tension of the alveoli
- b) prevent collapse of the alveoli
- c) prevent dehydration of the alveolar surface
- d) all of the above are correct

c   25. An individual with normal tidal volume is given morphine to relieve pain. Following administration of the drug, the rate of breathing decreases to 12 times per minute. The minute volume of respiration would be _____ liters per minute.
- a) 8
- b) 7.5
- c) 6.0
- d) 7.2
- e) none of the above is correct

e   26. The volume of the anatomical dead space is ___ ml.
- a) 15
- b) 50
- c) 100
- d) 500
- e) none of the above is correct

d   27.   Which one of the following is not true?
a) the volume of air that enters the alveoli from the atmosphere is more important than the volume of air that enters the respiratory passageways
b) carbon dioxide causes pulmonary blood vessels to constrict
c) alveolar airflow and bloodflow are usually closely matched
d) all of the above are true

d   28.   In the alveoli the partial pressure of carbon dioxide is about _____ mm. of Hg.
a) the same as that in the atmospheric air outside of the body
b) 100
c) 159
d) 40
e) it is dependent on cardiac output

d   29.   Within the tissues of the body where metabolic processes are taking place, the partial pressure of oxygen is about ____ mm. of Hg.
a) 159
b) 100
c) 45
d) 40
e) 0.3

e   30.   Which one of the following is not a factor that affects the binding of oxygen to hemoglobin?
a) partial pressure of oxygen
b) temperature
c) hydrogen ion concentration
d) carbon monoxide
e) all of the above are factors

a   31.   Most of the carbon dioxide transported by the blood is carried as:
a) bicarbonate ions
b) carbamino compounds
c) dissolved carbon dioxide gas
d) carboxyhemoglobin

d   32.   Which one of the following does not occur in the lungs?
a) carbon dioxide diffuses from the blood into the alveoli
b) within the erythrocyte hydrogen ions combine with bicarbonate ions
c) carbonic anhydrase breaks down carbonic acid
d) chloride ions enter the erythrocyte

a   33. In normal quiet breathing, expiration is brought about by:
        a)   inhibition of the inspiration
        b)   stimulation of sympathetic neurons
        c)   stimulation of expiratory muscles
        d)   stimulation of the diaphragms
        e)   none of the above is correct

c   34. Inspiratory neurons arising in the medulla can be stimulated by:
        a)   impulses from stretch receptors in the lungs
        b)   impulses from the pneumotaxic center
        c)   impulses from the apneustic center
        d)   all of the above are correct

e   35. Which one of the following is a cause of hypoxemic hypoxia?
        a)   emboli
        b)   deficiency of hemoglobin
        c)   heart failure
        d)   occluded arteries
        e)   none of the above is a cause of hypoxemic hypoxia

c   36. Cutting the nerves that go from the respiratory center in the medulla
        to the respiratory muscles will:
        a)   increase the breathing rate
        b)   decrease the breathing rate
        c)   completely stop breathing
        d)   do none of the above

e   37. Which one of the following will lead to an increase in alveolar
        ventilation?
        a)   decreasing the partial pressure of oxygen of inspired air by 25%
        b)   the partial pressure of carbon dioxide in arterial blood by 25%
        c)   increasing the hydrogen ion concentration of arterial blood
             by 25%
        d)   a blood clot that blocks the artery leading to the right lung
        e)   two of the above are correct

e   38. Which one of the following is not paired correctly?
        a)   apneustic center-stimulates inspiration
        b)   pneumotaxic center-inhibits inspiration
        c)   lung stretch receptors-inhibit respiration
        d)   all of the above are correct

b   39.   Which one of the following is not true of the chemoreceptors in the aortic arch and carotid artery?
- a)   they stimulate respiration
- b)   they have a partial pressure of oxygen almost identical to that in venous blood
- c)   they are sensitive to changes in oxygen due to moderate anemia
- d)   under certain conditions such as severe low blood pressure, they will increase their rates of discharge

c   40.   Which one of the following is not correct?
- a)   small increases in the partial pressure of carbon dioxide in the plasma have major effects on alveolar ventilation rates
- b)   the carotid and aortic bodies are less important than the central chemoreceptors in terms of their effect on respiratory rates
- c)   the chemical to which the central chemoreceptors respond directly is carbon dioxide
- d)   all of the above are true

NAME_____

SECTION _____

d   1. The mouth develops from:
      a)   the proctodeum
      b)   the foregut
      c)   the mesoderm
      d)   the stomodeum

c   2. The soft palate separates the oral cavity from the:
      a)   hard palate
      b)   oropharynx
      c)   nasopharynx
      d)   laryngopharynx

b   3. The cone-shaped structures distributed in V-shaped rows over the dorsal surface of the tongue are the:
      a)   taste buds
      b)   filiform papillae
      c)   vallate papillae
      d)   fungiform papillae

c   4. The teeth that are adapted for cutting are the:
      a)   bicuspids
      b)   molars
      c)   incisors
      d)   canines

b   5. Most of the tooth is composed of:
      a)   enamel
      b)   dentin
      c)   pulp
      d)   cementum

c   6. The gland that has a duct that crosses the masseter muscle and empties into the vestibule is known as the _____ gland.
      a)   bucal
      b)   sublingual
      c)   parotid
      d)   submandibular

d   7. Which one of the following listed is pierced thirdly by the point of a pin entering the esophagus from the outside and going through to the lumen?
      a)   tunica mucosa
      b)   tunica adventitia
      c)   tunica muscularis
      d)   tunica submucosa

c   8.   Which one of the following listed is pierced thirdly by the point of a pin passing from the lumen of the gut to the outside?
    a)   basement membrane
    b)   muscularis mucosae
    c)   lamina propria
    d)   epithelium

c   9.   The part of the stomach that bulges above the entrance of the esophagus is known as the:
    a)   body
    b)   pyloric region
    c)   fundus
    d)   greater omentum

d   10.   Longitudinal folds of the mucosa and the submucosa of the stomach are called:
    a)   gastric glands
    b)   fundic glands
    c)   cardiac glands
    d)   none of the above is correct

c   11.   The cells that secrete pepsinogen are the:
    a)   parietal cells
    b)   neck cells
    c)   chief cells
    d)   cells of the cardiac glands

a   12.   Pepsin initiates the chemical breakdown of:
    a)   proteins
    b)   complex carbohydrates
    c)   lipids
    d)   vitamins

a   13.   Which one of the following listed is passed thirdly by food going through the digestive tract?
    a)   ileum
    b)   cecum
    c)   duodenum
    d)   jejunum
    e)   colon

d   14.   Which one of the following produces mucus?
    a)   duodenal glands
    b)   cardiac glands
    c)   goblet cells
    d)   all of the above produce mucus

c   15. Which one of the following listed is passed thirdly by food?
   a)   sigmoid colon
   b)   ascending colon
   c)   descending colon
   d)   transverse colon

d   16. The portion of the gut in front of the sacrum is known as the:
   a)   cecum
   b)   sigmoid colon
   c)   appendix
   d)   rectum

d   17. Which one of the following is correct?
   a)   the internal anal sphincter is formed from the longitudinal muscles of the tunica muscularis
   b)   the external anal sphincter is under involuntary control
   c)   the internal anal sphincter is made of skeletal muscle
   d)   the internal anal sphincter is under involuntary control

b   18. The structures in the pancreas that secrete digestive enzymes are called:
   a)   islets
   b)   acini
   c)   lacteals
   d)   crypts

c   19. Which one of the following structures listed is passed thirdly by bile when the hepatopancreatic sphincter is contracted?
   a)   hepatic duct
   b)   bile canaliculi
   c)   common bile duct
   d)   cystic duct

d   20. The major process involved in bolum formation is:
   a)   deglutition
   b)   peristalsis
   c)   segmentation
   d)   mastication

a   21. Food is kept out of the nasopharynx by the:
   a)   soft palate
   b)   pharyngeal constrictor muscles
   c)   epiglottis
   d)   glottis

d   22.   Which one of the following is not correct?
- a)   food entering the stomach remains in a central position for a period of time
- b)   the pyloric sphincter is usually partly open
- c)   the gastric juices and food form a soup called chyme
- d)   a lot of food is forced past the pyloric sphincter with each peristaltic wave

d   23.   The major contractile activity in the small intestine is:
- a)   mass peristalsis
- b)   peristalsis
- c)   deglutition
- d)   segmentation

c   24.   Which one of the following is not correct?
- a)   segmentation and peristalsis occur within the colon
- b)   movement of material through the colon is slower than in the small intestine
- c)   the rectum is usually full
- d)   food in the stomach increases the motility of the colon

d   25.   Which one of the following structures listed is passed thirdly by a drop of blood?
- a)   hepatic artery
- b)   hepatic portal vein
- c)   sinusoid
- d)   central vein

c   26.   Which one of the following statements is true?
- a)   about 1100 ml. of blood flows through the liver per minute
- b)   proteins cannot diffuse out of the blood into the liver sinusoids
- c)   the liver sinusoids are lined by an endothelium
- d)   Kupffer cells in the liver make bile

b   27.   Plicae circulares are supported by and have a central core composed of:
- a)   lamina propria
- b)   submucosa
- c)   muscularis mucosa
- d)   muscularis externa

e   28.   A lacteal is found typically in each:
- a)   plica circularis
- b)   plica semilunaris
- c)   haustra
- d)   microvillus
- e)   none of the above is correct

d 29. Backflow of the intestinal contents from the cecum is prevented by the presence of:
   a) pocket valves
   b) parachute-type valves with tendinous cords
   c) a flutter valve within which a sphincter muscle exists
   d) a flutter valve with no sphincter

e 30. Which one of the following would inhibit gastric motility?
   a) stimulation of stretch receptors in the wall of the stomach
   b) gastrin
   c) secretin
   d) the enterogastric reflex
   e) two of the above inhibit gastric motility

e 31. Which one of the following would act to slow the motility of the small intestine?
   a) distension
   b) presence of hypertonic substances
   c) presence of excessively acidic substances
   d) gastroileal reflex
   e) none of the above is correct

d 32. Which one of the following is not a characteristic of the colon?
   a) haustra
   b) taeniae coli
   c) epiploic appendages
   d) plicae circulares
   e) all of the above are characteristic of the colon

a   1.   The anus develops from the:
           a)   proctodeum
           b)   hindgut
           c)   stomodeum
           d)   endoderm

b   2.   The space between the cheek and the teeth is known as the:
           a)   mouth
           b)   vestibule
           c)   oral cavity
           d)   fauces

a   3.   The dozen large structures forming an inverted V-shaped structure
           toward the back of the tongue are known as:
           a)   vallate papillae
           b)   fungiform papillae
           c)   filiform papillae
           d)   taste buds

d   4.   The teeth that are adapted for tearing are the:
           a)   incisors
           b)   molars
           c)   bicuspids
           d)   canines

a   5.   Which one of the following would be pierced thirdly by the point of
           a pin entering a tooth from the surface of the crown?
           a)   pulp
           b)   root canal
           c)   dentin
           d)   enamel

d   6.   The _____ gland opens via a single duct at the base of the
           frenulum.
           a)   buccal
           b)   parotid
           c)   sublingual
           d)   submandibular

a   7.   Which one of the following listed is pierced thirdly by the point of
           a pin passing from the lumen of the esophagus to the outside?
           a)   tunica muscularis
           b)   tunica adventitia
           c)   tunica mucosa
           d)   tunica submucosa

b   8.   Which one of the following structures listed is made up of or
           contains many of the same structures as the lamina propria?
           a)   tunica serosa
           b)   tunica submucosa
           c)   tunica muscularis
           d)   muscularis mucosa

a   9.   The opening from the esophagus into the stomach is known as the:
           a)   cardiac orifice
           b)   esophageal hiatus
           c)   pyloric sphincter
           d)   sphincter of Oddi

b   10.   The epithelium that lines the lumen of the stomach is:
           a)   simple squamous
           b)   simple columnar
           c)   stratified squamous
           d)   simple cuboidal

d   11.   The cells that secrete HCL are:
           a)   neck cells
           b)   zymogenic cells
           c)   chief cells
           d)   parietal cells

d   12.   Which one of the following is not a function of HCL?
           a)   kill bacteria
           b)   activate pepsin
           c)   aid in digestion of proteins
           d)   protect the gastric mucosa

a   13.   Which one of the following listed is passed thirdly by bile?
           a)   ileum
           b)   ampulla of Vater
           c)   duodenum
           d)   cecum

b   14.   Which one of the following does not serve to increase the surface
           area of the mucosa of the small intestine?
           a)   microvilli
           b)   plicae semilunares
           c)   villi
           d)   plicae circulares

d   15.   The three muscular bands that run the length of the colon are the:
           a)   haustra
           b)   epiploic appendages
           c)   plicae semilunares
           d)   taeniae coli

c   16. Which one of the following is not a correct statement about the
        rectum?
        a) it has no villi
        b) there are many mucous cells in the mucosa of the rectum
        c) it has taeniae coli
        d) the mucosal epithelium of the rectum is simple columnar
            epithelium

b   17. The mucosa of the anal canal form longitudinal folds called:
        a) rugae
        b) columns
        c) plicae
        d) haustra

d   18. The secretions of the pancreatic islets pass through the:
        a) common bile duct
        b) accessory pancreatic duct
        c) hepatopancreatic ampulla
        d) blood stream

a   19. Which one of the following structures listed is passed thirdly by
        bile when the hepatopancreatic sphincter is relaxed?
        a) common bile duct
        b) bile canaliculi
        c) hepatopancreatic ampulla
        d) cystic duct

b   20. Movement of food down the esophagus is by:
        a) segmentation
        b) peristalsis
        c) mass peristalsis
        d) deglutition

c   21. Food is kept out of the larynx by the:
        a) soft palate
        b) uvula
        c) epiglottis
        d) pharyngeal constrictor muscles

b   22. Mixing of chyme with digestive juices in the small intestine is done
        by:
        a) mastication
        b) segmentation
        c) peristalsis
        d) deglutition

a   23. Which one of the following is not true?
     a) the center for the defecation reflex is in the lumbar region of the spinal cord
     b) mass peristalsis initiates the defecation reflex
     c) the external anal sphincter is under voluntary control
     d) the defecation reflex is assisted by contracting the abdominal muscles

b   24. Which one of the following structures does not open directly into or out of a liver sinusoid?
     a) hepatic portal vein
     b) hepatic vein
     c) central vein
     d) hepatic artery

c   25. Which one of the following statements is correct?
     a) 400 ml. of blood arrive in the liver from the hepatic portal vein per minute
     b) Kupffer cells are found on the side of the liver cords away from the sinusoids
     c) the sinusoids contain a mixture of arterial and venous blood
     d) the oxygen content of the hepatic portal vein is high

d   26. Which one of the following is not a function of mucous membranes?
     a) protective
     b) secretory
     c) absorptive
     d) all of the above are functions of mucous membranes

e   27. Gastric glands are found in the _____ of the stomach.
     a) submucosa
     b) muscularis mucosa
     c) serosa
     d) epithelial layer
     e) none of the above is correct

a   28. The center of each microvillus contains:
     a) cytoplasm
     b) lamina propria, lacteal, vascular capillaries
     c) submucosa
     d) muscularis externa and serosa

d   29. Which one of the following would increase the movement of material into the cecum and large intestine?
     a) entrogastric reflex
     b) gastroileal reflex
     c) gastrin
     d) two of the above are correct
     e) all of the above are correct

a   30. Which one of the following does not inhibit gastric motility?
      a) gastrin
      b) secretin
      c) cholecystokinin
      d) enterogastric reflex

d   31. The result of rhythmic contractions that produce stationary constrictions in the wall of the intestine is called:
      a) peristalsis
      b) mass peristalsis
      c) deglutition
      d) segmentation
      e) none of the above is correct

a   32. The act of swallowing is divided into three stages. Which one of these is voluntary?
      a) only the first
      b) only the second
      c) the first and second
      d) the second and third
      e) all three stages are voluntary

d 1. Which one of the following is correct?
       a) most of the saliva comes from the parotid glands
       b) the submandibular glands produce only a serous secretion
       c) the parotid and buccal glands produce primarily a mucous
           secretion
       d) none of the above is correct

d 2. Which one of the following is not true of salivary amylase?
       a) it is an enzyme
       b) it has a stable range of pH 4-11
       c) it functions best at pH of 6.9
       d) it splits starch into polysaccharides
       e) all of the above are correct

a 3. Which one of the following is not a component of saliva?
       a) bradykinin
       b) kallikrein
       c) water
       d) electrolytes
       e) mucins

e 4. Which one of the following does not produce mucus?
       a) pyloric glands
       b) cardiac glands
       c) gastric glands
       d) goblet cells
       e) all of the above produce mucus

b 5. Which one of the following is not correct?
       a) the pH of the HCL secreted into the stomach is about 0.8
       b) the HCL of the stomach is produced by zymogenic cells
       c) the pH of the venous blood leaving the actively secreting
           stomach is higher than that of the arterial blood that flows to
           the stomach
       d) the HCL secreted into the stomach kills many bacteria that
           enter the digestive tract.
       e) all of the above are correct

e 6. The activation of pepsinogen is mediated by which of the following?
       a) gastrin
       b) enterogastrone
       c) HCL
       d) pepsin
       e) two of the above are correct

b   7.   Which one of the following is not true of the gastric phase of gastric secretion?
- a) a vagovagal reflex is involved
- b) a release of secretin occurs
- c) there is an increased flow of digestive juices from the stomach mucosa
- d) gastrin release occurs

b   8.   One action of gastrin in the stomach is to:
- a) stimulate release of secretagogues
- b) stimulate release of hydrochloric acid
- c) stimulate release of cholecystokinin
- d) none of the above

d   9.   Which one of these is not a function of secretin?
- a) stimulates bicarbonate-rich, watery pancreatic secretion
- b) stimulates duodenal glands
- c) stimulates the secretion of bile
- d) stimulates gastric acid secretion when gastrin is actively being produced

c   10.   The stomach is protected by all but which one of the following?
- a) mucous secretions
- b) rapid cell divisions of the mucosal epithelium
- c) gap junctions
- d) alkaline pH of the material that coats the mucosal epithelium

e   11.   Which one of the following would not initiate the enterogastric reflex?
- a) excessively acidic chyme in the duodenum
- b) hypertonic chyme in the duodenum
- c) hypotonic chyme in the duodenum
- d) distension of the duodenum
- e) all of the above initiate the enterogastric reflex

c   12.   Enzyme release by the pancreas is under control of which of these?
- a) secretin
- b) pepsin
- c) cholecystokinin
- d) gastric inhibitory peptide
- e) two of the above are correct

d   13.   Which one of the following is not true of pancreatic juice?
- a) isotonic
- b) basic pH
- c) contains bicarbonate ions
- d) contains insulin
- e) all of the above are true of pancreatic juice

c   14.   The enterohepatic circulation involves:
   a) bilirubin
   b) secretin
   c) bile salts
   d) bicarbonate ion
   e) kallikrein

e   15.   Which one of the following would decrease bile secretion?
   a) choleretics
   b) secretion
   c) increased hepatic blood flow
   d) stimulation of the vagus nerve
   e) none of the above is correct

a   16.   Which one of the following is true?
   a) the gall bladder can store about 60 ml. of bile
   b) the gall bladder does not release bile into the duodenum until several hours after a meal
   c) the concentration of electrolytes may increase five to ten times in the gall bladder
   d) none of the above is true

d   17.   Duodenal glands are also known as:
   a) lobules
   b) crypts (of Lieberkuhn)
   c) pits
   d) Brunner's glands

b   18.   Which one of the following is not true of intestinal juice?
   a) it has a pH of 6.5-7.5
   b) its enzymes are secretions of cells that cover most of each villus
   c) its production is stimulated by cholecystokinin
   d) its production is stimulated by local reflexes
   e) all of the above are correct

b   19.   A chronic inflammation of the liver that is progressive and diffuse is known as:
   a) hepatitis
   b) cirrhosis
   c) a gall stone attack
   d) jaundice

c   20.   Sucrose is composed of:
   a) two glucose molecules
   b) a glucose and a galactose molecule
   c) a fructose and a glucose molecule
   d) none of the above is correct

a   21. Which one of the following is not correct?
     a) final digestion of disaccharides to monosaccharides is done by some of the enzymes in pancreatic juice
     b) sugars are absorbed primarily in the duodenum
     c) glucose is absorbed by active transport
     d) fructose is absorbed by dialysis

a   22. Enterokinase:
     a) activates trypsinogen
     b) activates pepsinogen
     c) stimulates pancreatic enzyme release
     d) stimulates pancreatic bicarbonate release
     e) hydrolyzes proteins

b   23. What role does bile play in digestion?
     a) contains fat-splitting enzymes
     b) emulsification of fat
     c) slows peristalsis, thus allowing more time for digestion and absorption in the small intestine
     d) releases energy stored in fats
     e) none of the above is correct

b   24. Which one of the following splits peptide bonds linking aromatic amino acids such as tyrosine and phenylalanine?
     a) trypsin
     b) chymotrypsin
     c) carboxypeptidase
     d) enterokinase
     e) pepsin

d   25. Which one of the following is not correct?
     a) protein is 60-80% digested and absorbed by the time the chyme reaches the ileum
     b) amino acids and dipeptides are taken up by active transport
     c) the microvilli of the intestinal mucosa contain peptidase enzymes
     d) active protein splitting enzymes are secreted directly from the pancreas

e   26. Chylomicrons are mostly:
     a) phospholipids
     b) free fatty acids
     c) cholesterol
     d) protein
     e) triacylglycerols

b 27. Which one of the following would not be associated with micelles?
  a) bile salts
  b) vitamin B
  c) free fatty acids
  d) vitamin A
  e) cholesterol

d 28. Which one of the following is not absorbed by active transport in the duodenum?
  a) potassium
  b) magnesium
  c) phosphate
  d) chloride
  e) calcium

d 29. Which statement best describes the relationship between the gastric mucosa and vitamin B12? The mucosa:
  a) secretes the vitamin, which is absorbed further along in the ilium
  b) stores the vitamin
  c) converts the vitamin to other B vitamins
  d) secretes a substance that promotes the absorption of the vitamin
  e) none of the above is correct

e 30. Which one of the following is not correct?
  a) there are no villi in the large intestine
  b) the large intestine has many mucus secreting glands
  c) bicarbonate ions are actively secreted by the large intestine
  d) microbes in the large intestine produce vitamins
  e) all of the above are correct

c 1. Which one of the following is not correct?
       a)   about 2/3rds of the salivary output comes from the
            submandibular glands
       b)   the sublingual glands produce a mucous secretion
       c)   mucins are inactive salivary amylase proteins
       d)   the parotid glands produce serous secretions

c 2. Which one of the following is true of salivary amylase?
       a)   it has a narrow pH range at which it is stable
       b)   its pH optimum is around 8.2
       c)   it splits starch into disaccharides
       d)   it is able to continue working until the food enters the
            duodenum

b 3. Increased blood flow to salivary glands due to local vasodilation is
       due to the salivary glands secreting a substance known as:
       a)   mucin
       b)   kallikrein
       c)   bradykinin
       d)   none of the above is correct

a 4. Which one of the following produces mucus as well as considerable
       other secretions?
       a)   gastric glands
       b)   pyloric glands
       c)   cardiac glands
       d)   goblet cells
       e)   none of the above is correct

d 5. Which one of the following is correct?
       a)   the cells that produce HCL are known as chief cells
       b)   the pH of the acid secreted into the stomach is 1.6
       c)   the chloride ions of the HCL in the stomach are passively
            transported from the blood to the stomach lumen
       d)   none of the above is correct

e 6. The protein splitting enzyme in the stomach is:
       a)   trypsin
       b)   chymotrypsin
       c)   carbohypeptidase
       d)   amylase
       e)   none of the above is correct

c 7. Which one of the following is not true of the cephalic phase of
       gastric secretion?
       a) gastric secretion can be initiated by the sight, smell, or
           thought of food
       b) severing the vagus nerve stops secretion due to the cephalic
           phase
       c) fear or depression activate the cephalic phase
       d) the cephalic phase is basically a conditioned reflex

d 8. Which one of the following would not promote gastric secretion?
       a) vagovagal reflex
       b) secretagogues
       c) distension of the stomach by food
       d) high concentrations of hydrogen ions in the stomach

a 9. Which one of the following is not a function of cholecystokinin?
       a) stimulate gastric secretion when the stomach is full of food
       b) stimulate the contraction of the gall bladder
       c) stimulate the release of enzymes from the pancreas
       d) all of the above are functions

e 10. Which one of the following would not have an inhibitory effect on
        gastric secretion when the stomach contains food and gastric
        secretion is strongly stimulated?
        a) distension of the duodenum
        b) irritants in the duodenum
        c) enterogastric reflex
        d) secretin
        e) all of the above have an inhibitory effect on gastric secretion

e 11. The reason that the stomach mucosa is not digested by the enzymes
        within the lumen of that organ is because:
        a) the enzymes do not stay that long within the stomach
        b) the enzymes are destroyed when they come in contact with the
            mucosa
        c) the enzymes are produced in an inactive form within the stomach
            and are not activated until they reach the intestine
        d) the microvilli of the stomach mucosa are constantly being
            replaced every two weeks
        e) none of the above is correct

a 12. Which one of the following causes the release of a bicarbonate rich,
        watery secretion from the pancreas?
        a) secretin
        b) cholecystokinin
        c) pepsinogen
        d) none of the above

d   13. Which one of the following is not an exocrine pancreatic secretion?
         a)   carboxxpeptidase
         b)   lipases
         c)   ribonuclease
         d)   pepsin

d   14. Which one of the following is an example of a bile salt?
         a)   bilirubin
         b)   cholesterol
         c)   lecithin
         d)   none of the above is correct

c   15. Most of the bile salts that pass from the gall bladder to the
         duodenum are:
         a)   broken down during the digestive process
         b)   pass out with the feces
         c)   are recycled via the enterohepatic circulation
         d)   none of the above is correct

b   16. Which one of the following is not correct?
         a)   the gall bladder stores and concentrates bile produced by the
               liver
         b)   choleretics decrease bile production
         c)   bile secretion is chemically, hormonally, and neurally
               controlled
         d)   as much as 600 to 1000 ml. of bile may be produced per day
         e)   all of the above are correct

e   17. Which one of the following does not stimulate mucus secretion by
         duodenal glands?
         a)   vagus nerve
         b)   secretin
         c)   tactile stimulation of the mucosa
         d)   irritating stimulation of the mucosa
         e)   all of the above stimulate duodenal glands

a   18. The main stimulus for the production of intestinal juice is:
         a)   local reflexes
         b)   cholecystokinin
         c)   secretin
         d)   vagus nerve

b   19. An acute or chronic inflammation of the mucosa of the stomach or
         intestine is known as:
         a)   an ulcer
         b)   gastroenteritis
         c)   appendicitis
         d)   heart burn
         e)   none of the above is correct

d 20. Which one of the following is not a disaccharide?
    a)   maltose
    b)   sucrose
    c)   lactose
    d)   fructose

c 21. Which one of the following is not correct?
    a)   starch and sucrose are the major dietary carbohydrates
    b)   the microvilli of the duodenum contain disaccharidases
    c)   fructose is absorbed by active transport
    d)   sugar absorption is complete by the time the chyme reaches the ileum

b 22. Chymotrypsinogen is activated by :
    a)   enterokinase
    b)   trypsin
    c)   carboxypeptidase
    d)   pepsin
    e)   HCL

d 23. The first step in fat digestion is:
    a)   formation of micelles
    b)   breakdown of fatty acids by lipases
    c)   formation of monoacylglycerols
    d)   none of the above is correct

d 24. Which one of the following splits peptide bonds involving amino acids such as lysine and arginine?
    a)   chymotrypsin
    b)   carboxypeptidase
    c)   pepsin
    d)   trypsin
    e)   none of the above is correct

e 25. Which one of the following would not be an enzyme associated with intestine epithelial cells?
    a)   lipase
    b)   tripeptidases
    c)   aminopeptidases
    d)   disaccharidases
    e)   all of the above are associated with intestinal epithelial cells

c 26. Minute protein coated globules of triacylglycerols found in lacteals are known as:
    a)   chyme
    b)   micelles
    c)   chylomicrons
    d)   none of the above is correct

e   27. Epithelial cells of the small intestine are unable to synthesize
          which one of the following:
          a)   protein
          b)   cholesterol
          c)   glycerol
          d)   phospholipids
          e)   all of the above can be synthesized

b   28. Which one of the following is not absorbed by active transport in
          the small intestine?
          a)   potassium
          b)   water
          c)   sodium
          d)   chloride
          e)   calcium

e   29. Which one of the following is correct about vitamin B12?
          a)   it is absorbed actively
          b)   absorption occurs at specific sites in the ileum
          c)   an intrinsic factor produced in the stomach is necessary for
                vitamin B12 absorption
          d)   two of the above are correct
          e)   all of the above are correct

a   30. Which one of the following is not correct?
          a)   most of the water absorption takes place in the large intestine
          b)   vitamin K is produced by microbes in the large intestine
          c)   most of the secretion of the mucosa of the large intestine is
                mucus
          d)   sodium ions are actively absorbed in the large intestine
          e)   all of the above are correct

d   1.   The body can derive energy for organismal activity from:
         a)   amino acids
         b)   glucose
         c)   triacylglycerols
         d)   all of the above

a   2.   Which one of the following is not true of glycolysis?
         a)   it occurs in the mitochondria of the cell
         b)   a single molecule of glucose is converted to two molecules of
              lactic
         c)   a net gain of 2 ATP occurs
         d) $NAD+$ is reduced
         e)   no molecular oxygen is required

a   3.   When fructose-6-phosphate is converted to fructose 1.6-diphosphate
         during glycolysis:
         a)   ATP is converted to ADP
         b) $NAD+$ is reduced
         c) $NADH + H^+$ is converted to NAD^+
         d)   none of the above

b   4.   Which one of the following is not true of the Krebs cycle?
         a)   occurs in the mitochondria
         b)   2 molecules of NAD^+ are reduced
         c)   1 molecule of GTP is formed
         d)   1 molecule of FAD is reduced

d   5.   During the Krebs cycle, fumaric acid is converted into:
         a)   fumaric citric acid
         b)   oxaloacetic acid
         c)   a-ketoglutaric acid
         d)   malic acid
         e)   succinic acid

c   6.   Which one of the following is not true of the electron transport
         system?
         a)   it takes place within the mitochondria
         b)   when one NADH+H+ produced in the mitochondria passes its
              electrons through the electron transport system, 3 ATP
              molecules are produced
         c)   when one NADH+H+ produced in the cytoplasm during glycolysis
              passes its electrons through the electron transport system, 4
              ATP molecules are produced
         d)   oxygen is the final electron acceptor in the electron transport
              system

c 7. A gross total of _____ ATP molecules are made in the complete aerobic breakdown of a glucose molecule.
a)   40-42
b)   39-41
c)   36-38
d)   34-36
e)   none of the above is correct

b 8. A net total of _____ $FADH_2$ are produced during the complete aerobic breakdown of a glucose molecule.
a)   1
b)   2
c)   4
d)   6

e 9. The production of a single ATP from ADP traps approximately _____ calories of energy.
a)   200
b)   800
c)   2800
d)   6900
e)   7300

d 10. Cellular reactions are able to trap about _____% of the energy contained within a glucose molecule.
a)   15
b)   25
c)   35
d)   40
e)   more than 60

e 11. In the complete aerobic breakdown of a glucose molecule, the carbon atoms from the glucose ultimately end up in
a)   acetyl-coenzyme A
b)   citric acid
c)   the hydrogen transport system
d)   pyruvic acid
e)   none of the above

c 12. Which of the following is not true?
a)   amino acids may be converted to alpha keto acids
b)   some alpha keto acids may enter the Krebs cycle
c)   an alpha keto acid is a special kind of amino acid
d)   all of the above are true

b 13. The nitrogens from deaminated amino acids ultimately end up in
     a)   ammonia
     b)   urea
     c)   uric acid
     d)   keto acids

e 14. Which one of the following is not correct?
     a)   carbohydrates are readily converted to triacylglycerols
     b)   glucose can be used in the synthesis of triacylglycerols
     c)   amino acids can be converted to carbohydrates
     d)   fat stores are the major energy reserves for the body
     e)   all of the above are correct

e 15. Deamination of amino acids occurs in the
     a)   kidney
     b)   spleen
     c)   intestine
     d)   pancreas
     e)   liver

a 16. Which one of the following is not true of the absorptive state?
     a)   during this time the body uses fats as its major energy source
     b)   during this time the body stores carbohydrate as
         triacylglycerols
     c)   carbohydrates absorbed from the digestive tract are in the form
         of monosaccarides
     d)   most of the fat made in the liver is released for transport to
         adipose tissue
     e)   all of the above are true

c 17. Which one of the following is not true of proteins during the
    absorptive state?
     a)   some of the amino acids that result from protein digestion are
         stored as fat
     b)   many amino acids are converted to keto acids by the liver
     c)   excess amino acids in the body are stored only as newly made
         proteins
     d)   the liver deaminates many amino acids

a 18. Which one of the following is not true of fats during the absorptive
    state?
     a)   they move as triacylglycerols directly from the intestine to
         the liver via the hepatic portal vein
     b)   much of the fat is stored as adipose tissue fat
     c)   some fat may be used by the body as an energy source during the
         absorptive state
     d)   all of the above are true

e   19. Which one of the following is not correct?
         a)   during the postabsorptive state, glycogen is broken down to
              glucose
         b)   glycogen reserves in the liver are sufficient for approximately
              four hours
         c)   the breakdown of fats can provide additional supplies of
              glucose
         d)   the major blood glucose source during prolonged fasting is
              protein
         e)   muscle glycogen can be broken down directly to glucose

c   20. Which one of the following is true? During the postabsorptive state
         a)   the major energy source for most of the different kinds of
              tissue is glucose
         b)   the nervous system primarily takes up fatty acids and processes
              them through the Krebs cycle to obtain energy
         c)   the liver produces ketone bodies from fatty acids
         d)   the liver uses glucose as its only energy source

e   21. Which one of the following is not a function of the liver?
         a)   produces bile
         b)   synthesizes urea and ketone bodies
         c)   produces some plasma proteins
         d)   produces prothrombin
         e)   all of the above are functions of the liver

e   22. Calcium is not involved in which of the following?
         a)   blood clotting
         b)   muscle activity
         c)   bone formation
         d)   nerve activity
         e)   calcium is involved in all of the above

c   23. Which one of the following is a function of zinc in the body?
         a)   a component of hemoglobin
         b)   a constituent of amino acids
         c)   a component of carboxypeptidases
         d)   a component of urea

b   24. Which one of the following is not a fat-soluble vitamin?
         a)   A
         b) B_6
         c)   D
         d)   E
         e)   K

e   25. Which one of the following is not true of vitamins?
        a)   they function as coenzymes
        b)   they are not energy sources
        c)   they cannot generally be synthesized by the body
        d)   some may be made by bacteria in the intestine
        e)   all of the above are true

d   26. Which one of the following diseases is caused by deficiency of vitamin D?
        a)   dermatitis
        b)   scurvy
        c)   megaloblastic anemia
        d)   rickets
        e)   pellegra

e   27. The vitamin needed for the proper absorption of calcium from the gastrointestinal tract is
        a)   folic acid
        b)   thiamine
        c)   ascorbic acid
        d)   vitamin E
        e)   none of the above is correct

d   28. Basal metabolic rate can be determined by measuring the
        a)   amount of carbon dioxide produced by the body
        b)   the amount of oxygen produced by the body
        c)   the cardiac output
        d)   the amount of heat produced by the body
        e)   the respiratory rate

a   29. The heat lost from the hand when holding an ice cube is an example of heat loss by
        a)   conduction
        b)   convection
        c)   radiation
        d)   absorption

b   30. Which one of the following have the greatest specific dynamic action?
        a)   vitamins
        b)   proteins
        c)   glucose
        d)   starch
        e)   fats

b 31. Which one of the following is not true of body temperature?
  a) humans maintain a relatively constant body temperature
  b) the lowest temperatures occur in the evening
  c) the core temperature is usually warmer than the skin temperature
  d) the hypothalamus is involved in the integration of thermoregulatory activities
  e) all of the above are true

d 32. Which one of the following is not likely to be a symptom of heat exhaustion?
  a) little rise in body temperature
  b) cool, clammy skin
  c) hypotension
  d) all of the above are symptoms of heat exhaustion

a 33. Which one of the following is not true of fever?
  a) prostaglandin production is stimulated by aspirin
  b) pyrogens cause the hypothalamus to reset body temperature temporarily to a new higher level
  c) the effects of pyrogens are mediated by prostaglandins
  d) sweating and vasodilation are likely to occur when a fever breaks

d 34. At temperatures above about _____ degrees C, additional heat loss by the body as a result of sweating occurs.
  a) 15
  b) 21
  c) 26
  d) 28
  e) 34

e 35. Obesity often accompanies and contributes to which of the following diseases?
  a) high blood pressure
  b) atherosclerosis
  c) diabetes
  d) coronary artery disease
  e) all of the above

a   1.   The immediate source of energy for organismal activity is
         a)   ATP
         b)   glucose
         c)   protein
         d)   triglycerides
         e)   fatty acids

a   2.   Which one of the following is true of glycolysis?
         a)   it occurs in the cytoplasm of the cell
         b)   NAD is oxidized
         c)   molecular oxygen is required during glycolysis
         d)   a net gain of 4 ATP occurs

c   3.   When fructose-6-phosphate is converted to fructose 1.6-diphosphate
         during glycolysis
         a) NAD^+ is reduced
         b)   ATP is produced
         c)   ATP is utilized
         d) NAD^+ is oxidized
         e)   none of the above

b   4.   In one turn of the Krebs cycle, which one of the following does not
         occur?
         a)   three molecules of NAD have been reduced
         b)   two molecules of FAD have been oxidized
         c)   one molecule of GTP has been formed
         d)   two carbon dioxide molecules have been made

d   5.   When succinic acid is converted to fumaric acid during the Krebs
         cycle
         a)   ATP is utilized
         b) CO_2 is produced
         c) NAD^+ is reduced
         d)   FAD is reduced
         e)   none of the above

d   6.   Which one of the following is true of the electron transport system?
         a)   it occurs within the cytoplasm of the cell
         b)   when one NADH+H+ made in the Krebs cycle passes its electrons
              through the electron transport system 2 ATP molecules are
              produced
         c)   when one FADH+H+ made in the Krebs cycle passes its electrons
              through the electron transport system 3 ATP molecules are made
         d)   oxygen is the final electron acceptor in the electron transport
              system

e   7.   A total (gross) of _____ ATP molecules are made in the complete anaerobic breakdown of a glucose molecule
a)   40
b)   38
c)   36
d)   34
e)   none of the above is correct

e   8.   A net total of _____ ATP molecules are made in the electron transport system as a result of the complete aerobic breakdown of one glucose molecule
a)   40
b)   38
c)   30
d)   24
e)   none of the above is correct

d   9.   The net profit resulting from the aerobic breakdown of one mole of glucose is about _____ calories of energy.
a)   2880
b)   6900
c)   8000
d)   277,400
e)   686,000

d   10.   When glucose is completely broken down aerobically, about ____% of the energy goes off as heat.
a)   15
b)   35
c)   40
d)   60

e   11.   In the complete aerobic breakdown of a glucose molecule, the hydrogen atoms of the glucose ultimately end up in
a)   acetyl-coenzyme A
b)   citric acid
c)   pyruvic acid
d)   carbon dioxide
e)   none of the above is correct

b.   12.   Which one of the following is true?
a)   essential amino acids may be synthesized by the body
b)   deaminated amino acids may enter the Krebs cycle
c)   amino acids may be converted to triacylglycerols but not to carbohydrates
d)   an alpha keto acid is a special kind of fat

e   13. Urea is formed from
    a)   ketone bodies
    b)   keto acids
    c)   fatty acids
    d)   triglycerides
    e)   none of the above is correct

a   14. Which one of the following is not correct?
    a)   carbohydrates are not converted to fats
    b)   glucose can be used for fat synthesis
    c)   amino acids can be converted to alpha keto acids
    d)   fat stores are the major energy reserve for the body

e   15. Removal of urea from the body takes place in the
    a)   liver
    b)   spleen
    c)   intestine
    d)   pancreas
    e)   none of the above is correct

c   16. Which one of the following is not true of the absorptive state?
    a)   during this time the body uses glucose as its major energy source
    b)   during this time the body stores excess carbohydrate as fat
    c)   carbohydrates absorbed from the digestive tract are in the form of disaccharides
    d)   some of the triacylglycerols made in the liver are released for transport to adipose tissue
    e)   all of the above are true

b   17. Which one of the following is not true of proteins during the absorptive state?
    a)   the liver deaminates many amino acids
    b)   amino acids pass from the intestine via the lymphatic system
    c)   excess amino acids can ultimately be stored as fat
    d)   the liver converts amino acids to alpha keto acids

d   18. Which one of the following is not true of fats during the absorptive state?
    a)   absorbed triacylglycerols travel through the body as components of chylomicrons
    b)   much of the fat is stored in adipose tissue cells
    c)   some fat may be used as an energy source during the absorptive state
    d)   all of the above are true

b   19.   Which one of the following is not true?
- a)   during the postabsorptive state, glycogen is broken down to glucose
- b)   glucose sparing does not take place until several days of fasting have occurred
- c)   the breakdown of fats during the postabsorptive state can provide additional supplies of glucose
- d)   the major blood glucose source during prolonged fasting is protein
- e)   all of the above are true

a   20.   Which one of the following is not true? During the post-absorptive state . . .
- a)   the major energy source for most of the different kinds of tissue is glucose
- b)   the nervous system can eventually use ketone bodies as energy sources
- c)   the liver makes ketone bodies from fatty acids
- d)   muscle cells cannot breakdown glycogen into glucose and release it to the blood

e   21.   Which one of the following is not a function of the liver?
- a)   excretes bilirubin
- b)   stores glycogen
- c)   stores vitamins A and B
- d)   detoxifies toxins
- e)   all of the above are functions

a   22.   Which one of the following is not a function of iron in the body?
- a)   a constituent of vitamin B-12
- b)   oxygen transport
- c)   a component of myoglobin
- d)   a component of cytochromes
- e)   involved in ATP formation

c   23.   Cobalt is a component of which vitamin?
- a)   vitamin A
- b)   vitamin C
- c)   vitamin B_{12}
- d)   vitamin K
- e)   none of the above

b   24.   Which one of the following is not a fat soluble vitamin?
- a)   A
- b)   C
- c)   D
- d)   E
- e)   K

d 25. Which one of the following is true of vitamins?
    a)   they function as enzymes
    b)   they are important energy sources
    c)   in general they cannot be synthesized by the body
    d)   some may be made by bacteria in the gastrointestinal tract
    e)   none of the above is true

d 26. Which one of the following diseases is caused by thiamine deficiency?
    a)   dermatitis
    b)   scurvy
    c)   night blindness
    d)   beriberi
    e)   pellegra

c 27. The vitamin needed to prevent rickets is:
    a)   thiamine
    b)   riboflavin
    c)   vitamin D
    d)   vitamin E
    e)   vitamin B-12

d 28. Which one of the following is not a standard condition under which basal metabolic rate is measured?
    a)   patient is awake
    b)   room temperature is between 20 and 25 degrees C
    c)   patient has been in a postabsorptive state with nothing to eat for 12 hours
    d)   patient is performing a specific set of exercises
    e)   all of the above are standard conditions

a 29. The heat lost from the body when you emerge from a swimming pool on a windy day is heat lost by:
    a)   conduction and convection
    b)   absorption
    c)   radiation
    d)   osmosis

d 30. Which one of the following has the least specific dynamic action?
    a)   vitamins
    b)   proteins
    c)   fats
    d)   carbohydrates

c   31. Which one of the following is not true?
- a) the oral temperature is about 36.7 degrees
- b) the internal core temperature is usually warmer than the skin temperature
- c) there is no fluctuation in body temperature over a diurnal time period
- d) the hypothalamus plays a role in regulating body temperature

d   32. Which one of the following is not true of heat stroke?
- a) rapid rise in body temperature
- b) dry skin
- c) failure of heat regulating center
- d) not as serious as heat exhaustion

d   33. Which one of the following is not true of a fever?
- a) pyrogens reset the body temperature thermostat
- b) aspirin indirectly inhibits the production of pyrogens
- c) the shivering and chills of a fever are due to the fact that the body is trying to reach a new set point for its temperature
- d) all of the above are true

a   34. Which one of the following is not correct?
- a) at temperatures below 29 degrees C, there is no water loss from the body
- b) sweating can be an effective means of cooling the body when the humidity is low
- c) as much as 1.5 l of water per hour may be lost by sweating
- d) sweat contains urea, lactic acid, and potassium as well as sodium and chloride ions
- e) all of the above are correct

a   35. Obesity often accompanies and contributes to all but which one of the following diseases?
- a) polyneuritis
- b) diabetes
- c) stroke
- d) heart attack
- e) atherosclerosis

b   1.   Which one of the following is incorrect?
          a)   the kidneys are retroperitoneal
          b)   it is surgically impossible to operate on the kidneys without
               opening the peritoneal cavity
          c)   the right kidney is lower than the left
          d)   each kidney is capped by an adrenal gland

a   2.   Which one of the following is passed thirdly by a drop of urine?
          a)   renal pelvis
          b)   urinary bladder
          c)   papilla
          d)   major calyx

c   3.   Which one of the following statements is not correct about the renal
          hilus?
          a)   it is an indentation on the medial border of the kidney
          b)   renal arteries pass through it
          c)   the urethra passes through it
          d)   renal veins pass through it

d   4.   Renal corpuscles are located in the:
          a)   medulla
          b)   pelvis
          c)   renal columns
          d)   cortex

c   5.   The inner (visceral) layer of a glomerular capsule is made of:
          a)   simple squamous epithelium
          b)   endothelium
          c)   podocytes
          d)   pedicles

b   6.   Which one of the following listed is passed thirdly by water?
          a)   basal lamina
          b)   slit membrane
          c)   fenestrated endothelium
          d)   capsular space

a   7.   Microvilli are found in a
          a)   proximal convoluted tubule
          b)   thin segment of a loop of Henle
          c)   thick segment of a loop of Henle
          d)   distal convoluted tubule

b   8.   Which one of the following is false?
- a)   the loop of Henle passes into a pyramid
- b)   juxtamedullary golmeruli have shorter loops of Henle than glomeruli close to the surface of the kidney
- c)   the descending limb of the loop of Henle is made of simple squamous epithelium
- d)   the ascending limb of the loop of Henle is made of simple cuboidal epithelium

d   9.   The juxtaglomerular apparatus is made up of
- a)   an efferent arteriole and proximal convoluted tubule cells
- b)   an afferent arteriole endothelial cells and the macula lutea
- c)   tunica media cells of the efferent arteriole and juxta-glomerular cells
- d)   none of the above are correct

c   10.   Which one of the following is incorrect?
- a)   renal arteries carry about 20% of the total cardiac output
- b)   about 1100 ml. of blood per minute goes through the kidneys
- c)   the collecting tubules join the proximal convoluted tubules to the papilla
- d)   the kidney will not function if blood pressure drops below a minimum level

a   11.   Which one of the following structures listed is passed thirdly by a drop of blood?
- a)   glomerulus
- b)   arcuate artery
- c)   efferent arteriole
- d)   interlobar artery

c   12.   The movement of an essentially protein-free fluid from the blood into a glomerular capsule is known as glomerular _____.
- a)   reabsorption
- b)   secretion
- c)   filtration
- d)   discharge

b   13.   The peritubular capillaries converge into a _____ vein.
- a)   arcuate
- b)   interlobular
- c)   interlobar
- d)   renal

d   14. Which one of the following listed is pierced thirdly by the point of
        a pin going from the lumen of a ureter to the outside?
        a)   fibrous layer
        b)   mucosa
        c)   inner longitudinal muscle layer
        d)   circular muscle layer

c   15. Which one of the following statements about the urinary bladder is
        correct?
        a)   it is within the abdominal cavity
        b)   in females, it is immediately anterior to the rectum
        c)   in males, it is just behind the pubic symphysis
        d)   its internal urethral sphincter is made up of skeletal muscle
              tissue

a   16. In the male which one of the following structures is passed thirdly
        by a drop of urine?
        a)   cavernous urethra
        b)   prostatic urethra
        c)   membranous urethra
        d)   external urethral orifice

d   17. The disease caused by a bacterial infection of the pelvis of the
        kidney is known as:
        a)   cystitis
        b)   Bright's disease
        c)   renal calculi
        d)   pyelonephritis

d   18. The accumulation of the metabolic products of protein metabolism in
        the blood is known as:
        a)   ketonuria
        b)   proteinuria
        c)   glycosuria
        d)   uremia

b   19. About _____ % of the blood plasma that enters the kidneys is
        filtered from the glomerular capillaries.
        a)   5
        b)   15
        c)   30
        d)   45
        e)   65

c   20. If the glomerular blood pressure is 65 m. Hg, the capsular hydrostatic pressure 12 mm. Hg, and the colloidal osmotic pressure in the glomerulus averages 28 mm. Hg, what is the effective glomerular filtration pressure?
- a) 93 mm. Hg.
- b) 53 mm. Hg.
- c) 40 mm. Hg.
- d) 16 mm. Hg.
- e) none of the above is correct

a   21. The glomerular filtration barrier is _____ than a typical capillary.
- a) much more permeable
- b) much less permeable
- c) a little more permeable
- d) a little less permeable

c   22. When arterial pressures increase greatly, the glomerular filtration rate increases
- a) greatly
- b) moderately
- c) slightly
- d) not at all

b   23. An increase in arterial blood pressure from 100 to 120 mm. Hg will approximately _____ the output of urine.
- a) not significantly alter
- b) double
- c) triple
- d) increase six fold
- e) increase eight fold

b   24. Which one of the following is correct? Sympathetic stimulation of the kidneys
- a) dilates efferent arterioles
- b) results in a lowered glomerular capillary pressure
- c) results in an increased rate of blood flow into the glomeruli
- d) results in an increased glomerular filtration rate
- e) two or more of the above are correct

e   25. Which one of the following is not actively reabsorbed?
- a) glucose
- b) vitamins
- c) amino acids
- d) sodium
- e) urea

e   26. Which one of the following can enter renal tubules by tubular secretion?
      a) hydrogen ion
      b) potassium ion
      c) penicillin
      d) ammonia
      e) all of the above are correct

c   27. The concentration of the blood plasma is about _____ osmoles per liter.
      a) 75
      b) 150
      c) 300
      d) 600
      e) 1200

c   28. Which one of the following is correct?
      a) sodium and chloride ions move into the upper portions of the ascending limbs of the loops of Henle
      b) sodium and chloride ions move into the thin segments of the loops of Henle
      c) sodium and chloride ions move out of the collecting tubules
      d) sodium and chloride ions are carried upward toward the cortex by blood flowing through the vasa recta
      e) none of the above are correct

e   29. The countercurrent mechanism functions primarily in the
      a) vasa recta
      b) proximal convoluted tubule
      c) distal convoluted tubule
      d) loop of Henle
      e) two of the above are correct

c   30. Reabsorption takes place in:
      a) the afferent arteriole
      b) the renal vein
      c) the capillary net around the tubule
      d) the efferent arteriole
      e) the glomerulus

c   31. The term "clearance" refers to
      a) the amount of solute not reabsorbed
      b) the volume of plasma present in the glomerular filtrate
      c) the volume per minute of plasma from which a substance is entirely removed
      d) none of the above are correct

a 32. Fluid entering the distal convoluted tubule is likely to be _____ to the blood plasma.
   a) hypotonic
   b) isotonic
   c) hypertonic
   d) none of the above are correct

c 33. When ADH levels in the plasma are high, the collecting tubules are
   a) very permeable to sodium and chloride ion
   b) very permeable to vitamin D3 and calcium ion
   c) very permeable to water
   d) impermeable to sodium and chloride ions
   e) impermeable to water

a 34. Which one of the following structures are very permeable to urea?
   a) descending limb of the loop of Henle
   b) ascending limb of the loop of Henle
   c) first portions of the collecting tubules
   d) distal convoluted tubules

d 35. In the proximal convoluted tubule, reabsorption reduces the volume of the glomerular filtrate by about _____%.
   a) 30
   b) 45
   c) 55
   d) 65
   e) none of the above are correct

b 36. Which one of the following is not correct?
   a) the descending limb of the loop of Henle is relatively permeable to water
   b) at the bottom of the loop of Henle, the tubular fluid is hypotonic to plasma
   c) the tubular fluid in the ascending limb of the loop of Henle is impermeable to water
   d) just before the tubular fluid enters the distal convoluted tubule, it is isotonic to plasma
   e) all of the above are correct

c 37. The body must excrete about _____ milliosmoles of waste products each day.
   a) 100
   b) 300
   c) 600
   d) 1200
   e) 1400

a   38. Suppose that the concentration of urea in the plasma is 0.25 mg/ml. Suppose also that 50 ml of urine are formed in one hour, and that this urine has a urea concentration of 14 mg/ml. The plasma clearance of urea would be _____ ml/min.
  a)   46
  b)   15
  c)   42
  d)   none of the above are correct

b   39. Which one of the following does not develop from the ureteric buds?
  a)   calices
  b)   nephrons
  c)   collecting tubules
  d)   renal pelvis
  e)   ureter

b   40. Which one of the following would be pierced thirdly by the point of a pin entering a kidney from its ventral side?
  a)   anterior renal fascia
  b)   adipose capsule
  c)   renal capsule
  d)   parietal peritoneum
  e)   posterior renal fascia

NAME_____

SECTION _____

d 1. Which one of the following is pierced thirdly by a pin passing from the abdominal cavity through the kidney?
a) renal capsule
b) renal fascia
c) parietal peritoneum
d) adipose capsule

c 2. Which one of the following listed is passed thirdly by a drop of urine?
a) urethra
b) minor calyx
c) urinary bladder
d) ureter

c 3. Which one of the following listed is pierced thirdly by a pin entering the kidney from the outside?
a) base of pyramid
b) pelvis
c) papillae
d) cortex

b 4. The functional unit of the kidney is the
a) renal corpuscle
b) nephron
c) glomerulus
d) glomerular capsule

c 5. The epithelium of the glomerulus is composed of
a) podocytes
b) slit membranes
c) fenestrated endothelium
d) simple cuboidal epithelium

c 6. Which one of the following listed can pass through filtration membrane?
a) molecules 120-170 angstroms in diameter
b) plasma proteins
c) molecules less than 70 angstroms in diameter
d) molecules 80-120 angstroms in diameter

c 7. Cells lining the thick segment of the loop of Henle are composed of
a) simple squamous epithelium
b) simple columnar epithelium
c) simple cuboidal epithelium
d) stratified squamous epithelium

a   8.   The distal convoluted tubule contacts the
        a)   afferent arteriole
        b)   interlobular artery
        c)   interlobar artery
        d)   efferent arteriole

b   9.   Which one of the following is incorrect?
        a)   the juxtaglomerular apparatus secretes renin
        b)   the substance secreted by the juxtaglomerular apparatus is produced in response to high blood pressures
        c)   the product of the juxtaglomerular apparatus has an effect on angiotensinogen
        d)   the juxtaglomerular apparatus is made up of the macula densa and cells in the wall of the afferent arteriole

d   10.   Which one of the following is not true?
        a)   blood in the kidneys goes through two capillary beds
        b)   arterioles lead to and from a glomerulus
        c)   the kidneys can maintain a fairly constant blood pressure even if there are systemic pressure fluctuations
        d)   efferent arterioles lack smooth muscle in their walls

c   11.   Which one of the following structures listed is passed thirdly by a drop of blood?
        a)   interlobular artery
        b)   interlobar artery
        c)   arcuate vein
        d)   interlobular vein

c   12.   The uptake of substances from the lumen of the kidney tubules is known as
        a)   tubular secretion
        b)   tubular filtration
        c)   tubular reabsorption
        d)   does not occur

a   13.   The vasa recta supply
        a)   the loop of Henle
        b)   the proximal convoluted tubule
        c)   the distal convoluted tubule
        d)   the glomeruli

a   14.   Which one of the following structures listed is pierced thirdly by a pin entering a ureter from the abdominal cavity?
        a)   muscle layer
        b)   parietal peritoneum
        c)   mucosa
        d)   fibrous layer

a 15. Which one of the following statements is correct about the urinary
        bladder?
        a)   the external urethral sphincter is made of skeletal muscle
        b)   the ureters are open during micturition
        c)   when 600 ml of urine have accumulated, the sensory neurons of
              the urination reflex are stimulated
        d)   motor reflex impulses cause the muscle of the bladder wall to
              relax

c 16. Which one of the following is correct?
        a)   the membraneous urethra has openings from the ejaculatory ducts
        b)   the cavernous urethra receives the ducts of the prostate gland
        c)   the urethra passes through the corpus spongiosum
        d)   the urethra in the male is about 4 cm long

b 17. An inflammation of the urinary bladder accompanied by frequent and
        burning urination is known as
        a)   pyelonephritis
        b)   cystitis
        c)   Bright's disease
        d)   renal calculi

a 18. An inflammation of the kidney caused by an allergic reaction to the
        toxins produced by certain bacteria is known as
        a)   Bright's disease
        b)   pyelonephritis
        c)   cystitis
        d)   Dull's disease

e 19. Under normal circumstances the kidneys produce about _____ liters
        of glomerular filtrate per day.
        a)   15
        b)   50
        c)   80
        d)   100
        e)   180

c 20. Given the following conditions: glomerular blood hydrostatic
        pressure 75 mm. Hg, capsular hydrostatic pressure 15 mm. Hg, blood
        osmotic pressure 25 mm. Hg, the effective filtration pressure would
        be _____mm Hg.
        a)   0
        b)   25
        c)   35
        d)   40
        e)   65

a   21.   Which one of the following is not correct?
     a)   the components of the glomerular filtrate are at different
            concentrations than they are found in the plasma
     b)   the glomerulus, like other capillaries, is impermeable to
            proteins
     c)   a small amount of protein does appear in the glomerular fil-
            trate
     d)   essentially no protein appears in the urine in the urinary
            bladder

c   22.   When arterial pressures decrease greatly, the glomerular filtration
            rate decreases
     a)   greatly
     b)   moderately
     c)   slightly
     d)   not at all

d   23.   An increase in arterial blood pressure from 100 to 200 mm. Hg will _
            _____the output of urine.
     a)   not significantly alter
     b)   double
     c)   triple
     d)   increase six fold

c   24.   Which one of the following is not correct? Sympathetic stimulation
            of the kidneys
     a)   constricts the afferent arterioles
     b)   lowers the glomerular capillary pressure
     c)   constricts the efferent arterioles
     d)   lowers the glomerular filtration rate
     e)   all of the above are correct

b   25.   Which of the following is not absorbed passively?
     a)   water
     b)   glucose
     c)   lipid soluble substances
     d)   urea
     e)   all of the above are absorbed passively

e   26.   Which one of the following substances enters the tubules by tubular
            secretion?
     a)   amino acids
     b)   vitamins
     c)   sodium
     d)   calcium
     e)   hydrogen ion

c 27. The concentration of the urine can be as high as _____ milliosmoles per liter.
 a) 70
 b) 600
 c) 1200
 d) 3000

e 28. Which one of the following is not correct?
 a) sodium and chloride ions move out of the collecting tubules
 b) sodium and chloride ions move out of the upper portions of the ascending limbs of the loop of Henle.
 c) sodium and chloride ions are carried downward toward the medulla by the blood flowing in the vasa recta
 d) sodium and chloride ions move out of the thin segments of the loops of Henle
 e) all of the above are correct

e 29. The counter current mechanism functions in the vasa recta and the _____.
 a) loops of Henle
 b) collecting tubules
 c) proximal convoluted tubule
 d) distal convoluted tubule
 e) two of the above are correct

a 30. Most reabsorption of substances from the glomerular filtrate occurs in the
 a) proximal tubule
 b) loop of Henle
 c) collecting tubule
 d) Bowman's capsule
 e) distal tubule

e 31. The volume of plasma from which a substance is entirely removed per minute is known as
 a) effective renal plasma flow
 b) glomerular filtration rate
 c) hematocrit
 d) net filtration pressure
 e) none of the above are correct

c 32. Fluid leaving the distal convoluted tubule is likely to be_____ to blood plasma.
 a) hypotonic
 b) hypertonic
 c) isotonic
 d) none of the above are correct

a 33. When ADH levels in the plasma are low, the collecting tubules are
a) impermeable to water
b) very permeable to sodium and chloride ion
c) very permeable to vitamin D3 and calcium ion
d) impermeable to sodium and chloride ion
e) permeable to water

d 34. Which one of the following structures are quite impermeable to urea?
a) ascending limb of the loop of Henle
b) first portions of the collecting tubules
c) distal convoluted tubules
d) all of the above are impermeable

a 35. After the glomerular filtrate has passed down the proximal convoluted tubule only about _____ % is still moving down the nephron.
a) 35
b) 45
c) 55
d) 65

b 36. Which one of the following is not correct?
a) the tubular fluid at the bottom of the loop of Henle is hypertonic to blood plasma
b) the ascending limb of the loop of Henle is permeable to water
c) the descending limb of the loop of Henle is relatively permeable to water
d) the distal convoluted tubule is relatively impermeable to water
e) all of the above are correct

c 37. Regardless of water intake or availability, the minimum amount of urine that must be excreted each day to eliminate wastes is about _____ml.
a) 250
b) 300
c) 440
d) 600
e) 1200

c 38. Suppose that the concentration of urea in the plasma is 0.35 mg/ml. Suppose also that 75 ml of urine are formed in one hour, and that this urine has a urea concentration of 19 mg/ml. The plasma clearance of urea would then be _____ ml/min.
a) 23
b) 53
c) 58
d) none of the above is correct

d   39. Which one of the following is not a waste produce eliminated by the kidneys?
- a) urea
- b) uric acid
- c) creatine
- d) renin

d   40. The ureteric buds arise from the
- a) pronephros
- b) mesonephros
- c) metanephros
- d) mesonephric duct
- e) pronephric duct

NAME_____

SECTION _____

e   1.   The most abundant cation in extracellular fluid is:
a)   calcium
b)   potassium
c)   magnesium
d)   phosphorous
e)   none of the above is correct

c   2.   Which one of the following is not correct?
a)   the compositions of the plasma and interstitial fluid are very similar
b)   the plasma and interstitial fluid are separated only by capillary walls
c)   interstitial fluid has more protein than plasma
d)   there is less chloride within the plasma than within interstitial fluid

a   3.   Concerning intracellular fluid, which one of the following is not correct? There is more _____ in intracellular fluid than in any other fluid compartment.
a)   sodium
b)   phosphate
c)   protein
d)   sulphate
e)   potassium

c   4.   Plasma and interstitial fluid together make up about _____% of the total body weight.
a)   4
b)   16
c)   20
d)   40
e)   none of the above is correct

a   5.   Which one of the following comparisons between plasma and interstitial fluid is incorrect? There is _____ in interstitial fluid than in plasma.
a)   less chloride
b)   more organic acids
c)   less potassium
d)   more carbonic acid
e)   less calcium

e   6.   Which one of the following is not correct?
- a) ingestion of a closely regulated amount of sodium does not seem to be the major way by which internal sodium content of the body is regulated
- b) varying the glomerular filtration rate can alter the amount of sodium lost from the body
- c) varying the reabsorption rate in the tubules can alter the amount of sodium lost from the body
- d) aldosterone plays a major role in determining sodium reabsorption in the kidney tubules
- e) all of the above are true

d   7.   Which one of the following does not influence the glomerular filtration rate?
- a) arterial blood pressure
- b) diameter of glomerular afferent arterioles
- c) renin
- d) all of the above can influence the glomerular filtration rate

e   8.   Renin is released by the:
- a) lungs
- b) glomerulus
- c) Bowman's capsule
- d) liver
- e) none of the above is correct

d   9.   Angiotensinogen is made in the:
- a) lungs
- b) glomerulus
- c) Bowman's capsule
- d) liver
- e) none of the above is correct

a   10.   Which one of the following listed occurs thirdly?
- a) decrease in activity of sympathetic nerves to kidney
- b) increase in systemic blood pressure
- c) dilation of afferent arterioles
- d) increase in impulses from circulatory pressure receptors
- e) increase in glomerular filtration rate

e   11.   Which one of the following is not correct?
- a) regulation of sodium content of the body is closely linked with control of the extracellular fluid volume
- b) the factors involved in regulation of sodium content of the body are influenced by blood pressure
- c) sodium has a direct influence on extracellular volume
- d) loss of sodium tends to cause a decrease in plasma volume
- e) all of the above are correct

d   12.   The kind of receptors that are sensitive to the number of solute
          molecules present in the extracellular fluid are the:
          a)   baroreceptors
          b)   proprioceptors
          c)   chemoreceptors
          d)   osmoreceptors
          e)   none of the above is correct

a   13.   Which one of the following <u>listed</u> occurs thirdly?
          a)   decrease in activity of sympathetic neurons
          b)   increase in water excretion
          c)   dilation of afferent glomerular arterioles
          d)   increase in plasma volume
          e)   increase in systemic blood pressure

a   14.   Ammonia secretion by the renal tubules increases when:
          a)   the tubular fluid is very acid
          b)   the tubular fluid is very alkaline
          c)   the body is producing urea
          d)   there is not enough hydrogen ion
          e)   there is not enough potassium ion

e   15.   In an adult suffering from diabetes insipidus, a typical urine output
          per day would be _____ l.
          a)   1
          b)   1.5
          c)   2.5
          d)   3.5
          e)   none of the above is correct

d   16.   Hydrogen ions pass out of kidney tubule cells in exchange for
          a)   ammonia
          b)   chloride
          c)   phosphate
          d)   sodium
          e)   urea

c   17.   Which one of the following is not true?
          a)   lowering extracellular calcium concentration stimulates
               parathormone secretion
          b)   extracellular calcium concentration acts directly on the
               parathyroid glands without any intermediary hormones or nerves
          c)   parathormone secretion increases the calcium ion concentration
               found in the urine
          d)   calcitonin inhibits bone release of calcium

e   18.   ADH secretion:
    a)   is increased following the ingestion of large volumes of water
    b)   leads to an increased urine output
    c)   causes an increased permeability of the kidney tubules to water
    d)   is increased by increasing the osmolarity of the blood
    e)   two of the above are correct

e   19.   ADH is produced by the:
    a)   adrenal gland
    b)   kidney
    c)   lung
    d)   liver
    e)   none of the above is correct

d   20.   Potassium elimination is under the influence of:
    a)   ADH
    b)   PTH
    c)   calcitonin
    d)   aldosterone

a   21.   Which one of the following is not correct about parathyroid hormone?
    a)   it increases plasma calcium and phosphate concentration
    b)   it increases the movement of calcium and phosphate from bone into extracellular fluid
    c)   it stimulates osteoclastic activity
    d)   it decreases urinary excretion of calcium and increases excretion of phosphate
    e)   it is involved in the production of vitamin D_3 by the body

d   22.   Which one of the following does not act as a buffer?
    a)   intracellular phosphate complexes
    b)   plasma proteins
    c)   hemoglobin molecules
    d)   all of the above are buffers

c   23.   If the blood is alkaline:
    a)   more hydrogen ions are excreted and less bicarbonate ions are reabsorbed
    b)   more hydrogen ions are excreted and more bicarbonate ions are reabsorbed
    c)   less hydrogen ions are excreted and less bicarbonate ions are reabsorbed
    d)   less hydrogen ions are excreted and more bicarbonate ions are reabsorbed

c   24.   A normal pH value for arterial blood is:
        a)   7.1
        b)   7.2
        c)   7.4
        d)   7.6
        e)   7.8

e   25.   The amount of potassium secreted by the kidneys appears to be influenced in part by the potassium concentration of the extracellular fluid bathing the:
        a)   hypothalamus
        b)   pituitary gland
        c)   pancreas
        d)   parathyroid gland
        e)   adrenal glands

c   26.   Most of the water in the body is in the:
        a)   interstitial fluid
        b)   blood plasma and lymph
        c)   cells
        d)   cerebrospinal fluid
        e)   lumen of the urinary bladder and gut

c   27.   What is the effect of hemoglobin on acid-base balance? Hb:
        a)   releases hydroxide ions for neutralization of hydrogen ions
        b)   catalyzes the reaction leading to the formation of water and carbonic acid
        c)   picks up hydrogen ions and acts as a buffer
        d)   has nothing to do with acid-base balance

c   28.   Hyperventilation for an extended period results in:
        a)   metabolic alkalosis
        b)   metabolic acidosis
        c)   respiratory alkalosis
        d)   respiratory acidosis

d   29.   The kidney contributes to the maintenance of acid-base balance by excreting:
        a)   chloride ions
        b)   potassium ions
        c)   sodium ions
        d)   hydrogen ions
        e)   all of the above are correct

b 30. Extreme nervousness, muscular tetany, and convulsions are symptoms of:
   a) acidosis
   b) alkalosis
   c) hypertension
   d) none of the above is correct

c 31. An increase in the amount of carbon dioxide in the blood will:
   a) decrease the bicarbonate ion concentration of the blood
   b) raise the pH of the blood
   c) increase the hydrogen ion concentration of the blood
   d) none of the above is correct

d 32. Which one of the following is not correct? Within the proximal tubule:
   a) hydrogen ions are actively transported into the lumen
   b) sodium ions from the lumen enter the tubular cells
   c) carbonic anhydrase within the tubular cells facilitates the formation of carbonic acid from carbon dioxide and water
   d) all of the above are correct

e 33. Which one of the following is not correct?
   a) normally more bicarbonate ions are filtered into Bowman's capsule than hydrogen ions are secreted into the proximal tubule
   b) a loss of bicarbonate ions in the urine leads to a decrease in the hydrogen ion concentration of the plasma
   c) if the concentration of hydrogen ions in the internal environment falls, there is an increase in the amount of bicarbonate ion lost in the urine
   d) urinary excretion of bicarbonate ions leads to a decrease in the number of hydrogen ions tied up by buffer systems
   e) two of the above are not correct

b 34. Which one of the following takes place in the distal tubule?
   a) in the tubular cells, hydrogen ions combine with sodium ions
   b) hydrogen ions combine with phosphate compounds in the tubular lumen
   c) phosphate ions combine with bicarbonate ions in the tubular cells
   d) in the tubular lumen hydrogen ions combine with urea

a 35. Which one of the following listed occurs thirdly in the cells or lumen of the distal tubule?
   a) formation of ammonium ion
   b) deamination of glutamic acid
   c) formation of ammonium chloride
   d) diffusion of ammonia

a   36. Which one of the following <u>listed</u> occurs thirdly?
        a)   release of aldosterone
        b)   release of renin
        c)   sodium reabsorption from tubules
        d)   conversion of angiotensinogen to angiotensin

d   37. Which one of the following is not a major factor influencing water excretion?
        a)   osmotic pressure of extracellular fluid
        b)   blood pressure
        c)   plasma volume
        d)   all of the above are major factors

b   1.   The most abundant cation in intracellular fluid is:
          a)   sodium
          b)   potassium
          c)   magnesium
          d)   phosphorus
          e)   none of the above is correct

d   2.   Which one of the following is correct?
          a)   the compositions of plasma and interstitial fluid are very
               different
          b)   there is more chloride within the plasma than within interstitial
               fluid
          c)   interstitial fluid has more protein than plasma
          d)   the plasma and interstitial fluid are separated only by capillary
               walls

d   3.   Intracellular fluids are characterized by relatively high concentrations
          of:
          a)   chloride and potassium
          b)   sodium and chloride
          c)   sodium and phosphate
          d)   potassium and phosphate

b   4.   Intracellular fluid makes up about _____% of the total body weight.
          a)   56
          b)   40
          c)   20
          d)   16
          e)   4

e   5.   Which one of the following comparisons between plasma and interstitial
          fluid is correct? There is _____ in interstitial fluid than in
          plasma.
          a)   more calcium
          b)   less carbonic acid
          c)   more potassium
          d)   less organic acids
          e)   more chloride

d   6.   Which one of the following does not have an effect on the amount of
          sodium lost from the body?
          a)   varying the glomerular filtration rate
          b)   varying the reabsorption rate of sodium with the tubules
          c)   varying the concentration of aldosterone in the blood plasma
          d)   all of the above have an effect

e   7.   Which one of the following will not significantly influence the glomerular filtration rates?
     a)   arterial blood pressure
     b)   diameter of the afferent arterioles
     c)   activity of sympathetic neurons
     d)   renin
     e)   all of the above can influence the glomerular filtration rate

e   8.   The substance produced by the juxtaglomerular cells is:
     a)   calcitonin
     b)   angiotensin
     c)   aldosterone
     d)   ADH
     e)   none of the above

b   9.   Agiotensinogen is formed in the:
     a)   lungs
     b)   liver
     c)   glomerulus
     d)   Bowman's capsule
     e)   none of the above is correct

c   10.   Which one of the following listed occurs fourthly?
     a)   decrease in the activity of sympathetic nerves to the kidney
     b)   increase in systemic blood pressure
     c)   dilation of afferent arterioles
     d)   increase in impulses from circulatory pressure receptors
     e)   increase in glomerular filtration rate

b   11.   Which one of the following is not correct?
     a)   regulation of the sodium content of the body is closely linked with control of extracellular fluid volume
     b)   the factors involved in regulation of the sodium content of the body are not influenced by blood pressure
     c)   sodium has a direct influence on extracellular volume
     d)   loss of sodium tends to cause a decrease in plasma volume

e   12.   Osmoreceptors are sensitive to the number of water molecules present in which fluid?
     a)   plasma
     b)   intracellular fluid
     c)   interstitial fluid
     d)   extracellular fluid
     e)   none of the above is correct

c 13. Which one of the following <u>listed</u> occurs fourthly?
   a) decrease in the activity of sympathetic neurons
   b) increase in water excretion
   c) dilation of afferent glomerular arterioles
   d) increase in plasma volume
   e) increase in systemic blood pressure

b 14. Ammonia is secreted by the:
   a) proximal tubule
   b) distal tubule
   c) loop of Henle
   d) collecting tubules
   e) glomerulus

d 15. A condition in which the patient has a very large urine output of over 15 liters per day is:
   a) diabetes mellitus
   b) alkalosis
   c) acidosis
   d) none of the above is correct

e 16. Hydrogen ions pass out of kidney tubule cells in exchange for:
   a) water
   b) potassium ions
   c) bicarbonate ions
   d) ammonia
   e) none of the above is correct

d 17. Which one of the following is correct?
   a) increasing extracellular calcium concentration stimulates parathormone secretion
   b) calcitonin stimulates bone release of calcium
   c) extracellular potassium concentration stimulates parathyroid hormone secretion by acting directly on the parathyroid glands without any intermediary hormones or nerves
   d) parathyroid hormone secretion increases the calcium ion concentration of the plasma
   e) none of the above is correct

d 18. ADH secretion:
   a) decreases following the drinking of a large amount of water
   b) leads to a decrease in urine output
   c) causes an increased permeability of the kidney tubules to water
   d) all of the above are correct
   e) two of the above are correct

d 19. Which one of the following is released from the pars nervosa of the pituitary gland?
    a) aldosterone
    b) renin
    c) angiotensinogen
    d) ADH
    e) calcitonin

b 20. Calcium regulation is under the influence of:
    a) renin
    b) parathyroid hormone
    c) aldosterone
    d) ADH
    e) none of the above is correct

e 21. Which one of the following is not correct about the regulation of potassium?
    a) potassium is freely filtered from the plasma
    b) potassium can be actively reabsorbed by kidney tubules
    c) tubular secretion of potassium is enhanced by aldosterone
    d) when the potassium concentration of extracellular fluid is high, more potassium will be secreted by the tubular cells
    e) all of the above are correct

d 22. Which of the following are important buffers in the urine?
    a) phosphate complexes
    b) proteins
    c) ammonia
    d) two of the above are correct
    e) all of the above are correct

e 23. If the blood is too acidic, which one of the following would not occur?
    a) hydrogen ions would be excreted by the kidneys
    b) sodium ions would be taken up by the kidneys
    c) the person's rate of breathing would increase
    d) ammonia would be secreted by the cells of the kidney tubules
    e) all of the above would occur

b 24. A normal pH value for venous blood is:
    a) 7.25
    b) 7.35
    c) 7.45
    d) 7.55
    e) 7.75

d   25.   When hydrochloric acid is added to a buffer system made of carbonic acid and sodium bicarbonate, hydrogen ions will react with the:
          a)   carbonic acid to form carbon dioxide and water
          b)   sodium ion to form a buffer
          c)   carbonic acid to form a strong acid
          d)   bicarbonate ion to form carbonic acid

a   26.   Which one of the following is not true of calcitonin?
          a)   it is a hormone from the adrenal gland
          b)   its effect on plasma calcium levels is the opposite of parathyroid hormone
          c)   high interstitial calcium concentrations enhance its secretion
          d)   it is not considered to be the most important factor in calcium regulation within the body
          e)   all of the above are correct

c   27.   Which one of the following is the least amount in terms of body weight?
          a)   intracellular fluid
          b)   extracellular fluid
          c)   plasma
          d)   interstitial fluid

e   28.   Which one of the following is not generally recognized as being a blood buffer?
          a)   hemoglobin
          b)   plasma proteins
          c) $NaHCO3/H_2CO_3$
          d)   intracellular phosphate complexes
          e)   all of the above are blood buffers

d   29.   Diseases that interfere with the elimination of carbon dioxide by the lungs can result in:
          a)   metabolic alkalosis
          b)   metabolic acidosis
          c)   respiratory alkalosis
          d)   respiratory acidosis

e   30.   The kidney contributes to the maintenance of acid-base balance by:
          a)   excreting hydrogen ions in exchange for sodium ions
          b)   reabsorbing bicarbonate ions
          c)   excreting ammonia
          d)   excreting lactate
          e)   all of the above are correct

d   31. A patient who has severe diarrhea suffers a loss of water, bicarbonate
         ions, and sodium ions. Physiological compensation for these losses
         would include:
         a)   increased alveolar ventilation
         b)   increased hydrogen ion secretion by the kidney tubules
         c)   decreased output of urine
         d)   all of the above are correct

c   32. A decrease in the amount of carbon dioxide in the blood will _____
         of the blood.
         a)   decrease the hydrogen ion concentration
         b)   increase the bicarbonate ion concentration
         c)   raise the pH
         d)   none of the above is correct

c   33. Which one of the following is correct? Within the proximal tubule:
         a)   hydrogen ions are actively reabsorbed
         b)   sodium ions are actively secreted
         c)   carbonic anhydrase within the tubular cells facilitates the
              formation of carbonic acid from carbon dioxide and water
         d)   bicarbonate ions are secreted by the tubular cells
         e)   none of the above is correct

a   34. Which one of the following is correct?
         a)   normally more hydrogen ions are secreted than bicarbonate ions
              are absorbed in the proximal tubule
         b)   a loss of bicarbonate ions in the proximal tubule leads to a
              decrease in the hydrogen ion concentration of the plasma
         c)   if the concentration of hydrogen ions in the internal environment
              increases, there is an increase in the amount of bicarbonate ion
              lost in the urine
         d)   none of the above is correct

d   35. Which one of the following takes place in the distal tubule?
         a)   phosphates are actively secreted into the lumen by tubular cells
         b)   phosphates are actively reabsorbed
         c)   hydrogen ions combine with chloride ions, and as a result carbon
              dioxide enters the tubular cells
         d)   sodium ions are absorbed in exchange for hydrogen ions
         e)   none of the above take place in the distal tubule

b   36. Which one of the following listed occurs fourthly in the cells or lumen
         of the distal tubule?
         a)   diffusion of ammonia
         b)   formation of ammonium chloride
         c)   deamination of glutamic acid
         d)   formation of ammonium ions

331

c   37.   Which one of the following listed occurs fourthly?
       a)   release of aldosterone
       b)   release of renin
       c)   sodium reabsorption from tubules
       d)   conversion of angiotensinogen to angiotensin

d 1. Which one of the following structures is not an accessory sex organ?
   a) vas deferens
   b) oviduct
   c) epididymis
   d) ovary

c 2. Which one of the following is not derived from the mesonephric duct?
   a) efferent ductules
   b) epididymis
   c) urethra
   d) ductus deferens

a 3. Which one of the following is not derived from the paramesonephric ducts in the female?
   a) ureters
   b) uterus
   c) oviducts
   d) vagina

b 4. The structure located externally where the mesonephric and paramesonephric ducts open to the outside of the embryo is known as the:
   a) urethral groove
   b) genital tubercle
   c) urethral fold
   d) labioscrotal swelling

a 5. Which one of the following structures does not bound the perineum?
   a) superior ramus of the pubis
   b) coccyx
   c) ischiopubic rami
   d) sacrotuberous ligaments

a 6. Which one of the following structures listed is passed thirdly by a sperm?
   a) efferent ductules
   b) rete testis
   c) seminiferous tubules
   d) epididymis

e 7. Which one of the following is not required for spermatogenesis?
   a) Gn-RH
   b) testosterone
   c) FSH
   d) LH
   e) all of the above are required

d   8.   The principal cells that form the outer layer (closest to the basement membrane) of a seminiferous tubule are:
- a) spermatids
- b) secondary spermatocytes
- c) Sertoli cells
- d) spermatogonia

c   9.   Which one of the following structures is diploid?
- a) secondary spermatocytes
- b) spermatids
- c) primary spermatocytes
- d) sperm

c   10.   Most of the cytoplasm of a mature sperm is located in its:
- a) head
- b) neck
- c) middle piece
- d) tail

b   11.   The structure responsible for the wrinkled appearance of the scrotum is the:
- a) cremaster muscle
- b) dartos tunic
- c) tunica vaginalis
- d) gubernaculum

b   12.   Which one of the following statements about the epididymis is incorrect?
- a) it is lined with pseudostratified columnar epithelium
- b) the efferent ductules leave the tail of the epididymis
- c) it has smooth muscle in its wall
- d) it stores sperm during the final part of their maturation process

b   13.   Which one of the following statements about the seminal vesicles is incorrect?
- a) they secrete prostaglandins
- b) they pass their contents into the vas deferens
- c) their secretion contains fructose
- d) they are located lateral to the vas deferens on the posterior inferior surface of the urinary bladder

a   14.   Which one of the following statements about the prostate gland is false?
- a) its secretion enters the ejaculatory duct
- b) it lies directly in front of the rectum
- c) it secretes enzymes, lipids, and citric acid
- d) it completely surrounds the urethra just inferior to the urinary bladder

d   15.   Which one of the following statements is incorrect?
        a)   if the number of sperm per ejaculation is less than 60 million, the man is generally considered to be infertile
        b)   the pH of the semen is about 7.4
        c)   fructose is the major energy source for ejaculated spermatozoa
        d)   the pH of the vagina is alkaline

b   16.   Which one of the following does not aid in supporting the ovaries?
        a)   broad ligament
        b)   falciform ligament
        c)   mesovarium
        d)   suspensory ligament

c   17.   Which one of the following listed would be pierced thirdly by the point of a pin entering an ovary from the abdominal cavity?
        a)   tunical albuginea
        b)   medulla
        c)   cortex
        d)   germinal epithelium

d   18.   At birth, the reproductive cells of a young female are called:
        a)   oogonia
        b)   polar bodies
        c)   ova
        d)   primary oocytes

a   19.   The object expelled from the ovary at the moment of ovulation is known as a/an:
        a)   secondary oocyte
        b)   primary oocyte
        c)   ovum
        d)   oogonium

c   20.   A mature follicle is known as a _____ follicle.
        a)   primary
        b)   growing
        c)   Graafian
        d)   secondary

c   21.   If the egg is not fertilized, the fate of the ruptured follicle is to become a/an:
        a)   atretic follicle
        b)   corpus luteum
        c)   corpus albicans
        d)   corpus capitis

c 22. The finger-like projections around the opening of the upper end of the oviduct are known as:
a) ostia
b) mesosalpinx
c) fimbriae
d) villi

a 23. The dome-shaped region of the uterus above and between the points of entrance of the uterine tubes is known as the:
a) fundus
b) cervix
c) isthmus
d) body

d 24. The ovarian and round ligaments form a continuum that is homologous to the _____ of the male.
a) tunica vaginalis
b) scrotum
c) tunica albuginea
d) gubernaculum

c 25. Chief support for the uterus is provided by the:
a) broad ligament
b) round ligament
c) pelvic diaphragm
d) mesometrium

b 26. Which one of the following statements about the vagina is incorrect?
a) it has less smooth muscle than the wall of the uterus
b) the vaginal mucosa does not undergo changes during the menstrual cycle
c) the mucosa of the vagina is made up of a stratified squamous epithelium
d) the hymen is formed from a vascular fold of the vaginal mucosa

d 27. Which one of the following structures is homologous to the dorsal portion of the penis?
a) labia minora
b) labia majora
c) bulb of the vestibule
d) clitoris

a 28. Which one of the following listed is not part of the anterior triangle of the perineum?
a) anus
b) urethra
c) vagina
d) labia

c   29.   An embryo that is XXY would:
         a)   not develop normal male genitalia
         b)   not develop normal male secondary sex characteristics
         c)   not be able to produce sperm
         d)   be perfectly normal

c   30.   The disease characterized by a chancre, skin rash, and degeneration of
         the nervous system is caused by:
         a)   Herpes virus
         b)   <u>Neisseria</u>
         c)   <u>Treponema</u>
         d)   <u>Staphylococcus</u>

e   31.   Testosterone is secreted by the:
         a)   sustenticular cells
         b)   spermatogonia
         c)   tunica albuginea
         d)   rete testis
         e)   none of the above is correct

a   32.   Which one of the following is not correct?
         a)   the testes begin their development within the abdominal cavity
         b)   descent of the testes is initiated in part by hormonal interactions
         c)   the gubernaculum is believed to assist in the descent of the
              testes
         d)   the vaginal processes protrude through the inguinal canal and
              into the scrotum
         e)   the tunica vaginalis is derived from peritoneum

d   33.   The bulb of the penis is derived from the:
         a)   distal end of the corpus spongiosum
         b)   distal end of the corpora cavernosa
         c)   proximal end of the corpora cavernosa
         d)   proximal end of the corpus spongiosum
         e)   none of the above is correct

b   34.   The events that occur in the male at puberty are DIRECTLY the result
         of an increased production of:
         a)   Gn-RH
         b)   testosterone
         c)   FSH
         d)   LH
         e)   ICSH

a 35. The dilation of arterioles in the penis causes:
    a) the flaccid penis to become erect
    b) the erect penis to undergo detumescence
    c) ejaculation
    d) the retraction of the glans into the foreskin
    e) the drawing of the testes closer to the body

e 36. Failure of testicular descent is called:
    a) syngamy
    b) synapsis
    c) inguinal hernia
    d) eunuchidism
    e) none of the above is correct

e 37. The shortest region of the male urethra is the:
    a) prostatic region
    b) cavernous region
    c) serous region
    d) inguinal region
    e) membranous region

e 38. Which one of the following listed would be pierced thirdly by the point of a pin entering a uterine tube from the outside?
    a) serosa
    b) mucosa
    c) lamina propria
    d) longitudinal muscle layer
    e) circular muscle layer

c 39. Most of the wall of the uterus is made up of:
    a) endometrium
    b) lamina propria
    c) myometrium
    d) serosa

e 40. Which one of the following do not lie within or open into the vestibule?
    a) vagina
    b) greater vestibular glands
    c) urethra
    d) lesser vestibular glands
    e) labia majora

c 41. Milk would pass thirdly through which one of the following structures listed below?
    a) nipple
    b) glandular tissue
    c) ampulla
    d) lactiferous duct

c   42.   The events that occur during puberty in the female are DIRECTLY the result of an increased production of:
- a) progesterone
- b) Gn-RH
- c) estrogen
- d) FSH
- e) LH

e   43.   The noncellular membrane surrounding the ovum is known as the:
- a) corona radiata
- b) antrum
- c) theca interna
- d) cumulus oophorous
- e) none of the above is correct

e   44.   Which one of the following is not true? Estrogen:
- a) acts on the hypothalamus to inhibit or stimulate the production of Gn-RH depending on the level of estrogen
- b) acts on the pituitary to inhibit the release of FSH
- c) enhances the sensitivity of LH producing structures
- d) causes proliferation of the functional layer of the uterus
- e) does all of the above

e   45.   Which one of the following is not true of progesterone? Progesterone:
- a) is involved in causing ovulation
- b) is secreted by both the corpus luteum and the pregnant uterus
- c) inhibits the release of Gn-RH
- d) causes the endometrium to become glandular with spiral blood vessels
- e) does all of the above

b   46.   Which one of the following hormones would be present in low levels in the menopausal woman?
- a) FSH
- b) estrogen
- c) LH
- d) all of the above would be present in low levels

c   47.   The secretory phase of the menstrual cycle is characterized by all but which one of the following items?
- a) the glands of the endometrium secrete a fluid rich in glycogen
- b) the arteries in the endometrium become enlarged and spiraled
- c) it lasts from day 20 to day 28
- d) spontaneous contractions of the myometrium decrease

c   48. Which one of the following statements about hormone levels is false?
a)   lowering levels of corpus luteum hormones induce ischemia of blood vessels in the endometrium
b)   increasing levels of estrogen stimulate growth and proliferation of new endometrial tissue
c)   there is a low decrease in LH secretion by the adenohypophysis
d)   during the proliferative phase, progesterone levels slowly increase due to the activity of the Graafian follicle

a   1.   Which one of the following structures listed is not an accessory sex organ?
- a) testes
- b) seminal vesicles
- c) vagina
- d) prostate gland

c   2.   What is the fate of the paramesonephric ducts in the male?
- a) they become the epididymis
- b) they become the ductus deferens
- c) they degenerate
- d) they become the ureters

d   3.   Which one of the following is not derived from the mesonephric ducts in the female?
- a) vagina
- b) oviduct
- c) uterus
- d) none of the above is derived from the mesonephric ducts

c   4.   The genital tubercle in the female embryo develops into the:
- a) labia minora
- b) labia majora
- c) clitoris
- d) urethra

c   5.   Which one of the following is incorrect?
- a) the perineum is bonded by the pubic arch anteriorly
- b) the perineum is bonded by the coccyx posteriorly
- c) the anterior triangle of the perineum contains the anus
- d) a transverse line passing between the ischial tuberosities divides the perineum into two triangles

c   6.   Which one of the following structures listed is passed thirdly by a sperm?
- a) tubulus rectus
- b) epididymis
- c) rete testis
- d) seminiferous tubule

c   7.   Interstitial endocrinocytes secrete:
- a) FSH
- b) ICSH
- c) testosterone
- d) LH

b   8.   The cells that protect and support the developing spermatozoa are known as:
a)   intestinal endocrinocytes
b)   Sertoli cells
c)   spermatogonia
d)   secondary spermatocytes

d   9.   Which one of the following structures is haploid?
a)   sustenticular cells
b)   spermatogonia
c)   primary spermatocytes
d)   secondary spermatocytes

c   10.   The principal structures in the neck of the sperm are the:
a)   acrosomes
b)   mitochondria
c)   centrioles
d)   Golgi apparatus

a   11.   The structure mainly responsible for regulating the temperature of the testes to some extent is the:
a)   cremaster muscle
b)   dartos tunic
c)   tunica vaginalis
d)   gubernaculum

c   12.   Which one of the following statements about the ductus deferens is incorrect?
a)   its lumen is lined with pseudostratified columnar epithelium
b)   there are 3 layers of smooth muscle in its wall
c)   it receives fluid from the prostate gland
d)   it has stereocilia

d   13.   Secretions of seminal vesicles include all but which one of the following?
a)   fructose
b)   prostaglandin
c)   pH
d)   sperm

c   14.   Which one of the following statements is false?
a)   within the penis, the corpus spongiosum is ventral
b)   within the penis, the corpora cavernosa are dorsal
c)   the corpora cavernosa at their distal ends anchor the penis to the ischial and pubic bones
d)   the proximal end of the corpus spongiosum forms the bulb of the penis

b   15. Which one of the following statements is false?
- a) each ejaculation has a volume of about 2 ml.
- b) the roles of most of the substances in the semen have been determined
- c) many sperm must be present in order for fertilization to occur
- d) the pH of the vagina is normally acidic

e   16. Which one of the following does not aid in supporting the ovary?
- a) suspensory ligament
- b) ovarian ligament
- c) mesovarium
- d) broad ligament
- e) all of the above support the ovary

a   17. Which one of the following listed would be pierced secondly by the point of a pin entering an ovary from the abdominal cavity?
- a) tunica albuginea
- b) medulla
- c) cortex
- d) germinal epithelium

c   18. Which one of the following listed occurs thirdly during oogenesis?
- a) ovum
- b) oogonium
- c) secondary oocyte
- d) primary oocyte

d   19. Prophase of the first meiotic division of oogenesis takes:
- a) about an hour
- b) 15 days
- c) a month
- d) years

b   20. If the egg is fertilized, the fate of the ruptured follicle is to become ultimately a corpus _____.
- a) follicle
- b) albicans
- c) luteum
- d) capitis

d   21. Which one of the following structures listed is pierced secondly by the point of a pin entering a uterine tube from the abdominal cavity?
- a) mucosa
- b) serosa
- c) circular muscle layer
- d) longitudinal muscle layer
- e) lamina propria

b   22.   The portion of the uterus that joins with the vagina is known as the:
           a)   isthmus
           b)   cervix
           c)   body
           d)   fundus

c   23.   Which one of the following structures does not aid in anchoring the
           uterus in place?
           a)   mesometrium
           b)   broad ligament
           c)   umbilical ligaments
           d)   round ligaments

a   24.   Which one of the structures listed is pierced thirdly by the point of a
           pin entering the uterus from the abdominal cavity?
           a)   endometrium
           b)   myometrium
           c)   functional layer
           d)   serosa

b   25.   Which one of the following listed occurs thirdly?
           a)   corpus luteum
           b)   mature follicle
           c)   corpus albicans
           d)   primary follicle
           e)   developing follicle

a   26.   Which one of the following structures is homologous to the ventral
           portion of the penis?
           a)   bulb of the vestibule
           b)   labia minora
           c)   clitoris
           d)   labia majora

d   27.   The "clinical perineum" is the area:
           a)   around the vestibule
           b)   within the anterior triangle
           c)   within the posterior triangle
           d)   between the vagina and the anus

d   28.   Which one of the following would be a normal male?
           a)   XO
           b)   XX
           c)   YO
           d)   XY

b   29.   Gonorrhea is caused by:
    a)   Herpes virus
    b)   <u>Neisseria</u>
    c)   <u>Treponema</u>
    d)   <u>Staphylococcus</u>

d   30.   Which one of the following would be passed thirdly by an egg?
    a)   body of uterus
    b)   coelom
    c)   uterine tube
    d)   fimbria
    e)   ovary

a   31.   Assume a pin is entering the testis through the scrotal wall. Listed below are the layers through which this pin would pass. Select the third layer which would be pierced by the pin.
    a)   tunica vaginalis
    b)   cremaster muscle fibers
    c)   dartos muscle fibers
    d)   tunica albuginea
    e)   seminiferous tubules

e   32.   Ovulation is the result of the sudden and dramatic release of large quantities of:
    a)   oxytocin
    b)   ADH
    c)   FSH
    d)   estrogen
    e)   none of the above is correct

b   33.   The constriction of arterioles in the penis causes:
    a)   the flaccid penis to become erect
    b)   the erect penis to undergo detumescence
    c)   ejaculation
    d)   the retraction of the glans into the foreskin
    e)   the drawing of the testes closer to the body

b   34.   The longest region of the male urethra is the:
    a)   prostatic region
    b)   cavernous region
    c)   inguinal region
    d)   membranous region

d   35.   Which one of the following is not correct?
    a)   it is possible for an ovary to descend into the labia majora
    b)   the ligaments do not provide much support for the uterus
    c)   a prolapsed uterus is one that protrudes into the vagina
    d)   the round and ovarian ligaments have no equivalents in the male
    e)   all of the above are correct

a   36. Milk would pass fourthly through which one of the following
          structures listed below?
          a)   nipple
          b)   glandular acini (alveoli)
          c)   ampulla
          d)   lactiferous duct

e   37. The events that occur during puberty in the female are DIRECTLY the
          result of an increased production of:
          a)   progesterone
          b)   Gn-RH
          c)   FSH
          d)   LH
          e)   none of the above is correct

e   38. The events that occur in the male at puberty are DIRECTLY the result
          of an increased production of:
          a)   Gn-RH
          b)   ICSH
          c)   FSH
          d)   LH
          e)   none of the above is correct

c   39. If the testes fail to descend, the principal medical complication is:
          a)   an empty scrotum
          b)   failure of development of secondary sex characteristics
          c)   sterility
          d)   a small penis
          e)   loss of sex drive

e   40. Fertilization usually takes place in the:
          a)   abdominal cavity
          b)   vagina
          c)   fundus
          d)   body
          e)   oviduct

c   41. The mesosalpinx is part of the:
          a)   suspensory ligament
          b)   ovarian ligament
          c)   broad ligament
          d)   round ligament
          e)   umbilical ligament

d   42.   Which one of the following statements about the vagina is incorrect?
- a) the vagina has less smooth muscle than the wall of the uterus
- b) its mucosa contain rugae
- c) its mucosa proliferates during the menstrual cycle in a similar manner to the endometrium of the uterus
- d) the epithelium of the mucosa is simple squamous
- e) all of the above are correct

d   43.   The male and female reproductive systems have many homologous structures. Which one of the following pairs is not a correct set of male-female homologues?
- a) ovary-testes
- b) labia majora-scrotal sac
- c) clitoris-dorsal portion of penis
- d) oviduct-vas deferens
- e) all of the above are male-female homologues

a   44.   Which one of the following is not correct? Estrogen:
- a) causes uterine gland secretion
- b) is responsible for the characteristic body fat distribution of women
- c) enhances the sensitivity of LH producing structures
- d) acts on the pituitary to inhibit the release of FSH
- e) at high levels stimulates the production of Gn-RH

e   45.   Which one of the following is not correct? Progesterone:
- a) aids in inhibiting the release of Gn-RH
- b) aids in inhibiting the release of FSH and LH
- c) stimulates the development of the endometrium during its secretory phase
- d) causes blood vessels in the wall of the uterus to become enlarged and spiraled
- e) all of the above are correct

d   46.   The cessation of the menstrual cycle in women over 50 years of age is known as:
- a) menarche
- b) the female climacteric
- c) amenorrhea
- d) menopause
- e) dysmenorrhea

c   47.   The menstrual phase is characterized by all but which one of the following?
- a) low estrogen and progesterone levels
- b) sloughing of the outer layer of the endometrium
- c) low levels of FSH-RH
- d) this phase lasts from 3-6 days

a   48.   Which one of the following statements about hormone levels is false?
         a)   estrogens are responsible for the differentiation and further development of new tissue during the proliferative phase of the menstrual cycle
         b)   LH and FSH are inhibited by rising levels of estrogen and progesterone
         c)   estrogen levels peak around the 12th day
         d)   high levels of LH induce ovulation

NAME_____

SECTION _____

b   1.   Which one of the following statements is false?
          a)   the fertilized egg is called a zygote
          b)   it takes several hours for the egg to reach the uterus
          c)   it is possible for the fertilized egg to implant in the mesenteries
          d)   the fertilized egg will usually implant in the endometrium of the
                uterus

d   2.   Which one of the following statements is false?
          a)   the hyaluronidase in the acrosome of each sperm breaks down the
                corona radiata
          b)   the sperm must undergo capacitation before they can fertilize an
                ovum
          c)   the sex of the embryo is determined at the time of fertilization
          d)   after the fusion of the pronuclei, the ovum completes the second
                meiotic division

d   3.   The embryo develops from the:
          a)   trophoblast
          b)   blastocoel
          c)   entire blastocyst
          d)   inner cell mass

c   4.   Which one of the following statements is false?
          a)   when the blastocyst is in the uterus, it initially gets its nutrients
                from the uterine fluids
          b)   implantation takes place 7 days after ovulation
          c)   the zona pellucida becomes the trophectoderm
          d)   estrogen and progesterone have prepared the endometrium for
                implantation

d   5.   Which one of the following nutrient sources is used thirdly by a
          developing embryo?
          a)   placenta
          b)   cytoplasm of ovum
          c)   uterine fluids
          d)   eroded cells of the endometrium

c   6.   The corpus luteum is stimulated to produce its hormone during
          pregnancy by the hormone _____.
          a)   LH
          b)   FSH
          c)   HCG
          d)   estrogen and progesterone

c   7.   Which one of the following develop from the trophectoderm?
         a)   embryo
         b)   allantois
         c)   chorion
         d)   amnion

d   8.   The placenta secretes:
         a)   FSH
         b)   LH
         c)   HCG
         d)   none of the above is correct

c   9.   The portion of the uterus which forms the maternal portion of the
         placenta is the:
         a)   myometrium
         b)   endothelium
         c)   decidua basalis
         d)   decidua capsularis

d   10. Which one of the following vessels removes fetal blood from the
         placenta?
         a)   uterine arteries
         b)   uterine veins
         c)   umbilical arteries
         d)   umbilical vein

d   11. Which one of the following structures is not derived from ectoderm?
         a)   skin
         b)   nervous system
         c)   cutaneous sense organs
         d)   peritoneum

c   12. Which one of the embryonic membranes is the source of the germ cells
         that migrate to the gonads?
         a)   amnion
         b)   chorion
         c)   yolk sac
         d)   allantois

a   13. The embryonic membrane that encloses the fluid that protects the
         embryo against injury and allows it to move freely is the:
         a)   amnion
         b)   chorion
         c)   yolk sac
         d)   allantois

c   14.   Which one of the following is correct? The umbilical cord contains:
         a)   2 arteries, 2 veins, and 1 allantoic duct
         b)   1 artery, 2 veins, and 1 allantoic duct
         c)   2 arteries, 1 vein, and 1 allantoic duct
         d)   none of the above is correct

d   15.   The foramen ovale connects the:
         a)   pulmonary trunk to the aorta
         b)   right atrium to the left ventricle
         c)   the right ventricle to the left ventricle
         d)   none of the above is correct

a   16.   Which one of the following fetal vessels does not carry unoxygenated blood?
         a)   umbilical vein
         b)   aorta
         c)   umbilical artery
         d)   inferior vena cava

b   17.   The arteries that took blood to the placenta in the fetus become the
         _____ in the adult.
         a)   ligamentum teres
         b)   umbilical ligaments
         c)   fossa ovalis
         d)   ligamentum venosum

d   18.   Which one of the following statements is false regarding changes in circulation at birth?
         a)   after birth there is a lowered resistance to blood flow through the lungs
         b)   pressure in the right atrium decreases
         c)   pressure in the left atrium increases
         d)   pressures in the aorta are lower than in the pulmonary trunk

b   19.   The ovum remains capable of being fertilized for about ___ hours after ovulation.
         a)   6
         b)   18
         c)   48
         d)   72

e   20.   During pregnancy high levels of estrogen and progesterone are necessary to:
         a)   inhibit secretion of FSH
         b)   prevent the further development of any follicles
         c)   maintain the endometrium
         d)   inhibit production of LH
         e)   all of the above are correct

a   21.   Which one of the following is not true of the allantois?
- a)   it is an outpouching of the midgut of the embryo
- b)   it is derived from mesoderm
- c)   it is in the umbilical cord
- d)   in most animals it serves as a storage site for wastes
- e)   the blood vessels that supply the allantois are the umbilical vessels which carry the fetal blood supply to and from the placenta

b   22.   Which one of the following is not derived from mesoderm?
- a)   muscle
- b)   epidermis of the skin
- c)   blood
- d)   cartilage and bone
- e)   epithelium of the coelom and kidneys

e   23.   Which one of the following is not derived from endoderm?
- a)   epithelium of the digestive tract
- b)   glands of the digestive tract
- c)   epithelium of the urinary bladder
- d)   epithelium of the larynx and trachea
- e)   all of the above are derived from endoderm

b   24.   By _____ weeks, it is possible to determine the gender of the fetus, which by this time also has most of the major organs developed.
- a)   5
- b)   8
- c)   18
- d)   28

a   25.   Which one of the following is not characteristic of monozygotic twins?
- a)   are of different sexes
- b)   may have separate placentas
- c)   have identical genes
- d)   come from an inner cell mass that separated into two masses
- e)   all of the above are characteristic of monozygotic twins

e   26.   Which one of the following is not correct?
- a)   in advance labor oxytocin levels in maternal and fetal blood are high
- b)   oxytocin stimulates prostaglandin production by the placenta
- c)   prostaglandins stimulate smooth muscle to contract
- d)   estrogen and uterine distension can increase the number of oxytocin receptors within the uterus and placenta
- e)   all of the above are correct

d   27.   When the placenta implants low in the uterus and completely or partially overlaps the internal os, the condition is called:
     a)   toxemia
     b)   ectopic pregnancy
     c)   cryptorchidism
     d)   placenta previa
     e)   prolapsed uterus

c   28.   At puberty the hormone _____ stimulates the growth and branching of the duct system within the breast.
     a)   prolactin
     b)   oxytocin
     c)   estrogen
     d)   progesterone
     e)   chorionic somatomammotropin

b   29.   The major hormone responsible for milk production is:
     a)   oxytocin
     b)   prolactin
     c)   chorionic somatomammotropin
     d)   estrogen
     e)   progesterone

c   30.   Which one of the following listed occurs thirdly in the milk let-down reflex?
     a)   contraction of myoepithelial cells
     b)   sensory nerve impulses to the hypothalamus
     c)   release of oxytocin
     d)   suckling of the baby
     e)   milk let-down

e   31.   A unit of genetic information responsible for a particular biochemical or structural characteristic of an individual is known as a:
     a)   chromosome
     b)   locus
     c)   DNA molecule
     d)   trait
     e)   gene

d   32.   Egg or sperm cells contain:
     a)   the diploid chromosome number
     b)   46 chromosomes
     c)   two sex chromosomes
     d)   22 autosomes
     e)   none of the above is correct

a 33. A person who has the genotype XO has the condition known as _____ syndrome.
   a) Turner's
   b) Down's
   c) Kleinfelter's
   d) Spence's
   e) Mason's

c 34. Consider that the gene for tongue rolling is dominant. If a person who is heterozygous dominant for this trait mates with a person of the opposite sex who is also heterozygous dominant for this trait, the probability that their child will be a tongue roller too is:
   a) 25%
   b) 50%
   c) 75%
   d) cannot be determined from the data available

b 35. The actual expression of the genetic information within a person is her or his:
   a) genotype
   b) phenotype
   c) pedigree
   d) genetic potential
   e) none of the above is correct

e 36. Which of the following are genetic diseases linked to the X chromosome?
   a) sickle cell anemia
   b) galactosemia
   c) color blindness
   d) hemophilia
   e) two of the above are correct

c 37. A genotype that contains two identical genes for a particular trait is:
   a) dominant
   b) recessive
   c) homozygous
   d) heterozygous

a   1.   Which one of the following statements is false?
        a)   ciliary currents in the oviduct help move the sperm upwards toward the egg
        b)   fertilization generally takes place in the upper region of the oviduct
        c)   after ovulation the egg is in the peritoneal cavity
        d)   the sphere of follicle cells surrounding the egg is called the corona radiata

c   2.   Which one of the following statements is false?
        a)   the sperm reach the upper portion of the uterine tubes within several minutes
        b)   prostaglandins may assist the sperm to move up the uterine tubes
        c)   the sperm are capable of fertilizing an ovum as soon as they are deposited in the vagina
        d)   the acrosome of the sperm contains hyaluronidase

c   3.   The embryo that reaches the uterus is most accurately called:
        a)   an ovum
        b)   a zygote
        c)   a blastocyst
        d)   a fertilized egg

d   4.   Implantation of the embryo in the wall of the uterus takes place:
        a)   immediately after the embryo reaches the uterus
        b)   about seven days after ovulation
        c)   when the woman is in approximately the 21st day of her menstrual cycle
        d)   two of the above are correct

b   5.   The endometrium is digested by enzymes released from the:
        a)   inner cell mass
        b)   trophoblast
        c)   chorionic villi
        d)   blastocoel

a   6.   Maintaining the endometrium is the role of:
        a)   estrogen and progesterone
        b)   estrogen and FSH
        c)   progesterone and FSH
        d)   LH and FSH
        e)   none of the above is correct

d   7.   The structure that maintains the endometrium after the third month of pregnancy is the:
a) corpus luteum
b) chorion
c) ovary
d) placenta

a   8.   A human embryo is called a fetus after:
a) two months
b) one month
c) two weeks
d) three weeks

d   9.   The chorion develops from the:
a) endometrium
b) decidua basalis
c) decidua capsularis
d) trophectoderm and extra-embryonic mesoderm

a   10.   Which one of the following vessels removes maternal blood from the placenta?
a) uterine veins
b) uterine arteries
c) umbilical veins
d) umbilical arteries

b   11.   Which one of the following structures is not derived from mesoderm?
a) muscles
b) pancreas
c) blood
d) bones

b   12.   Which one of the following membranes surrounds the entire embryo and the other three embryonic membranes?
a) amnion
b) chorion
c) allantois
d) yolk sac

d   13.   The blood vessels that supply the embryonic membrane known as the _____ also provide the fetal blood supply to the placenta.
a) amnion
b) chorion
c) yolk sac
d) allantois

b   14.   The vessel that acts as a bypass of the liver of the developing fetus
          is the:
          a)   ductus arteriosus
          b)   ductus venosus
          c)   umbilical vein
          d)   umbilical artery

d   15.   The bypass known as the ductus arteriosus connects the:
          a)   right atrium to the left atrium
          b)   the umbilical vein to the inferior vena cava
          c)   the umbilical vein to the hepatic portal vein
          d)   none of the above is correct

b   16.   Which one of the following fetal vessels carries only oxygenated
          blood?
          a)   umbilical arteries
          b)   umbilical vein
          c)   aorta
          d)   inferior vena cava

d   17.   The umbilical vein of the fetus becomes the ____ in the adult.
          a)   ligamentum arteriosum
          b)   ductus venosus
          c)   ligamentum venosum
          d)   ligamentum teres

c   18.   Which one of the following statements is false regarding changes in
          circulation at birth?
          a)   after birth, the build-up in carbon dioxide and decrease in
               oxygen cause the infant to take its first breath
          b)   expansion of the lungs decreases their resistance to blood flow
          c)   pressures decrease in the left atrium
          d)   pressures in the pulmonary arteries are reduced

c   19.   The sperm remain capable of fertilizing an egg for about _____
          hours after intercourse.
          a)   10-12
          b)   12-24
          c)   24-72
          d)   more than 72

e   20.   Which one of the following is not correct about the yolk sac?
          a)   in the human it serves no nutritive function
          b)   it extends into the umbilical cord
          c)   it is an outpouching of the midgut:
          d)   the gut of the embryo develops from this membrane
          e)   all of the above are correct

b   21.   Which one of the following is not derived from endoderm?
- a) liver
- b) epithelium of the mouth
- c) epithelium of the thyroid glands
- d) epithelium of the urethra
- e) all of the above are derived from endoderm

d   22.   Which one of the following is not derived from ectoderm?
- a) lens of the eye
- b) receptor cells of sense organs
- c) adrenal medulla
- d) epithelium of the gonads
- e) enamal of the teeth

e   23.   By _____ weeks, the fingernails are present, and the respiratory, circulatory, and nervous systems are established.
- a) 5
- b) 8
- c) 16
- d) 20
- e) 28

b   24.   Which one of the following is not characteristic of dizygotic twins?
- a) can be of different sexes
- b) may or may not have separate placentas
- c) come from the fertilization of two different ova
- d) can vary as much as any siblings
- e) all of the above are characteristic of dizygotic twins

b   25.   Which one of the following is true?
- a) labor means the same thing as parturition
- b) the afterbirth and the placenta are the same thing
- c) menstrual cycles resume about the end of the 5th week if the mother is not nursing
- d) uterine contractions move towards the fundus

e   26.   Which one of the following does not play a role in parturition?
- a) oxytocin
- b) prostaglandins
- c) a neuroendocrine reflex from the cervix to the hypothalamus
- d) positive feedback
- e) all of the above play a role in parturition

b   27.   Implantation of the egg at a site outside of the uterus is known as:
- a) cryptorchidism
- b) ectopic pregnancy
- c) toxemia
- d) placenta previa
- e) none of the above is correct

a 28. Which one of the following hormones does not contribute to the development of the breasts during pregnancy?
   a) oxytocin
   b) prolactin
   c) estrogen
   d) progesterone
   e) chorionic somatomammotropin

c 29. The reason that there is little milk production during pregnancy is due to the influence of:
   a) FSH and estrogen
   b) estrogen and oxytocin
   c) progesterone and estrogen
   d) oxytocin and HCG
   e) none of the above is correct

a 30. Which one of the following occurs fourthly in the milk let-down reflex?
   a) contraction of myoepithelial cells
   b) sensory nerve impulses to the hypothalamus
   c) release of oxytocin
   d) suckling of the baby
   e) milk let-down

d 31. The place on a chromosome where a unit of genetic information is located is known as a:
   a) trait
   b) DNA molecule
   c) gene
   d) locus
   e) allele

b 32. All cells but egg or sperm cells contain:
   a) the haploid chromosome number
   b) 23 pairs of chromosmes
   c) 22 autosomes and one sex chromosome
   d) none of the above is correct

c 33. A person who has the genotype XXY has the condition known as ____ ____syndrome.
   a) Turner's
   b) Down's
   c) Kleinfelter's
   d) Spence's
   e) Mason's

d   34.   Considering that the gene for red hair is recessive to dark hair, what are the chances that a couple, both of whom have dark hair, will have a red-haired child if each had a red-haired parent?
- a)   100%
- b)   75%
- c)   50%
- d)   25%
- e)   none

d   35.   Referring to question #34 above, what is the genotype of the couple, each of whom had a red-haired parent?
- a)   dark hair
- b)   red hair
- c)   homozygous
- d)   heterozygous

e   36.   Which of the following are genetic diseases not linked to the X chromosome?
- a)   color blindness
- b)   hemophilia
- c)   sickle cell anemia
- d)   galactosemia
- e)   two of the above are correct

d   37.   A genotype that contains two different genes for a particular trait is:
- a)   dominant
- b)   recessive
- c)   homozygous
- d)   heterozygous

Bone Features Superior Extremity

a.	acromion process
b.	axillary border
c.	coracoid process
d.	capitulum
e.	coronoid process
ab.	deltoid tuberosity
ac.	glenoid fossa
ad.	greater tubercle
ae.	head
bc.	inferior angle
bd.	infraspinous fossa
be.	lateral epicondyle
cd.	lesser tubercle
ce.	medial epicondyle
de.	olecranon fossa
abc.	olecranon process
abd.	radial tuberosity
abe.	spine
acd.	subscapular fossa
ace.	superior border
ade.	supraspinous fossa
bcd.	styloid process
bce.	trochlea
bde.	trochlear notch
cde.	vertebral border

Bones

a.	calcaneus
b.	carpal
c.	clavicle
d.	femur
e.	fibula
ab.	gladiolus (body)
ac.	humerus
ad.	ilium
ae.	ischium
bc.	manubrium
bd.	metacarpal
be.	metatarsal
cd.	patella
ce.	phalange
de.	pubis
abc.	radius
abd.	rib
acd.	scapula
ace.	talus
ade.	tarsal
bce.	tibia
bce.	ulna
bde.	xiphoid process

Bone Features Inferior Extremity

a.	acetabulum
b.	anterior inferior iliac spine
c.	anterior superior iliac spine
d.	articulating surface for head of fibula
e.	greater sciatic notch
ab.	greater trochanter
ac.	head
ad.	iliac crest
ae.	iliac fossa
bc.	ischial tuberosity
bd.	inferior ramus of pubis
be.	intercondylar eminence
cd.	lateral condyle
ce.	lateral epicondyle
de.	lateral malleolus
abc.	lesser trochanter
abd.	linea aspera
abe.	medial condyle
acd.	medial epicondyle
ace.	medial malleolus
ade.	neck
acd.	obturator foramen
bce.	pelvic inlet
bde.	posterior inferior iliac spine
cde.	posterior superior iliac spine
abcd.	pubic symphysis
abce.	superior ramus of pubis
abde.	surface for articulation with talus
acde.	surface for articulation with tibia
bcde.	tibial tuberosity

Rib Features

a.	costal groove
b.	head
c.	inferior margin
d.	neck
e.	shaft

Vertebrae

ab.	atlas
ac.	axis
ad.	cervical
ae.	coccyx
bc.	lumbar
bd.	sacral (sacrum)
be.	thoracic

Vertebral Features

a.	auricular surface
b.	body
c.	demifacet
d.	costotransverse facet
e.	dens
ab.	inferior articulating process
ac.	inferior articulating surface
ad.	inferior vertebral notch
ae.	lamina
bc.	lateral mass
bd.	pedicle
be.	sacral foramen
cd.	spinous process
ce.	superior articulating process
de.	superior articulating surface
abc.	superior vertebral notch
abd.	transverse foramen
ace.	transverse process
acd.	vertebral foramen

Skull Features

a. anterior cranial fossa
b. carotid canal
c. coronal suture
d. cribriform plate
e. crista galli
ab. external auditory meatus
ac. foramen magnum
ad. internal auditory meatus
ae. jugular foramen
bc. lambdoidal suture
bd. mastoid process
be. middle cranial fossa
cd. middle nasal concha
ce. occipital condyles
de. optic canal (foramen)
abc. posterior cranial fossa
abd. sagittal suture
abe. sella turcica (hypophyseal fossa)
acd. squamosal suture
ace. superior nasal concha
ade. zygomatic arch

Cranial Bones

a. ethmoid
b. frontal
c. parietal
d. occipital
e. sphenoid
ab. temporal

Facial Bones

ac. inferior nasal concha
ad. lacrimal
ae. mandible
bc. maxillary
bd. nasal
be. palatine
cd. vomer
ce. zygomatic

Views

a. anterior
b. posterior
c. lateral
d. medial
e. superior
ab. inferior

Joints

a. condyloid joint
b. gomphosis
c. hinge joint
d. hyaline synchondrosis
ab. multiaxial joint
ac. nonaxial joint
ad. pivot joint
ae. saddle joint
bc. sutures
bd. symphysis
be. syndesmosis

FINAL EXAMINATION

CHAPTER 1 - INTRODUCTION

d   1. The organs of the abdominopelvic cavity are covered with:
    a) visceral pleura
    b) parietal pleura
    c) parietal peritoneum
    d) visceral peritoneum
    e) no correct answer

b   2. The thoracic cavity includes all but which one of the following?
    a) pericardial cavity
    b) vertebral canal
    c) mediastinum
    d) pleural cavities
    e) all of the above are in the thoracic cavity

c   3. In the anatomical position, all but which one of the following are true?
    a) toes directed forward
    b) eyes directed forward
    c) body supine
    d) thumbs lateral
    e) all of the above are true

e   4. The term distal means:
    a) toward the attached end of a limb
    b) away from the head
    c) toward the origin of a structure
    d) toward the mid-line of the body
    e) none of the above

d   5. The study of the body as visible to the unaided eye (no microscope) is spoken of as what kind of anatomy?
    a) systemic
    b) regional
    c) histological
    d) gross
    e) radiological

c   6. The chin is _____ to the nose.
    a) anterior
    b) posterior
    c) inferior
    d) dorsal
    e) lateral

c 7. The plane that divides the body or its parts into superior and inferior portions is called the _____.
a) sagittal
b) frontal
c) horizontal
d) coronal
e) no correct answer

b 8. The mechanism by which an organism maintains itself in a condition that is relatively stable and constant is:
a) metabolism
b) homeostasis
c) growth and reproduction
d) positive feedback mechanisms
e) none of the above is correct

b 9. Which region of the abdomen is superior and lateral to the hypogastric region?
a) inguinal region
b) hypochondriac region
c) umbilical
d) hypergastric
e) no correct answer

c 10. The movement of a solvent across a semi permeable membrane in response to a concentration difference across the membrane is called:
a) diffusion
b) dialysis
c) osmosis
d) facilitated transport
e) no correct answer

d 11. The portion of the thoracic cavity between the two pleural cavities is known as the:
a) pericardial cavity
b) dorsal cavity
c) pleural space
d) mediastinum

d 12. The maintenance of constant internal environmental conditions within the body is referred to as:
a) anabolism
b) catabolism
c) negative feedback
d) homeostasis
e) none of the above is correct

e   13. Which one of the following listed is the third largest in size (starting with the largest)?
- a) organ
- b) organ system
- c) organelle
- d) cell
- e) tissue

b   14. The membrane that covers the lungs is known as the:
- a) parietal peritoneum
- b) visceral pleura
- c) parietal pleura
- d) visceral peritoneum
- e) visceral pericardium

e   15. A longitudinal section that passes through the midline of the body is called the _____ section.
- a) parasagittal
- b) frontal
- c) cross
- d) coronal
- e) midsagittal

FINAL EXAMINATION

CHAPTER 2 - CHEMICAL AND PHYSICAL BASIS OF LIFE

d   1. The mass of an atom is determined primarily by the:
    a)   number of protons
    b)   number of electrons
    c)   number of electrons and protons
    d)   number of protons and neutrons
    e)   number of neutrons

c   2. An atom has the same number of:
    a)   protons and neutrons
    b)   electrons and neutrons
    c)   protons and electrons
    d)   none of the above

d   3. If a certain atom has 13 neutrons and 11 protons in its nucleus, how many electrons would be in the outermost orbit?
    a)   11
    b)   8
    c)   2
    d)   1
    e)   none of the above

a   4. Isotopes of an atom:
    a)   have different numbers of neutrons
    b)   are almost always radioactive
    c)   have the same atomic weight
    d)   none of the above

c   5. A gamma ray:
    a)   has essentially the same mass as an electron
    b)   has the same charge as a helium nucleus
    c)   is a form of electromagnetic radiation
    d)   is also called a negatron
    e)   none of the above

e   6. Which one of the following is not a cation?
    a)   sodium
    b)   hydrogen
    c)   calcium
    d)   magnesium
    e)   none of the above are cations

c   7. Covalent bonding differs from ionic bonding in that covalent bonding involves:
- a) transfer of electrons from one atom to another
- b) bonding between cations
- c) sharing of one of more pairs of electrons
- d) none of the above is correct

c   8. Which one of the following terms includes all the others?
- a) glucose
- b) disaccharide
- c) carbohydrate
- d) polysaccharide
- e) monosaccharide

e   9. Fats consist of:
- a) two amino acids linked together
- b) carbon, hydrogen, oxygen, and nitrogen
- c) phosphates and fatty acids
- d) two glycerols and one fatty acid linked together
- e) none of the above is correct

b   10. A peptide bond joins:
- a) the amino groups of two amino acids
- b) the amino group of one and the acid group of another amino acid
- c) the acid groups of two amino acids
- d) the amino group of one and the peptide group of another amino acid
- e) none of the above is correct

b   11. The secondary structure of a protein is maintained largely by:
- a) peptide bonds
- b) hydrogen bonds
- c) nitrogen bonds
- d) R-group interactions
- e) none of the above is correct

e   12. RNA differs from DNA in that RNA contains the base:
- a) adenine
- b) thymine
- c) guanine
- d) cytosine
- e) uracil

d   13. Which one of the following contains all the others?
- a) sugar
- b) phosphate
- c) nitrogen-containing base
- d) nucleotide

d   14. Which one of the following does not influence the rate of an enzyme-catalyzed reaction?
a)   changes in pH
b)   increases in temperature
c)   changes in the amount of enzyme present
d)   all of the above can influence the rate of an enzyme-catalyzed reaction

b   15. A solution with which of the following pH values would be considered the most acidic?
a)   7
b)   3
c)   5
d)   9
e)   12

c   16. Which mixture contains the largest solute particles?
a)   true solution
b)   colloid
c)   suspension
d)   solvent

d   17. With regard to enzyme-catalyzed reactions, many vitamins act as
a)   non-protein, but organic catalysts
b)   inorganic catalysts
c)   specific substrates
d)   coenzymes
e)   enzymes cooperating with other enzymes

a   18. The process of selective separation of solutes in solution by taking advantage of their different diffusibilities through a selectively permeable membrane is called:
a)   dialysis
b)   osmosis
c)   diffusion
d)   active transport
e)   none of the above is correct

e   19. Which one of the following is not a function of water in the body?
a)   a solvent
b)   a heat absorber
c)   a transporting agent
d)   a source of hydrogen ions
e)   all of the above are functions of water

b   20. Which one of the following is not an example of a colloid?
- a) milk
- b) a mixture of sodium chloride and water
- c) a mixture of sand and water
- d) all of the above are examples of colloids

a   21. If a solution consisting of sucrose dissolved in water is separated by a semipermeable membrane from pure water, which of the following will occur?
- a) water molecules will move from the pure water into the sugar solution
- b) sugar molecules will move from the sugar solution into the pure water
- c) both water molecules and sugar molecules will move from the sugar solution into the pure water
- d) there will be no movement of either sugar molecules or water molecules across the membrane in either direction
- e) none of the above

b   22. One mole of glucose
- a) occupies the same volume as one mole of sucrose
- b) contains the same number of molecules as one mole of sucrose
- c) has the same weight as one mole of sucrose
- d) contains the same number of protons as one mole of sucrose
- e) none of the above

FINAL EXAMINATION

CHAPTER 3 - THE CELL

b   1. Which of the following is a movement of materials across a membrane
        that involves the expenditure of cellular energy?
        a)   filtration
        b)   active transport
        c)   dialysis
        d)   osmosis
        e)   diffusion

c   2. Which term describes the process of a cell taking in solid material
        from the external environment?
        a)   exocytosis
        b)   ectocytosis
        c)   erythrocytosis
        d)   endocytosis
        e)   none of the above is correct

e   3. The internal structure of cilia most nearly resembles that of:
        a)   mitochondria
        b)   lysosomes
        c)   Golgi bodies
        d)   centromeres
        e)   centrioles

b   4. If the normal human chromosome number is 46, then at prophase II of
        meiosis and cytokinesis there would be _____ chromosomes in a
        cell.
        a)   46
        b)   23
        c)   92
        d)   none of the above

c   5. During mitosis, the chromatids of the chromosomes become attached
        to:
        a)   astral fibers
        b)   microfibers
        c)   spindle fibers
        d)   nuclear fibers
        e)   none of the above

c   6. If a small part of one side of a DNA molecule has the base sequence
        A T C T G C, then the corresponding RNA base sequence would be:
        a)   ATCTGC
        b)   AUGTCG
        c)   UAGACG
        d)   CGUGTU

c   7. Cytoplasmic organelles which are the major organelles involved in producing energy for the cell are called:
a)   lysosomes
b)   Golgi bodies
c)   mitochondria
d)   ribosomes
e)   endoplasmic reticulum

d   8. Which of these organelles is most directly concerned with protein synthesis?
a)   centriole
b)   mitochondria
c)   cilia
d)   ribosomes
e)   vacuoles

e   9. Which of these cellular organelles is composed of cytoplasmic canals?
a)   mitochondria
b)   chromosomes
c)   lysosomes
d)   centrioles
e)   none of the above

a   10. The lysosome of the cell contains:
a)   digestive enzymes
b)   enzymes for organic synthesis
c)   hormones
d)   ribosomes
e)   mitochondria

d   11. Chromosomes are widely separated, having moved two-thirds the distance from the equatorial plate during _____ of mitosis and cytokinesis.
a)   interphase
b)   prophase
c)   metaphase
d)   anaphase
e)   telophase

a   12. If the total number of chromosomes in a cell is six, then there will be _____ chromosomes in each daughter cell after mitosis and cytokinesis.
a)   six
b)   three
c)   twelve
d)   none of these

b   13. Which cell organelle forms small secretory vesicles?
a)   lysosome
b)   Golgi apparatus
c)   nucleolus
d)   centriole
e)   mitochondria

e   14. The nucleolus is:
a)   the storehouse of genetic information within the cell
b)   the structure that contains the nucleus
c)   a cytoplasmic structure that contains DNA
d)   a structure that is bound by a membrane
e)   none of the above is correct

c   15. Which structure would not be evident in a cell at metaphase of mitosis?
a)   chromatids
b)   centromeres
c)   nuclear envelope
d)   spindle fibers
e)   all of the above would be evident

d   16. Which one of the following is not true of facilitated diffusion?
a)   uses carrier molecules
b)   moves molecules across the cell membrane that by themselves might pass the membrane slowly or not at all
c)   is characterized by specificity
d)   can move material into cells but not out of cells
e)   all of the above are true of facilitated diffusion

b   17. The fact that only a limited number of molecules can be transported across a cell membrane by mediated transport in a given amount of time implies:
a)   competition
b)   saturation
c)   specificity
d)   reciprocal molecular configuration

a   18. Which one of the following substances can generally pass through the lipid portion of the cell membrane with relatively little difficulty?
a)   nonpolar molecules
b)   positively charged ions
c)   large polar molecules
d)   none of the above is correct

d 19. Which one of the following is not a mediated transport process?
    a)   facilitated diffusion
    b)   active transport
    c)   dialysis
    d)   all of the above are mediated transport processes

b 20. Which one of the following does not have a similar structure to all the others listed?
    a)   cilia
    b)   microvilli
    c)   centrioles
    d)   basal bodies

d 21. Which one of the following do not occur as cellular inclusions?
    a)   hemoglobin
    b)   melanine
    c)   glycogen
    d)   all of the above can occur as cellular inclusions

c 22. Which one of the following stages occurs first in mitosis and cytokinesis?
    a)   metaphase
    b)   telophase
    c)   prophase
    d)   anaphase

c 23. The phase of interphase that immediately precedes mitosis is the:
    a)   S phase
    b) G_1 phase
    c) G_2 phase
    d)   M phase
    e)   cytokinesis phase

b 24. Which one of the following serves to carry instructions from nuclear DNA to sites of synthesis in the cytoplasm?
    a)   transfer RNA
    b)   messenger RNA
    c)   ribosomal RNA
    d)   endoplasmic reticulum
    e)   instructions from nuclear DNA do not reach the cytoplasm.

CHAPTER 4 - TISSUES

b   1.   Which one of the following is a property of <u>all</u> living cells?
    a)   contractility
    b)   reproduction
    c)   conductivity
    d)   metabolism
    e)   irritability

e   2.   Which one of the following listed is <u>not</u> true of mucous membranes?
    a)   they may be absorptive structures
    b)   the nature of their secretions can vary depending upon their location in the body
    c)   they always have a layer of connective tissue directly below the surface epithelium
    d)   they are topologically continuous with the outside of the body
    e)   all of the above are true of mucous membranes

b   3.   Which of these is <u>not</u> a serous membrane?
    a)   pleura
    b)   lining of esophagus
    c)   lining of abdominal cavity
    d)   membrane over surface of heart

b   4.   If a tissue has no blood supply but has lacunae, and a matrix with many fine collagen fibers, it would be _____.
    a)   compact bone
    b)   hyaline cartilage
    c)   elastic cartilage
    d)   fibrous cartilage
    e)   spongy bone

a   5.   The layers of matrix which characterize compact bone are called:
    a)   lamellae
    b)   lacunae
    c)   lunula
    d)   lamecula
    e)   lunacy

c   6.   In a living bone, what is contained in a lacuna?
    a)   matrix
    b)   calcium salts
    c)   bone cell (osteocyte)
    d)   fat
    e)   none of the above

c   7.   Which one of these features is <u>not</u> characteristic of cardiac muscle?
        a)   branching fibers
        b)   striations
        c)   elliptical, peripherally-located nuclei
        d)   longitudinal myofibrils
        e)   intercalated disks

a   8.   Skeletal muscle can be histologically characterized by all but which
         one of the following?
        a)   intercalated disks
        b)   striations
        c)   elliptical, flattened nuclei at periphery of cell
        d)   multinucleated
        e)   all the above characterize skeletal muscle

b   9.   Which of the following is <u>incorrect</u> about areolar connective tissue?
        a)   it contains various connective tissue cells, including fibroblasts
        b)   it always has its fibrous elements arranged in parallel fashion
        c)   it fills spaces between various organs
        d)   it forms the hypodermis of the skin
        e)   it forms adipose tissue with the accumulation of fat

c   10. When a secretion is formed by the tips of the individual cells pinching
         off and forming part of the secretion, the gland is spoken of as:
        a)   endocrine
        b)   holocrine
        c)   apocrine
        d)   merocrine
        e)   none of the above

e   11. Which one of the following is not true of epithelial tissues?
        a)   cover body surfaces
        b)   line body cavities
        c)   line ducts
        d)   line blood vessels
        e)   all of the above are correct

c   12. How does epidermal epithelium differ from the epithelium of the oral
         cavity (mouth)?
        a)   epidermal epithelium is simple columnar while oral epithelium is
             stratified squamous
        b)   epidermal epithelium is stratified squamous, while oral
             epithelium is ciliated
        c)   epidermal epithelium is keratinized, while oral epithelium is not
        d)   oral epithelium is merely wetter, because of saliva
        e)   none of the above

d   13. Which one of the following is not a way of classifying epithelia?
     a)   the shape of the cells
     b)   the arrangement of the cells into one or more layers
     c)   the kinds of decorations that may be present on the cells
     d)   the kinds of fibers present
     e)   all of the above are ways of classifying epithelia

c   14. Which one of the following is not true of simple squamous epithelium?
     a)   is found in regions where diffusion occurs
     b)   is found in regions where filtration occurs
     c)   is found in regions where an effective barrier is needed
     d)   is part of a serous membrane
     e)   all of the above are true of simple squamous epithelium

c   15. In spite of its stratified appearance, ciliated pseudostratified columnar epithelium is considered a simple epithelium. Why?
     a)   because all cells reach the free surface
     b)   because cells do not duplicate themselves
     c)   because all cells are in contact with the basement membrane
     d)   because it is a moist membrane
     e)   none of the above are valid explanations

e   16. Functions of epithelial tissues include all but which one of the following?
     a)   protection
     b)   secretion
     c)   absorption
     d)   excretion
     e)   conduction

d   17. Which one of the following listed is the deepest part of a mucous membrane?
     a)   lamina propria
     b)   epithelium
     c)   basement membrane
     d)   muscularis mucosa

e   18. In tissue repair, granulation tissue is formed by:
     a)   dividing epithelial cells
     b)   scar tissue
     c)   the blood clot
     d)   regrowth of capillaries
     e)   none of the above is correct

b   19.   Which one of the following is not true of the reticuloendothelial system?
    a)   its cells are phagocytic
    b)   its fibers are short, thin and unbranched
    c)   some of its cells are involved in the immune processes of the body
    d)   the spleen and tonsils are part of this system

d   20.   Mucous acini are easily distinguished from serous acini, because mucous acini have:
    a)   dark-staining cytoplasm
    b)   a large central nucleus
    c)   a small central lumen
    d)   a lack of obvious granules in the cytoplasm

a   21.   Which one of the following is not true of epithelia?
    a)   vascular
    b)   closely packed cells
    c)   few deposits between cells
    d)   one side in contact with a "free surface"
    e)   all of the above are true of epithelia

c   22.   Which one of the following is not shared between hyaline cartilage and bone?
    a)   collagen fibers
    b)   lacunae
    c)   interconnections between cells
    d)   supporting tissue

c   23.   A striated border of a cell is created by:
    a)   thickened endoplasmic reticulum at the free surface of the cell
    b)   congregation of mitochondria at the free surface of the cell
    c)   the presence of microvilli
    d)   the accumulation of secretion granules awaiting release
    e)   none of the above are correct

d   24.   Which one of the following is not a location of hyaline cartilage?
    a)   ends of long bones
    b)   trachea
    c)   costal cartilages
    d)   ear

b   25.   Which one of the following is not true of collagen fibers?
    a)   strong
    b)   elastic
    c)   unbranched
    d)   often white in color

e   26.   Which of the following may be found holding adjacent epithelial cells
           together?
           a)   gap junctions
           b)   tight junctions
           c)   desmosomes
           d)   intermediate junctions
           e)   all of the above are associated with epithelia

b   27.   The epithelial membrane that lines the mouth, esophagus, stomach, and
           intestines is called:
           a)   serous membrane
           b)   mucous membrane
           c)   dense irregular connective tissue
           d)   adipose tissue
           e)   there is no one general term for the membrane that lines all of
                 these regions.

FINAL EXAMINATION

CHAPTER 5 - THE SKIN

e   1.   Which one of the following is not a function of the skin?
         a)   protection
         b)   temperature regulation
         c)   excretion of water and salts
         d)   sensation
         e)   all of the above are functions of the skin

d   2.   The dermis of the skin is largely composed of _____, except in the
         papillary area and around hair follicles.
         a)   keratinized stratified squamous epithelium
         b)   non-keratinized stratified squamous epithelium
         c)   dense fibrous connective tissue, regularly arranged
         d)   dense fibrous connective tissue, irregularly arranged
         e)   adipose tissue

b   3.   The pigment of the skin is:
         a)   hyaline
         b)   melanin
         c)   keratin
         d)   lucidum
         e)   sebum

c   4.   The epidermis is an example of which of these epithelial varieties?
         a)   simple squamous epithelium
         b)   non-keratinized stratified squamous epithelium
         c)   keratinized stratified squamous epithelium
         d)   transitional epithelium
         e)   the stratum corneum

d   5.   Which of the following is not true of sebaceous glands?
         a)   they may be associated with hair follicles
         b)   their mode of secretion is holocrine
         c)   they secrete sebum
         d)   they are endocrine glands

e   6.   Which of the following items is passed through thirdly by a pin
         entering from the outside?
         a)   stratum lucidum
         b)   stratum corneum
         c)   stratum germinativum
         d)   dermis
         e)   stratum granulosum

379

c   7.   Sweat glands are classified as:
         a)   simple tubular glands
         b)   branched tubular glands
         c)   coiled tubular glands
         d)   holocrine glands
         e)   simple alveolar glands

c   8.   The nail matrix is found in which of the following areas?
         a)   eponychium
         b)   nail bed
         c)   nail plate
         d)   cuticle
         e)   lunula

d   9.   The characteristic fingerprint patterns are a result of unique
         development of the:
         a)   dermal reticular layer
         b)   stratum spinous
         c)   stratum germinativum
         d)   dermal papillary layer
         e)   corneum

e   10. Which one of the following is not part of the integumentary system?
         a)   hair
         b)   nails
         c)   sweat glands
         d)   sebaceous glands
         e)   hypodermis

b   11. Which one of the following <u>does</u> <u>not</u> apply to a burn?
         a)   disrupts the body's homeostasis
         b)   is a pathological condition
         c)   can result in large losses of tissue fluid
         d)   can result in losses of electrolytes
         e)   all of the above are true of burns

d   12. The skin aids in regulating calcium and phosphate levels in the body
         by producing vitamin _____.
         a)   A
         b)   B
         c)   C
         d)   D
         e)   none of the above is correct

b   13. A nail consists of:
   a) dense fibrocartilage tissue
   b) flattened keratinized cells
   c) a thin plate of bone
   d) modified tooth enamel
   e) non cellular secretion of fibrocytes

a   14. Which one of the layers of the hair shaft is pierced thirdly by the point of a pin entering the hair at right angles to the long axis of the hair shaft?
   a) medulla
   b) cortex
   c) cuticle
   d) dermal papilla
   e) hair bulb

c   15. The relatively transparent and narrow band of cells in the epidermis is the:
   a) dermis
   b) stratum corneum
   c) stratum lucidum
   d) stratum germinativum
   e) stratum granulosum

c   16. The mitotically active cells that cause hair growth are in the:
   a) medulla
   b) cortex
   c) matrix
   d) cuticle

CHAPTER 6 - SKELETAL SYSTEM

a   1.   Which item is <u>not</u> part of the upper (superior) extremity?
     a)   tarsals
     b)   phalanges
     c)   humerus
     d)   radius
     e)   ulna

e   2.   All thoracic vertebrae are easily identified by the presence of:
     a)   a large body
     b)   a bifurcated spinous process
     c)   mammillary processes
     d)   a horizontal, blunt-tipped spinous process
     e)   rib articular facets

e   3.   Which of the following is NOT a cranial bone?
     a)   ethmoid
     b)   frontal
     c)   sphenoid
     d)   temporal
     e)   mandible

b   4.   The acetabulum is most closely related to the:
     a)   humerus
     b)   frontal
     c)   sphenoid
     d)   temporal
     e)   mandible

b   5.   The greater trochanter is on the:
     a)   os coxa
     b)   mandible
     c)   humerus
     d)   femur
     e)   tibia

a   6.   Which of these bones is not part of the axial skeleton?
     a)   clavicle
     b)   mandible
     c)   rib
     d)   sternum
     e)   two of the above

d   7.   Which of these is not part of the temporal bone?
     a)   external auditory meatus
     b)   styloid process
     c)   mastoid process
     d)   pterygoid process
     e)   zygomatic process

c   8.   An achondroplastic dwarf:
         a)   is a short individual of normal proportions
         b)   is the result of insufficient growth hormone before puberty
         c)   failed to produce a functional cartilage model during his
               embryonic period
         d)   produces excess growth hormone after puberty
         e)   none of the above is correct

b   9.   On which of the following can you find the glenoid fossa?
         a)   temporal bone
         b)   scapula
         c)   femur
         d)   os coxa
         e)   sternum

c   10.   Which of the following occurs thirdly in the formation of a long bone?
         a)   perichondrium invades diaphysis carrying osteoblasts
         b)   growth of the epiphyseal plate
         c)   formation of trabeculae
         d)   erosion of cartilage in the diaphysis
         e)   formation of ossification center in the epiphysis

d   11.   Spongy bone:
         a)   covers the outer surface of most bones directly beneath the
               periosteum
         b)   forms the articulating surfaces of bones
         c)   fills the marrow cavity of a long bone
         d)   can be found in the central area of the ends of long bones
         e)   none of the above

a   12.   Growth in the <u>length</u> of a long bone occurs at which of these
         locations?
         a)   at the end of the diaphysis at the epiphyseal plate
         b)   at the inner end of the epiphysis next to the epiphyseal plate
         c)   at the end of the epiphysis beneath the articular cartilage
         d)   from the center of the diaphysis
         e)   from beneath the periosteum

b   13.   The blood clot which forms in the area of a fracture is replaced by a
         callus of non-bone material which temporarily unites the broken pieces
         of bone. This non-bony material is:
         a)   hyaline cartilage
         b)   fibrocartilage
         c)   dense fibrous connective tissue
         d)   granulation tissue
         e)   reticular fibers

c   14.   In the formation of long bones, secondary ossification centers develop within the:
- a)   marrow cavity of the diaphysis
- b)   compact bone of the diaphysis
- c)   proximal and distal epiphyses of hyaline cartilage
- d)   proximal and distal epiphyses of cancellous bone
- e)   procallus

a   15.   A compound fracture is one in which:
- a)   the bone has pierced the skin and has become exposed to infection
- b)   the bone is broken in several places
- c)   the bone has many fragments with sharp ends
- d)   the region of its fracture requires a splint

a   16.   The nutrient artery supplies blood to the:
- a)   marrow cavity and its contents
- b)   Volkmann canals
- c)   Haversian canals
- d)   canaliculi
- e)   epiphyseal disks

b   17.   The general effect of calcitonin release by the thyroid gland is to:
- a)   increase calcium levels in the blood
- b)   decrease calcium levels in the blood
- c)   activate vitamin D conversion to the active form in the kidneys
- d)   produce no change in calcium levels

b   18.   The canaliculi of bones serve to:
- a)   convey capillaries to bone cells
- b)   provide spaces for thin cytoplasmic extensions of osteocytes to reach through the matrix
- c)   hold artery, nerve, and vein of each Haversian system
- d)   repair bony breaks
- e)   none of the above is correct

d   19.   The reabsorption of previously laid down bone is done by:
- a)   fibrocytes
- b)   osteoblasts
- c)   chondrocytes
- d)   osteoclasts
- e)   osteoprogenitor cells

b   20.   The hormone that increases the reabsorptive activity of specific bone
           cells and increases blood calcium levels is:
           a)   calcitonin
           b)   parathormone
           c)   pituitary hormone
           d)   estrogen
           e)   thyroxin

a   21.   Which one of the following is not correct about the organic framework
           of bone?
           a)   it allows the bone to withstand compressive forces
           b)   it contains collagen fibers
           c)   it provides tensile strength
           d)   all of the above are correct

c   22.   A fracture that splinters the bone into many fragments is a:
           a)   compound fracture
           b)   depressed fracture
           c)   comminuted fracture
           d)   impacted fracture
           e)   none of the above is correct

e   23.   The osteoblastic enzyme thought to play a major role in the release of
           organically bound blood phosphate, allowing highly insoluble
           hydroxyapatite to form in bone, is:
           a)   parathormone
           b)   pituitary hormone
           c)   catalase
           d)   7-dehydrocholesterol
           e)   none of the above is correct

FINAL EXAMINATION

CHAPTER 7 - ARTICULATIONS

a   1.   Which one of the following is not correct about synovial fluid?
        a)   it is a clear viscous fluid secreted by articular capsules
        b)   it serves to lubricate joint surfaces
        c)   it can cause joint swelling when secreted in excess amounts
        d)   it serves to nourish articular cartilages

b   2.   Abduction means to:
        a)   increase the angle between two bones moving in an anterior-posterior (parasagittal) plane
        b)   increase the angle between two bones moving in a frontal plane
        c)   reduce the angle between two bones moving in an anterior--posterior plane
        d)   reduce the angle between two bones moving in a frontal plane
        e)   none of the above

e   3.   The joint between the metacarpals and phalanges can best be classified as a/an:
        a)   pivot joint
        b)   plane joint
        c)   hinge joint
        d)   uniaxial joint
        e)   condyloid joint

c   4.   Which of these is an uniaxial hinge joint?
        a)   metacarpal-phalangeal joint
        b)   radio-carpal joint
        c)   interphalangeal joint
        d)   sacro-iliac joint
        e)   none of the above

b   5.   Which of these is classified as a pivot joint?
        a)   occipital-atlas
        b)   radius-humerus
        c)   tibia-fibular with tarsal bones
        d)   knee joint
        e)   none of the above is a pivot joint

b   6.   What specific joint variety exists between the fifth rib and the sternum?
        a)   fibrous joint
        b)   fibrous synchondrosis
        c)   hyaline synchondrosis
        d)   synostosis
        e)   condyloid diarthrotic joint

d   7.   Which of these joints can be considered a symphysis (fibrous synchondrosis)?
a)   between articulatory processes of vertebrae
b)   between ribs and sternum
c)   between distal end of tibia and fibula
d)   none of these is a symphysis

b   8.   Condyloid and saddle joints are both examples of:
a)   multiaxial joints
b)   biaxial joints
c)   uniaxial joints
d)   condyloid and saddle joints belong to different categories

a   9.   Which of these bones does <u>not</u> articulate with the humerus by means of a synovial joint?
a)   clavicle
b)   radius
c)   ulna
d)   scapula
e)   two of the above

c   10. Sacs of fibrous connective tissue lined with a synovial membrane and commonly located at joints to prevent friction are called:
a)   vesicles
b)   lacunae
c)   bursae
d)   sinuses
e)   cushions

e   11. Extending or protruding a body part (mandible or scapula) is known as:
a)   extension
b)   elevation
c)   circumduction
d)   pronation
e)   protraction

d   12. What is the synarthrotic joint variety in which the fibers are relatively long and are generally referred to as ligaments?
a)   suture
b)   symphysis
c)   synchondrosis
d)   syndesmosis
e)   none of the above is correct

c   13.   Which one of the following is not a circular motion?
         a)   supination
         b)   pronation
         c)   inversion
         d)   circumduction
         e)   all of the above are circular motions

b   14.   The joint between the skull and the atlas is an example of a _____ joint.
         a)   hinge
         b)   condyloid
         c)   gliding
         d)   saddle
         e)   none of the above is correct

e   15.   Many different types of inflammation of joints fall under the general term:
         a)   gout
         b)   degenerative osteoarthropathy
         c)   bursitis
         d)   sprains
         e)   arthritis

e   16.   Which one of the following is not an angular motion?
         a)   flexion
         b)   extension
         c)   abduction
         d)   adduction
         e)   elevation

d   17.   The carpometacarpal joint of the thumb is an example of a _____ joint.
         a)   hinge
         b)   condyloid
         c)   gliding
         d)   saddle
         e)   none of the above is correct

a   18.   Which one of the following is not a fibrous joint?
         a)   symphysis
         b)   suture
         c)   syndesmosis
         d)   synostosis

b   19. The joint between the carpals is known as a _____ joint.
        a)   hinge
        b)   gliding
        c)   condyloid
        d)   multiaxial
        e)   none of the above is correct

d   20. Which of the following is associated with the shoulder joint?
        a)   iliofemoral ligament
        b)   articular capsule
        c)   glenoid labrum
        d)   both b and c

a   21. Which of the following is associated with the knee joint?
        a)   medial and lateral menisci
        b)   acetabular labrum
        c)   ligamentum teres
        d)   glenohumeral ligament

FINAL EXAMINATION

CHAPTER 8 - MUSCLE PHYSIOLOGY

d 1. The substance that directly provides the energy for muscle cell
contraction is:
a) ADP
b) creatine phosphate
c) glucose
d) ATP
e) glycogen

d 2. The specific action of free calcium ion in the cell which allows the
contractile action of the muscle cell to occur is:
a) to participate in cellular depolarization
b) to increase the permeability of the T tubules
c) to attach to the cross-bridges
d) to cause the tropomyosin to sink into the groove between the
actin filaments and thus block its inhibitory function
e) to catalyze the removal of phosphate from ATP

d 3. What is the best definition of the origin of a muscle?
a) an attachment of muscle to bone
b) the proximal attachment of a muscle to a bone
c) the distal attachment of a muscle to a bone
d) the attachment of a muscle to a bone which is relatively fixed in
position
e) the attachment of a muscle to a bone which is relatively movable

a 4. A muscle which opposes the action of the muscle performing the
specific action is called the:
a) antagonist
b) synergist
c) fixator
d) prime mover
e) agonist

a 5. Which of these is a bipennate muscle?
a) rectus femoris
b) semitendinosus
c) soleus
d) temporalis
e) sternocleidomastoid

c 6. Which of these descriptions best accounts for the H zone?
a) thin filaments only (no thick filaments)
b) thick and thin filaments
c) thick filaments only (no thin filaments)
d) intermolecular bonds between adjacent filaments
e) no filaments present at all

c   7.   During muscle contraction which of these does not become shorter?
        a)   H zone
        b)   I band
        c)   A band
        d)   sarcomere
        e)   all of the above <u>do</u> become shorter

a   8.   The A band of a skeletal muscle cell is composed of:
        a)   overlapping thin and thick filaments
        b)   thin filaments only
        c)   thick filaments only
        d)   intercalated disks
        e)   motor end plates

c   9.   Thin filaments are composed of all but which one of the following?
        a)   F-actin
        b)   troponin
        c)   myosin
        d)   tropomyosin
        e)   G-actin

c   10. Which muscle fiber arrangement lends itself to a great range of
        motion, but rather limited power?
        a)   radiate
        b)   bipennate
        c)   longitudinal
        d)   multipennate

d   11. Most of the high energy phosphates available to ADP for resynthesis
        of ATP during the first ten seconds of vigorous muscular exercise are
        provided by:
        a)   aerobic metabolic pathways
        b)   anaerobic metabolic pathways
        c)   stored ATP
        d)   the breakdown of creatine phosphate

b   12. The attachment of ATP to which of the following frees the cross-
        bridge after contraction, breaking the "rigor complex".
        a)   transverse tubule
        b)   cross-bridge
        c)   tropomyosin
        d)   troponin
        e)   G-actin

e   13.   A cross-section of a myofibril shows that each thick filament is surrounded by _____ thin filaments.
- a)   2
- b)   3
- c)   4
- d)   5
- e)   6

a   14.   The inclusion of more muscle fibers into the contractile state in order to lift a heavy weight is termed:
- a)   synchronous motor unit summation
- b)   wave summation
- c)   tetanus
- d)   recruitment
- e)   isometric contraction

b   15.   The variety of muscle contraction seen in lifting a carton of milk from the table by a normal individual is:
- a)   hypotonic
- b)   isotonic
- c)   hypertonic
- d)   hypometric
- e)   isometric

d   16.   Which one of the following listed is the third largest in size (starting with the largest)?
- a)   filamentous actin
- b)   muscle fiber
- c)   sarcomere
- d)   myofibril
- e)   muscle

a   17.   Which one of the following is not characteristic of smooth muscle?
- a)   multinucleate
- b)   poorly developed sarcoplasmic reticulum
- c)   may exhibit inherent rhythmicity
- d)   indistinct cell membranes

b   18.   The specific function of the T tubules is to:
- a)   secrete acetylcholine
- b)   transmit action potentials
- c)   store calcium ion
- d)   transmit ATP to the myofilaments

e   19.   Which one of the following is not characteristic of the thick filament?
           a)   it is composed of myosin
           b)   it contains a bare zone
           c)   it contains cross-bridges
           d)   its molecules face in opposite directions from the center of the
                 filament
           e)   all of the above are correct

b   20.   Which one of the following is not a metabolic by-product of muscles
           during strenuous exercise?
           a)   ADP
           b)   creatine phosphate
           c)   lactic acid
           d)   heat
           e)   all of the above are by-product

d   21.   A sustained muscle contraction in response to repeated maximal stimuli
           is a:
           a)   twitch
           b)   multiple motor unit summation
           c)   treppe
           d)   complete tetanus

e   22.   During a muscle contraction, calcium ions bind specifically to:
           a)   tropomyosin
           b)   myosin
           c)   actin
           d)   lateral sacs
           e)   none of the above is correct

a   23.   The formation of the <u>active-complex</u> (actomyosin-ATP) occurs during
           the:
           a)   latent period
           b)   period of contraction
           c)   period of relaxation
           d)   all of these periods

c   24.   If a series of stimuli is strong enough to bring about excitation of the
           motor units of a skeletal muscle and each stimulus is delivered before
           the muscle has a chance to completely relax, and each successive
           contraction is stronger than the preceding one, the series of
           increasing responses is termed:
           a)   multiple motor unit summation
           b)   wave summation
           c)   asynchronous motor unit summation
           d)   isometric contraction
           e)   none of the above

d   25. The effect of acetylcholine on a muscle cell is to:
a)   restore calcium ions to their storage areas
b)   cause cross-bridges to move in a ratchet-like fashion
c)   bind cross-bridges to their active sites
d)   cause a change in the permeability of the cell membrane at the neuromuscular junction
e)   split ATP to ADP thus freeing energy for muscle action

d   26. Which one of the following is not a characteristic result of high intensity, short duration exercise?
a)   hypertrophy of low-oxidative fibers
b)   increase in muscle mass
c)   increase in muscle strength
d)   transformation of low-oxidative fibers into high-oxidative fibers

d   27. Which one of the following is not true of single unit smooth muscle?
a)   cells in contact with each other via gap junctions
b)   rhythmic, spontaneous contractions
c)   located in intestine, uterus, and small blood vessels
d)   cells require individual stimulation
e)   all of the above are correct

CHAPTER 9 - MUSCLE ANATOMY

a   1.   The insertion of the biceps femoris is on the:
          a)   head of fibula
          b)   tibial tuberosity
          c)   ischial tuberosity and linea aspera
          d)   linea aspera only
          e)   supra-acetabular ridge

a   2.   The common action of the component muscles of the quadriceps groups
          is to:
          a)   extend the leg
          b)   rotate the thigh
          c)   flex the thigh
          d)   plantar flex the foot
          e)   flex the leg

b   3.   The muscle primarily responsible for straightening at the hip is the:
          a)   sartorius
          b)   gluteus maximus
          c)   vastus medialis
          d)   rectus femoris
          e)   gastrocnemius

d   4.   The insertion of the gastrocnemius is:
          a)   pubic symphysis
          b)   medial and lateral epicondyles of femur
          c)   posterior surface of tibia and fibula
          d)   on the calcaneus of the foot
          e)   none of the above

c   5.   The brachioradialis has its origin:
          a)   on the inferior 2/3 of humerus, anteriorly
          b)   on the head of radius
          c)   supracondylar ridge of humerus
          d)   olecranon process of ulna
          e)   coronoid process of ulna

d   6.   Which of these is not an attachment of the pectoralis major muscle?
          a)   sternum
          b)   medial 2/3 of clavicle
          c)   lateral lip of intertubercular (bicipital) groove of humerus
          d)   coracoid process of scapula
          e)   all of the above are attachments of the pectoralis

a   7.   The origin of the deltoid is the:
        a)   lateral third of clavicle, acromion process and spine of scapula
        b)   deltoid tuberosity of humerus
        c)   greater tubercle of humerus
        d)   coracoid process, acromion process and subscapular fossa of scapula
        e)   outer surface of ribs 3-5

d   8.   Which one of these is not an attachment of the triceps brachii?
        a)   proximal end of posterior surface of humerus
        b)   axillary border of scapula (infraglenoid tuberosity)
        c)   olecranon process
        d)   anterior portion of humerus, distally
        e)   all of the above correctly state the attachments of the triceps brachii

d   9.   Which one of these is not attached to the scapula?
        a)   deltoid
        b)   triceps brachii
        c)   biceps brachii
        d)   pectoralis major
        e)   trapezius

c   10.   The muscle which is used in turning the head from side to side is the:
        a)   trapezius
        b)   deltoid
        c)   sternocleidomastoideus
        d)   temporalis
        e)   masseter

d   11.   Name the muscle responsible for puckering the lips.
        a)   frontalis
        b)   sternocleidomastoid
        c)   levator palpebra superioris
        d)   orbicularis oris
        e)   none of the above

d   12.   Which of these muscles is antagonistic to the temporalis?
        a)   medial pterygoid
        b)   lateral pterygoid
        c)   masseter
        d)   digastric
        e)   none of the above

b   13. If a pin were passed through the abdominal wall which of the layers
          listed would be passed through third?
          a)   skin and fascia
          b)   external oblique
          c)   transverse abdominis
          d)   internal oblique
          e)   parietal peritoneum

b   14. The fascia at the lower margin of the external abdominal oblique
          muscle forms a band called the:
          a)   pelvic diaphragm
          b)   inguinal ligament
          c)   urogenital diaphragm
          d)   perineum

CHAPTER 10 - THE NERVOUS SYSTEM

b   1.   A bundle of nerve fibers in the peripheral nervous system is called:
    a)   a neuron
    b)   a nerve
    c)   a ganglion
    d)   a tract
    e)   an axolemma

d   2.   The "neuroglia" refers to:
    a)   afferent neurons
    b)   efferent neurons
    c)   internuncial neurons
    d)   supportive, connective tissue-like cells of the nervous system
    e)   none of the above

c   3.   The "node of Ranvier" is an:
    a)   indentation of the nerve fiber
    b)   area of an axon lacking satellite cells
    c)   interruption of the myelin sheath
    d)   area making synaptic contact
    e)   none of the above

a   4.   The neuron process termed the axon normally:
    a)   conducts impulses away from the cell body only
    b)   conducts impulses toward the cell body only
    c)   conducts impulses toward the central nervous system (CNS) only
    d)   conducts impulses away from the CNS only
    e)   does not conduct impulses at all

c   5.   Efferent nerves:
    a)   are sensory nerves
    b)   carry impulses toward the central nervous system
    c)   carry impulses away from the central nervous system
    d)   a and b are correct
    e)   a and c are correct

a,e   6.   Receptors that are sensitive to pressure or stretch can be classified as: (two correct answers)
    a)   mechanoreceptors
    b)   thermoreceptors
    c)   chemoreceptors
    d)   photoreceptors
    e)   proprioceptors

a   7.   When impulses from receptors for deep somatic pain and visceral pain are incorrectly interpreted by the brain as having come from the surface of the body, we call that pain ____.
         a)   referred
         b)   phantom
         c)   psychosomatic
         d)   visceral
         e)   no correct answer

b,d 8.   Which two of the following are not true of bipolar neurons?
         a)   sensory
         b)   associated with receptors for heat or cold
         c)   found in organs of special sense
         d)   their axons run in rami of spinal nerves
         e)   have two processes

e   9.   The structure which carries out the response is called a/an:
         a)   motor
         b)   efferent
         c)   synapse
         d)   receptor
         e)   effector

c   10. A mixed nerve is:
         a)   one that goes to both the skin surface and the viscera
         b)   one that has its pathway mixed with other nerves
         c)   one that carries both sensory and motor fibers
         d)   a nerve that carries large and small motor fibers

d   11. Individual nerve fibers are surrounded by which one of these while they are part of a nerve?
         a)   endomysium
         b)   perimysium
         c)   epineurium
         d)   endoneurium
         e)   perineurium

e   12. In the PNS the myelin sheath is created by:
         a)   oligodendrocytes
         b)   secretions of fatty substances by the axon
         c)   neurofibrils
         d)   secretion of phospholipids by the Schwann cell
         e)   a wrapping of the Schwann cell around the nerve fiber

c   13.   What is the relationship of the neurilemma to the Schwann cell?
   a)   the neurilemma lies just outside the Schwann cell
   b)   the neurilemma is the same thing as the endoneurium and surrounds the Schwann cell
   c)   the neurilemma is the outermost cell membrane of the Schwann cell
   d)   the Schwann cell lies just outside the neurilemma. The neurilemma is the cell membrane of the nerve fiber

b   14.   The role of oligodendrocytes is to:
   a)   secrete acetylcholine
   b)   form the myelin sheath of axons in the CNS
   c)   form part of the blood brain barrier
   d)   line the cavities of the CNS
   e)   none of the above is correct

d   15.   Which one of these is not true of dendrites?
   a)   contain Nissel granules
   b)   conduct impulses toward the nerve cell body
   c)   are highly branched
   d)   secrete acetylcholine
   e)   all of the above are correct

c   16.   The portion of the neuron that is the "receptive" part is the:
   a)   axon
   b)   nerve cell body
   c)   dendrite
   d)   collateral branch

c   17.   The part of the body that receives sensory input and formulates responses to this input is the:
   a)   receptor
   b)   effector
   c)   central nervous system
   d)   peripheral nervous system
   e)   autonomic nervous system

d   18.   Which one of the following is not a type of <u>cell</u> found in the nervous system?
   a)   neuron
   b)   Schwann cell
   c)   glial cell
   d)   nerve
   e)   all of the above are types of cells found in the nervous system

a   19. Which one of the following is not found in a cell body?
         a)   axoplasm
         b)   nucleus
         c)   neurofibrils
         d)   ribosomes
         e)   Nissel bodies

e   20. Which one of the following is pierced thirdly by a pin going through
         an internodal area?
         a)   axolemma
         b)   endoneurium
         c)   neurilemma (Schwann cell)
         d)   axoplasm
         e)   myelin

e   21. Which one of the following is not a characteristic of the neuron that
         has a part of itself modified as a pressure receptor?
         a)   sensory
         b)   psuedounipolar
         c)   afferent
         d)   has its axon passing through the dorsal root
         e)   all of the above are correct

a   22. Which one of the following is not true of receptors?
         a)   they may be modified axons
         b)   they may be individual cells
         c)   they are usually most sensitive to one kind of stimulus
         d)   they act as transducers

d   23. Which one of the following is not a cutaneous sense?
         a)   heat
         b)   cold
         c)   touch
         d)   taste
         e)   pressure

c   24. Which one of the following is not true of axons that are myelinated in
         the CNS?
         a)   appear white in color
         b)   have glial cells around them
         c)   have a neurilemma
         d)   have an endoneurium

e   25. Which one of the following organ systems serves primarily as a means
         of internal communication in the body?
         a)   digestive
         b)   urinary
         c)   respiratory
         d)   circulatory
         e)   none of the above is correct

e   26. Which one of the following is not true of the ANS?
         a)   part of the peripheral nervous system
         b)   transmits impulses to glands
         c)   is motor only
         d)   generally cannot be consciously controlled
         e)   transmits impulses to skeletal muscle

e   27. The specialized portion of the nerve cell at the neuromuscular junction
         is called the:
         a)   muscle spindle
         b)   node of Ranvier
         c)   synaptic cleft
         d)   tubule
         e)   none of the above is correct

d   28. Which one of the following is not true of the neurons whose axon
         ends at the myoneural junction?
         a)   multipolar
         b)   motor
         c)   efferent
         d)   all of the above are correct

e   29. The receptors for light touch sensation are the:
         a)   Pacinian corpuscles
         b)   brushes of Ruffini
         c)   end bulbs of Krause
         d)   free nerve endings
         e)   none of the above is correct

d   30. Fibers are to nerves as cell bodies are to:
         a)   the CNS
         b)   axons
         c)   dendrites
         d)   neurofibrils

c   31. A nerve is:
         a)   a neuron
         b)   composed of sensory axons and motor dendrites
         c)   composed of the long fibers of a number of neurons
         d)   a part of the central nervous system
         e)   none of the above is correct

CHAPTER 11 - NEURONS, SYNAPSES, AND RECEPTORS

e   1.   Upon strong stimulation of a nerve fiber:
         a)   Na$^+$ ions move in
         b)   the nerve fiber becomes depolarized
         c)   an impulse is propagated
         d)   the nerve cell membrane becomes permeable to sodium
         e)   all of the above

c   2.   A synapse is of <u>importance</u> because:
         a)   it is the gap between two nerve cells
         b)   impulses can pass in both directions across a synapse
         c)   it acts as a one-way valve for nerve impulses
         d)   none of the above correctly states the real importance of the
              synapse

a   3.   The positive charge on the surface of living cells is due to the
         accumulation of which one of the following cations?
         a)   sodium
         b)   calcium
         c)   potassium
         d)   magnesium

e   4.   Even though a neuron is stimulated at a threshold level or above, it
         does not always react or fire. What mechanism would most likely
         suppress that excitation? There may be:
         a)   resistance at the myoneural junction
         b)   the cell membrane may exhibit hyperexcitability
         c)   not enough transmitter substance available
         d)   fatigue
         e)   inhibitory fibers produced an inhibitory postsynaptic potential

c   5.   The depolarization of a myelinated neuron occurs:
         a)   along the entire length of the axolemma
         b)   in the myelinated regions only
         c)   in the nodes of Ranvier only
         d)   occurs in dendrites only
         e)   occurs in axons only

a   6.   Which one of these is most related to the concept of "local
         excitatory state"?
         a)   partial depolarization
         b)   refractory period
         c)   hyperpolarization
         d)   all-or-nothing law
         e)   supramaximal stimulus

a   7.   Which one of the following is true of adaptation?
- a) the responsiveness of the receptor decreases with time
- b) the same number of action potentials are being formed in the sensory neuron
- c) it causes sensations to diminish in intensity when the stimulus is removed
- d) pain is an example of a sensation that shows adaptation

c   8.   Which one of these statements about nerve fibers is not correct?
- a) nerve impulses can be conducted in either direction along a nerve fiber
- b) a nerve impulse is as strong at one end of a fiber as at the other
- c) nerve fibers can fatigue readily if given a rapid series of stimuli
- d) impulse frequency is limited mostly by the duration of the refractory period

b   9.   Which one of the following is true of the absolute refractory period?
- a) another action potential can be evoked if a stronger stimulus is used
- b) the duration of this period corresponds to the time of sodium permeability changes
- c) refractory periods have no effect on the number of action potentials that can be generated in a given time period
- d) the duration of this period is quite long

a 10.   Which one of the following is not a way of discriminating differing stimulus intensities?
- a) varying the size of the action potential in a sensory neuron
- b) varying the number of sensory neurons stimulated
- c) varying the number of impulses generated by a stimulus depending on its strength
- d) varying the speed of the action potential depending on the strength of the stimulus

d 11.   Which one of the following is not true of generator potentials?
- a) a large enough generator potential is able to trigger action potentials in a sensory neuron
- b) generator potentials do not increase in magnitude with increasing rates of stimuli
- c) generator potentials may increase in magnitude with the removal of a stimulus
- d) a strong stimulus will cause the same sized generator potential as a weak stimulus

b   12. Which of the following is not true of neurons and action potentials?
    a) action potentials are an all-or-nothing phenomenon
    b) large diameter axons conduct action potentials at a slower rate than smaller diameter axons
    c) the speed of conduction is faster in myelinated axons
    d) action potentials in unmyelinated axons do not show saltatory conduction

e   13. If a stimulus of sufficient intensity is applied to a neuron, the local depolarization of the cell may reach a critical level or threshold and trigger a _____ potential.
    a) IPSP
    b) EPSP
    c) generator
    d) local
    e) action

e   14. When many nerve impulses arrive in a short time period at a synapse between a single stimulatory presynaptic cell and a postsynaptic cell, this is known as:
    a) convergence
    b) divergence
    c) spatial summation
    d) facilitation
    e) none of the above are correct

e   15. To cross a synaptic cleft, which one of the following listed is required?
    a) generator potential
    b) saltatory conduction
    c) an IPSP
    d) an EPSP
    e) neurohumor molecules

d   16. Nerve impulses in many different stimulatory presynaptic cells traveling to one postsynaptic cell may arrive there at about the same time. Which one of the following is not related to that phenomenon?
    a) convergence
    b) spatial summation
    c) facilitation
    d) all of the above are related

c   17. The release of acetylcholine at the neuromuscular junction causes a local hypopolarization on the muscle cell called a:
    a) generator potential
    b) local potential
    c) end plate potential
    d) IPSP
    e) EPSP

a 18. Generator potentials:
    a)   can summate
    b)   have long absolute refractory periods
    c)   have short absolute refractory periods
    d)   are of shorter duration than an action potential
    e)   are an all-or-nothing phenomenon

b 19. The "local potential" of a membrane:
    a)   can be as great as the action potential
    b)   is proportional to the magnitude of the stimulus
    c)   is an all-or-nothing phenomenon
    d)   is a propagated event, i.e., it travels along the length of the cell from the site of stimulation
    e)   is an example of a hyperpolarization of the cell membrane

b 20. When a neuron reaches threshold and an action potential is generated:
    a)   potassium ions rapidly enter the cell
    b)   sodium ions rapidly enter the cell
    c)   the permeability to both potassium and sodium decreases dramatically
    d)   sodium ions rapidly leave the cell
    e)   the cell hyperpolarizes

CHAPTER 12 - THE CENTRAL NERVOUS SYSTEM

c   1.   Destruction of the right precentral gyrus of the cerebrum results in loss of:
- a)   sensations from the left side of the body
- b)   sensations from the right side of the body
- c)   voluntary movements of left side of the body
- d)   voluntary movements of the right side of the body
- e)   loss of reasoning and logical behavior

b   2.   Gray matter of the CNS consists of:
- a)   myelinated neurons only
- b)   cell bodies and unmyelinated neurons
- c)   pia matter and arachnoid
- d)   post-ganglionic parasympathetic fibers
- e)   nerve tracts

d   3.   A raised area of the cerebrum could be termed a:
- a)   genu
- b)   folium
- c)   sulcus
- d)   gyrus
- e)   gymnast

a   4.   The connections between gyri of a single hemisphere are called:
- a)   association tracts
- b)   commissural tracts
- c)   internuncial tracts
- d)   projection tracts
- e)   none of the above

c   5.   Which of these is not part of the mesencephalon?
- a)   cerebral aqueduct
- b)   superior colliculi
- c)   thalamus
- d)   cerebral peduncle
- e)   all of the above are associated with the mesencephalon

e   6.   The vital centers for control of heart, respiration and blood pressure are located in the:
- a)   pons
- b)   cerebrum
- c)   cerebellum
- d)   midbrain
- e)   medulla

a   7.   Which of the following word sequences is <u>correct,</u> as related to the circulation of cerebral spinal fluid?
- a)   lateral ventricle, interventricular foramen (of Monro), third ventricle, cerebral aqueduct, fourth ventricle
- b)   lateral ventricle, cerebral aqueduct, third ventricle, interventricular foramen (of Monro), fourth ventricle
- c)   lateral ventricle, foramen of Luschka, third ventricle, fourth ventricle, aqueduct of Sylvius
- d)   lateral ventricle, foramen of Luschka, cerebral aqueduct, fourth ventricle, third ventricle

a   8.   The pia matter is the:
- a)   innermost meninx of the brain
- b)   middle meninx of the brain
- c)   outermost meninx of the brain
- d)   double meninx of the brain
- e)   none of the above

d   9.   Conscious awareness of heat, cold, and pressure are localized in the ____ of the cerebrum.
- a)   frontal lobe
- b)   occipital lobe
- c)   temporal lobe
- d)   parietal lobe
- e)   island of Reil

c   10.   The part of the body that receives sensory input and formulates responses to this input is the:
- a)   receptor
- b)   effector
- c)   central nervous system
- d)   peripheral nervous system
- e)   autonomic nervous system

d   11.   A group of cell bodies within the CNS with a specific function is called a:
- a)   ganglion
- b)   nerve
- c)   center
- d)   nucleus
- e)   none of the above is correct

b   12.   Voluntary motor impulses reach skeletal muscle from the cerebral cortex by way of _____ neuron path.
- a)   one
- b)   two
- c)   three
- d)   four
- e)   ten or more

c   13. The nerve cell bodies of efferent nerve fibers of spinal nerves are
        located in:
        a)   the dorsal gray column of the cord
        b)   the lateral gray column of the cord
        c)   the ventral gray column of the cord
        d)   the ventral root
        e)   the dorsal root ganglion

d   14. Which of these is most superior (with respect to vertical axis of the
        body)?
        a)   cauda equina
        b)   lumbar enlargement
        c)   conus medullaris
        d)   cervical enlargement

a   15. Which of these statements is not true about second order neurons of
        most somesthetic pathways?
        a)   their cell bodies lie in the dorsal root ganglion
        b)   the fiber decussates
        c)   the fibers end in the thalamus
        d)   the ascending fibers lie in the white matter

d   16. The proper term for neurons joining the sensory and motor neuron in
        the CNS is the:
        a)   afferent neuron
        b)   efferent neuron
        c)   commissural neuron
        d)   internuncial neuron
        e)   none of the above

b   17. The fasciculus cuneatus lies in the:
        a)   gray matter of the cord
        b)   dorsal white column
        c)   ventral white column
        d)   lateral white column

a   18. The part of the brain concerned with water balance, appetite, and
        regulating body temperature is the:
        a)   hypothalamus
        b)   cerebral cortex
        c)   thalamus
        d)   basal ganglia
        e)   cerebellum

c 19. The limbic system of the brain is a collection of structures that are particularly important in:
   a) controlling breathing
   b) coordination of muscular activity
   c) emotional behavior
   d) maintaining wakefulness
   e) relaying sensory impulses to the cerebral cortex

c 20. Within the spinal cord, descending tracts:
   a) connect sensory fibers to motor fibers
   b) relay sensory projection fibers to the receptor
   c) are composed of motor fibers arising in the brain or higher cord levels
   d) transmit impulses to the brain
   e) carry impulses from muscle spindles

a 21. When the following five features are arranged in sequential order, the third item in the passage along a monosynaptic reflex pathway is a(n):
   a) synapse
   b) efferent neuron
   c) afferent neuron
   d) receptor
   e) effector

d 22. The reticular formation of the brain is essential in:
   a) controlling breathing
   b) coordination of muscular activity
   c) emotional behavior
   d) maintaining wakefulness
   e) relaying sensory impulses to the cerebral cortex

b 23. In the brain the sensation of sound is perceived in the:
   a) frontal lobe
   b) temporal lobe
   c) parietal lobe
   d) occipital lobe
   e) insula

c 24. Tracing the pathway of voluntary motor impulses from the cerebral cortex, which of the named areas listed below occurs third?
   a) internal capsule
   b) precentral gyrus
   c) cerebral peduncle
   d) spinal cord
   e) decussation in the medulla

a 25. Lower motor neuron damage normally produces:
       a) flaccid paralysis
       b) spastic paralysis
       c) loss of coordination
       d) inadequate sensory interpretation
       e) ataxia

a 26. A patient has suffered a cerebral hemorrhage that has caused injury
       and non-functioning of the primary motor area of his right cerebral
       cortex. As a result:
       a) he cannot voluntarily move his left arm or hand or his left leg
           or foot
       b) he feels no sensations on the left side of his body
       c) reflexes cannot be elicited on the left side of his body
       d) all of the above are correct

c 27. The major control systems for balance and posture in the body as well
       as muscular coordination are located in the:
       a) pons
       b) medulla
       c) cerebellum
       d) hypothalamus
       e) cerebrum

e 28. The two cerebral hemispheres are divided by the:
       a) internal capsule
       b) central sulcus
       c) island of Reil
       d) tentorium cerebelli
       e) none of the above are correct

c 29. Which of the following is a true statement concerning long-term
       memory?
       a) it is due to the same type of neural activity that is responsible
           for short-term memory
       b) it depends on the activation of reverberating circuits of neurons
       c) it is thought to be the result of changes in the structure or
           biochemistry of neurons or synapses
       d) all of the above are true

d 30. Which of the following is a true statement concerning the cerebellum?
       a) it appears to play an important role in providing the CNS with
           the ability to predict the future position of a body part during a
           movement
       b) it is important in maintaining body equilibrium
       c) it contains the areas concerned with language
       d) both a and b are true

CHAPTER 13 - THE PERIPHERAL NERVOUS SYSTEM

e 1. Which cranial nerve would tell you your coffee is "hot"?
    a) optic
    b) abducens
    c) olfactory
    d) facial
    e) trigeminal

d 2. The IV cranial nerve:
    a) controls the lateral rectus muscle of the eye
    b) is the sensory nerve from the face and controls muscles of mastication
    c) causes the eye to look medially when activated
    d) controls the superior oblique eye muscle
    e) carries sensation from the nasal epithelium

d 3. The trigeminal nerve:
    a) is sensory from the face
    b) is motor to the facial muscles and carries taste sensations from the anterior 2/3 of the tongue
    c) is motor to the muscles of mastication
    d) is two of the above
    e) is none of the above

c 4. Taste buds on the anterior 2/3 of the tongue are innervated by the _ ____cranial nerve.
    a) III
    b) V
    c) VII
    d) IX
    e) XII

b 5. The cranial nerve that controls most of the viscera of the body is the:
    a) glossopharyngeal
    b) vagus
    c) spinal accessory
    d) hypoglossal
    e) trigeminal

e 6. The nerve controlling the trapezius and sternocleidomastoid muscles is the:
    a) hypoglossal
    b) trigeminal
    c) trochlear
    d) vagus
    e) spinal accessory

c   7. An anastomosing network of nerves in the peripheral nervous system is
        known as a:
        a)   ganglion
        b)   nucleus
        c)   plexus
        d)   segment
        e)   cauda equina

c   8. Which of the following branches of spinal nerves passes posteriorly
        from the intervertebral foramina to supply the skin and muscles of the
        back?
        a)   dorsal roots
        b)   ventral roots
        c)   dorsal rami
        d)   splanchnic nerves
        e)   pelvic nerves

b   9. Cervical spinal nerve 4 emerges between cervical vertebrae:
        a)   2 and 3
        b)   3 and 4
        c)   4 and 5
        d)   5 and 6
        e)   there is no cervical nerve 4

b   10. If the ventral root of a spinal nerve was cut, what would be the
        result in the region supplied by that nerve?   There would be
        complete loss of:
        a)   sensation
        b)   movement
        c)   sensation and movement
        d)   sensation, movement, and autonomic activity

a   11. Which one of the following listed is the function of the trochlear
        nerve?
        a)   motor to superior oblique muscle
        b)   motor to inferior oblique muscle
        c)   motor to lateral rectus muscle
        d)   motor to medial rectus muscle
        e)   none of the above is correct

c   12. The bulge of the spinal cord called the cervical (brachial) enlargement
        is due to the presence of:
        a)   a ventricle in the central canal
        b)   a structure responsible for the production of cerebrospinal fluid
        c)   greater amounts of gray matter
        d)   greater amounts of white matter
        e)   accumulations of glial cells

c   13. Which cranial nerve supplies sensory fibers to the taste buds of the
        posterior portion of the tongue?
        a) trigeminal
        b) facial
        c) glossopharyngeal
        d) vagus
        e) hypoglossal

d   14. Loss of sense of balance might be due to injury of which cranial
        nerve?
        a) II
        b) IV
        c) VI
        d) VIII
        e) X

b   15. The dorsal ramus of a spinal nerve:
        a) contains sensory neurons only
        b) supplies the skin and muscle to either side of the vertebral
            spinous processes
        c) connects the spinal nerve to the paravertebral ganglion
        d) conveys motor impulses only
        e) possesses a ganglion

b   16. Which one of the following statements, concerning peripheral nerve
        regeneration, is false?
        a) when a peripheral nerve is severed, the portion of the nerve
            beyond the injury undergoes degenerative changes
        b) the neuronal process degenerates, leaving a hollow cylinder of
            myelin and Schwann cells
        c) sprouts form on the stump of the damaged neurons, some of
            which find their way into the connective tissue tubes
        d) neurons regenerate at the rate of 1-4 mm a day
        e) all of the above statements are correct

d   17. There are _____ pairs of thoracic spinal nerves.
        a) 5
        b) 7
        c) 8
        d) 12
        e) none of the above is correct

c   18. A horizontal band of skin supplied by a ramus of a spinal nerve is
        known as a:
        a) myotome
        b) scleratome
        c) dermatome
        d) motor unit
        e) none of the above is correct

e   19.   Which one of the following is a branch of the lateral cord of the brachial plexus?
     a)   axillary nerve
     b)   radial nerve
     c)   ulnar nerve
     d)   median nerve
     e)   none of the above are correct

a   20.   Which one of the following is not part of the lumbar plexus?
     a)   sciatic nerve
     b)   femoral nerve
     c)   saphenous nerve
     d)   anterior femoral cutaneous nerve
     e)   all of the above are part of the lumbar plexus

d   21.   Which of the following is a branch of the sacral plexus?
     a)   sciatic nerve
     b)   common peroneal nerve
     c)   tibial nerve
     d)   all of the above

d   22.   Which of the following is a branch of the cervical plexus?
     a)   phrenic nerve
     b)   lesser occipital nerve
     c)   supraclavicular nerve
     d)   all of the above

FINAL EXAMINATION

CHAPTER 14 - THE AUTONOMIC NERVOUS SYSTEM

e   1. Which of the following characteristics of the autonomic nervous system is also characteristic of the somatic nervous system?
   a) there are two peripheral motor nerve cells between the central nervous system and organ innervated
   b) the autonomic nervous system acts upon muscles of the viscera and arteries as well as on glands
   c) it is concerned with non-willful actions only
   d) it is entirely a motor system
   e) none of the above is characteristic of both systems

d   2. Which of the following is not under control of the autonomic nervous system?
   a) diameter of blood vessels
   b) movements of the intestine
   c) secretion of exocrine glands
   d) reflex control of skeletal muscle
   e) all of the above are controlled by the autonomic nervous system

c   3. The thoraco-lumbar outflow of the autonomic nervous system is called the:
   a) craniosacral division
   b) somatic division
   c) sympathetic division
   d) parasympathetic division
   e) none of the above

b   4. What specific neurohumor is released by post-ganglionic parasympathetic neurons?
   a) norepinephrine
   b) acetylcholine
   c) atropine
   d) GABA
   e) nicotine

a   5. Terminal ganglia are:
   a) found in parasympathetic division
   b) found in the sympathetic division
   c) found in the dorsal root
   d) found alongside the vertebral bodies
   e) just anterior to the aorta

d   6. The gray communicating ramus of a spinal nerve conveys:
   a) somatic afferents
   b) preganglionic sympathetic efferents
   c) preganglionic parasympathetic fibers
   d) post-ganglionic sympathetic efferents
   e) none of the above

e   7. Stimulation of the sympathetic system results in:
        a) contraction of the pupil
        b) slowing of heart rate
        c) constriction of bronchial tubes
        d) increased peristalsis in the digestive tract
        e) contraction of arrector pili muscles

c   8. What forms the lump or swelling (along the sympathetic trunk) called
        the paravertebral ganglion?
        a) receptor cells
        b) nerve cell bodies or preganglionic sympathetic neurons
        c) nerve cell bodies of postganglionic sympathetic neurons
        d) nerve cell bodies of sensory neurons
        e) none of the above

b   9. Preganglionic sympathetic axons, located below the diaphragm in the
        abdominopelvic cavity, form pathways called:
        a) collateral ganglia
        b) splanchnic nerves
        c) pelvic nerves
        d) ventral rami
        e) none of the above is correct

b 10. Autonomic fibers travelling with the somatic motor and afferent fibers
        of the vagus nerve are:
        a) sympathetic
        b) parasympathetic
        c) absent
        d) none of the above is correct

a 11. Most preganglionic parasympathetic fibers release:
        a) acetylcholine
        b) cholinesterase
        c) norepinephrine
        d) parasympathin
        e) none of the above is correct

e 12. The cell bodies of preganglionic sympathetic neurons are located in
        the:
        a) paravertebral ganglia
        b) collateral ganglia
        c) in the brain or sacral cord
        d) in the terminal ganglia
        e) none of the above are correct

a   13. What sort of fiber forms the white ramus communicans?
          a)   sympathetic preganglionic fibers
          b)   sympathetic postganglionic fibers
          c)   parasympathetic preganglionic fibers
          d)   parasympathetic postganglionic fibers
          e)   fibers of the CNS

a   14. Select the neuron that may be adrenergic.
          a)   sympathetic postganglionic neurons
          b)   preganglionic parasympathetic neurons
          c)   parasympathetic postganglionic neurons
          d)   motor neurons of the CNS
          e)   post ganglionic sympathetic fibers to sweat glands

c   15. What is meant by "double innervation" with respect to the ANS?
          a)   there are two neurons between the CNS and the effector
          b)   there are two different kinds of neurohumors utilized
          c)   effectors are generally controlled by both divisions of the ANS
          d)   the term has no relevance to the ANS

c   16. The preganglionic fibers of both the sympathetic and sacral
          parasympathetic divisions of the ANS exit the spinal cord via the:
          a)   dorsal root
          b)   ventral gray column
          c)   lateral gray column
          d)   dorsal gray column
          e)   ventral root

d   17. Postganglionic sympathetic neurons generally stimulate which of the
          following receptors?
          a)   nicotinic
          b)   alpha$_1$
          c)   beta$_2$
          d)   both b and c

c   18. Postganglionic parasympathetic neurons generally stimulate which of
          the following receptors?
          a)   alpha$_1$
          b)   muscarinic
          c)   beta$_1$
          d)   all of the above

418

FINAL EXAMINATION

CHAPTER 15 - THE SPECIAL SENSES

a   1.   Focusing of light rays involves which of the following?
          a)   change in shape of lens
          b)   change in relative position of the lens
          c)   change in shape of the cornea
          d)   change in the density of the aqueous humor
          e)   none of the above

d   2.   When the curvature of the cornea is not equal in all meridians, the defect is termed:
          a)   emmetropia
          b)   myopia
          c)   nystagmus
          d)   astigmatism
          e)   glaucoma

d   3.   The fibrous tunic consists of two parts, the:
          a)   sclera and choroid
          b)   cornea and conjunctiva
          c)   choroid and iris
          d)   sclera and cornea
          e)   orbit and sclera

a   4.   At which point in the visual system does the greatest bending of the light rays occur? As light:
          a)   enters the cornea
          b)   leaves the cornea to enter the aqueous humor
          c)   enters the lens
          d)   leaves the lens
          e)   the amount of bending is identical in each of the above

c   5.   The aqueous humor is produced by the:
          a)   canal of Schlemm
          b)   vitreous body
          c)   ciliary processes
          d)   iris
          e)   uvula

d   6.   Which of the following is passed through thirdly by an impulse from the eye?
          a)   optic radiation
          b)   optic nerve
          c)   lateral geniculate body
          d)   optic tract
          e)   optic chiasma

c 7. The images of objects being closely examined fall on the:
       a) optic disk
       b) ora serrata
       c) fovea centralis
       d) blind spot
       e) sclera

b 8. The cornea is a continuation of the:
       a) retina
       b) sclera
       c) choroid
       d) iris
       e) optic disk

b 9. In the accommodation reflex for close-up vision, what adjustments are made?
       a) the muscles of the ciliary body relax, the lens becomes less convex and the sphincter of the pupil relaxes
       b) the ciliary muscles contract, the lens becomes more convex, and the pupil constricts
       c) the ciliary muscles contract, tightening the suspensory ligaments, the lens flattens, and the pupil becomes dark adapted
       d) the ciliary muscles contract
       e) the extrinsic muscles contract, the lens does not change, but the radial muscles relax

d 10. Of the structures listed, which is passed through thirdly by a light beam?
       a) vitreous humor
       b) cornea
       c) aqueous humor
       d) lens
       e) retina

b 11. The sense of hearing is made possible by sound vibrations:
       a) directly stimulating the tectorial membrane
       b) causing the hair cells of the organ of Corti to be distorted by the tectorial membrane
       c) stimulation of the tectorial membrane by hair cells "tickling" it, when the latter are caused to vibrate
       d) causing the vestibular membrane to vibrate
       e) none of the above

c 12. According to Helmholtz's resonance theory, the pitch of a sound is determined by:
    a) the frequency of action potentials arriving in the CNS
    b) the amplitude of the action potentials arriving in the CNS
    c) the specific location or the basilar membrane which oscillates in response to the sound
    d) the number of neurons discharging
    e) the mode of response of the macula acoustica

e 13. The bony cochlea is divided into two canals connected by the helicotrema. This partitioning is by the:
    a) oval window
    b) crista galli
    c) crista ampullaris
    d) cochlear septum
    e) osseous spiral lamina and the basilar and vestibular membranes

c 14. The macula acoustica is concerned with:
    a) hearing
    b) vision
    c) static equilibrium
    d) equilibrium
    e) none of the above

d 15. Which of the following sequences correctly traces the sound wave across the middle ear?
    a) oval window, stapes, malleus, incus, ear drum
    b) oval window, malleus, incus, stapes, ear drum
    c) ear drum, stapes, malleus, incus, oval window
    d) ear drum, malleus, incus, stapes, oval window
    e) round window, incus, stapes, malleus, oval window

c 16. The Eustachian tube is associated with the equalization of air pressure in the:
    a) pharynx
    b) external ear
    c) middle ear
    d) inner ear
    e) none of these

e 17. The middle ear cavity is filled with:
    a) endolymph only
    b) perilymph only
    c) endolymph and perilymph separated by a membrane
    d) blood
    e) air

d   18. The scala vestibuli of the inner ear is separated from the cochlear duct by the:
- a) oval window
- b) tectorial membrane
- c) basilar membrane
- d) vestibular membrane
- e) the scala vestibuli is connected to the cochlear duct by the helicotrema

d   19. The loudness of a sound is physically determined by:
- a) the number of wavelengths as expressed in cycles per second
- b) the speed at which the sound wave travels through the air
- c) the density of the sound conducting medium (steel, water, wood, air, etc.)
- d) the height of its wavelength
- e) none of the above is correct

c   20. Within which of the following hearing frequency ranges is the human ear most sensitive?
- a) 0-20Hz
- b) 20-1000Hz
- c) 1000-4000Hz
- d) 3000-16000Hz
- e) 16000-20000Hz

b   21. The function of the crista is to detect
- a) sound waves
- b) rotational acceleration or deceleration
- c) linear acceleration or deceleration
- d) taste sensations
- e) odors

b   22. Which of these is not a taste modality?
- a) sweet
- b) acrid
- c) sour
- d) bitter
- e) salt

FINAL EXAMINATION

CHAPTER 16 - THE ENDOCRINE SYSTEM

c   1.   An increase in thyrotrophin (TSH) in the blood of a mammal causes:
         a)   proliferation of cells in the thyroid gland
         b)   accumulation of thyroglobulin in follicles
         c)   release of thyroxine by the thyroid gland into the blood
         d)   flow of nerve impulses into the thyroid follicles
         e)   cessation of thyroxine secretion

b   2.   An increased amount of ADH leads to:
         a)   an increased amount of urine
         b)   a decreased amount of urine
         c)   no change in the amount of urine
         d)   kidney failure
         e)   increased renin production

c   3.   The anterior lobe of the pituitary is controlled by:
         a)   nerve fibers whose cell bodies lie in the hypothalamus and which
              extend down through the pituitary stalk
         b)   inhibition by the target gland's hormone directly acting on the
              pituitary
         c)   releasing factors/hormones
         d)   changes in the blood supply to the lobe
         e)   none of the above

a,b,d 4. Which three of the following are correct with regards to calcium
         hemostasis?
         a)   lowering extracellular calcium concentration stimulates
              parathormone secretion
         b)   extracellular calcium concentration acts directly on the
              parathyroid glands without any intermediary hormones or nerves
         c)   parathormone secretion increases the calcium ion concentration
              found in the urine
         d)   calcitonin inhibits bone release of calcium

d   5.   The action of glucagon is to:
         a)   lower blood calcium ion levels
         b)   cause uterine contractions
         c)   evoke vasoconstriction
         d)   decrease liver glycogen levels

a   6.   The hormone insulin promotes:
         a)   formation of glycogen from glucose
         b)   formation of glucose from glycogen
         c)   the action of adrenalin
         d)   deamination

a 7. The primary stimulus for the secretion of aldosterone is:
a) rising potassium concentration in the adrenal cortex
b) falling potassium concentration in the adrenal cortex
c) rising sodium concentration in the adrenal cortex
d) falling sodium concentration in the adrenal cortex
e) none of the above

d 8. Secretions from this gland could cause a person to have temporary high blood pressure, rapid heartbeat, dilated pupils, momentary muscular power, and be generally alert in response to stress.
a) adrenal cortex
b) thyroid
c) anterior pituitary
d) adrenal medulla
e) pancreas

c 9. The role of cyclic AMP in endocrinology is becoming recognized as critical. What exactly is the role of this chemical factor?
a) it supplies cellular energy so that hormonally induced actions can occur
b) it is the messenger substance which causes the release of hormones from their binding sites in the cells
c) it is the final activator which causes the cell to carry out the functions for which it was genetically designed
d) it is responsible for release of adenylate cyclase when the hormone binds to the cellular receptor site

a 10. Generally speaking, a hormone:
a) is carried by blood throughout the body but is effective in only certain cells
b) is carried by blood throughout the body and is ineffective in all cells
c) is carried by ducts throughout the body but is effective in only certain cells
d) is carried by ducts throughout the body and is effective in all cells

c 11. Hormones whose primary site of action are OTHER endocrine glands are called:
a) precursor hormones
b) ambiotic hormones
c) tropic hormones
d) pheromones
e) synergistic hormones

a   12. Loss of sodium and retention of potassium are typical symptoms of:
    a)   reduced adrenal cortex function
    b)   increased adrenal cortex function
    c)   reduced adrenal medulla function
    d)   increased adrenal medulla function
    e)   increased secretion of ACTH

c   13. The role of adenylate cyclase in hormone activity is to:
    a)   attach to the hormone to trigger its action
    b)   catalyze release of the hormone from its receptor site
    c)   convert ATP to cyclic AMP
    d)   convert ATP to ADP

a   14. Hormones released by the pars nervosa (posterior lobe) of the pituitary
        are actually synthesized by cells in the:
    a)   hypothalamus and transported axonally to the pars nervosa
    b)   pars distalis and transported axonally to the pars nervosa
    c)   hypothalamus and transported by the hypothalamicohypophyseal
         portal veins to the pars nervosa
    d)   the pars nervosa

e   15. The pars distalis produces all but which one of the following
        hormones?
    a)   GH
    b)   FSH
    c)   LH
    d)   prolactin
    e)   all of the above are produced by the pars distalis

a   16. Which one of the following stimulates the adrenal cortex to produce
        glucocorticoids?
    a)   ACTH
    b)   GH
    c)   TSH
    d)   FSH
    e)   none of the above is correct

d   17. Releasing or inhibiting hormones are produced entirely by:
    a)   the target organs
    b)   the anterior lobe of the pituitary
    c)   the posterior lobe of the pituitary
    d)   the hypothalamus

a   18. Parathormone is a hormone which:
    a)   regulates calcium and phosphorous utilization by the body
    b)   causes involuntary muscular twitchings when present in excess
    c)   causes bones to become thickened when present in excess
    d)   all of the above are correct

d   19.   Which one of the following is not true of adrenaline?
   a)   it stimulates the rate of heart beat
   b)   it stimulates the conversion of glycogen to glucose
   c)   it causes the capillaries in the skin and digestive organs to decrease in diameter
   d)   it is produced in the cortex of the adrenal glands

b   20.   What is the probable relationship between somatomedins and GH?
   a)   somatomedins stimulate GH secretion
   b)   somatomedins are released in response to GH secretion
   c)   somatomedins bind to GH to activate them
   d)   GH modifies somatomedins so they can bind with actively growing tissue

b   21.   The release of melatonin is inhibited by light. The effect of light is to:
   a)   excite the pituitary
   b)   cause releasing factors to be secreted
   c)   inhibit the gonads
   d)   none of the above is correct

c 1.   Oxygen is transported in blood most efficiently by:
         a)   being dissolved in the plasma
         b)   chemical conversion to H_2O
         c)   forming a combination with hemoglobin
         d)   combining with carbon dioxide to form bicarbonate

c 2.   Which is the most abundant type of white blood cell?
         a)   lymphocyte
         b)   basophil
         c)   neutrophil
         d)   platelet
         e)   none of the above

e 3.   Which specific white blood cell can develop into a macrophage?
         a)   eosinophil
         b)   neutrophil
         c)   leucocyte
         d)   basophil
         e)   monocyte

a 4.   The liquid portion of the blood is called:
         a)   plasma
         b)   serum
         c)   tissue fluid
         d)   interstitial fluid
         e)   lymph

d 5.   The hematocrit expresses the:
         a)   concentration of hemoglobin in the blood
         b)   number of erythrocytes in the blood
         c)   viscosity of the blood
         d)   proportion of erythrocytes in a sample of blood
         e)   oxygen carrying capacity of the blood

d 6.   Which one of the following statements is correct?
         a)   each globin portion of hemoglobin contains an iron atom
         b)   the protein portion of the blood is the heme group
         c)   there are 6 heme groups in each hemoglobin molecule
         d)   oxygen combined with hemoglobin forms oxyhemoglobin
         e)   two of the above statements (a-d only) are correct

c 7.   Most of the iron of the body is found in:
         a)   ferritin
         b)   hemosiderin
         c)   hemoglobin
         d)   transferrin
         e)   the plasma as free ionized iron

b   8.   Which structures are normally released from the bone marrow into the blood stream?
- a)   fully mature erythrocytes
- b)   reticulocytes
- c)   normoblasts
- d)   hemocytoblasts
- e)   polychromatophilic erythroblasts

d   9.   Which one of the following is incorrect?
- a)   renal erythropoietic factor converts a precursor substance found in the plasma into erythropoietin
- b)   erythropoietin production is controlled, negative feedback fashion, by the level of circulating oxygen
- c)   as oxygen concentration in the blood increases, the production of erythropoietin declines
- d)   all of the above are correct

e   10.   When hemoglobin is broken down by the spleen, the iron is removed from the heme molecule and the remainder of the heme molecule is first converted into a molecule of:
- a)   transferrin
- b)   apoferritin
- c)   ferritin
- d)   bilirubin
- e)   biliverdin

c   11.   The role of transferrin in erythropoiesis is that it:
- a)   stimulates erythropoiesis
- b)   links the heme and globin portions of the hemoglobin molecule
- c)   holds iron in loose association while the iron is transported in the plasma to storage or use sites
- d)   transfers the hemoglobin into the developing erythrocyte
- e)   is an intracellular storage form of iron

c   12.   The normal synthesis of prothrombin by the liver specifically requires the presence of:
- a)   calcium ions
- b)   thrombin
- c)   vitamin K
- d)   vitamin B-12
- e)   vitamin C

c   13.   The leukocyte which is believed to ingest and destroy antigen-antibody complexes is the:
- a)   neutrophil
- b)   monocyte
- c)   eosinophil
- d)   basophil
- e)   lymphocyte

a   14.   Which one of the following is not a granulocyte?
- a)   monocyte
- b)   neutrophil
- c)   basophil
- d)   eosinophil

b   15.   The leukocyte most resembling the mast cell of connective tissues is the:
- a)   monocyte
- b)   basophil
- c)   eosinophil
- d)   lymphocyte

b   16.   Blood platelets are:
- a)   the extruded nuclei of erythrocytes
- b)   fragments of megakaryocytes, released into the blood stream
- c)   phagocytic cells
- d)   whole cells containing granules of histamine and heparin
- e)   mats of fibrin formed on the walls of blood vessels due to local injury

a   17.   Platelets release certain chemicals involved in hemostasis. Which one of the following is the chemical that causes the surfaces of platelets to become "sticky" and thus form a clump of platelets which plugs the injured area?
- a)   ADP
- b)   heparin
- c)   serotonin
- d)   prothrombin
- e)   factor X

c   18.   Which of the following structures contain a nucleus?
- a)   platelets
- b)   reticulocytes
- c)   monocytes
- d)   erythrocytes

d   19.   In the extrinsic clotting mechanism, plasmin converts a
- a)   fibrinogen to fibrin
- b)   inactive factor X to active factor X
- c)   prothrombin to thrombin
- d)   none of the above

a   20.   A clot which floats within the blood stream is known as a/an:
- a) embolus
- b) aneurysm
- c) thrombocyte
- d) thrombus
- e) none of the above is correct

d   21.   Clotting does not occur within normal blood vessels. Which one of the following does not prevent this clotting?
- a) lining of the vessels is smooth, preventing contact activation of the intrinsic mechanism
- b) heparin, secreted by the mast cells and basophils, accelerates the activity of antithrombin III
- c) plasmin destroys some clotting factors
- d) calcium ions are normally in such low concentration in the plasma that clotting cannot occur until more are released from platelets
- e) negatively charged protein molecules are absorbed onto the endothelial cell surfaces and repel negatively charged clotting factors and platelets

CHAPTER 18 - THE HEART

d   1.   Of the structures named, which is passed through third by an action potential spreading through the heart?
     a)   atrial myocardium
     b)   Purkinje fibers
     c)   sinoatrial node
     d)   atrioventricular node
     e)   ventricular myocardium

c   2.   That part of the circulation involved with pumping blood to and from the lungs is known as the:
     a)   systemic circulation
     b)   respiratory circulation
     c)   pulmonary circulation
     d)   metabolic circulation
     e)   aerobic circulation

e   3.   Blood returning from the pulmonary branch of the circulatory system enters this chamber of the heart.
     a)   atrium
     b)   vena cava
     c)   right ventricle
     d)   left ventricle
     e)   none of the above

c   4.   The major venous channel returning blood from the heart muscle is the:
     a)   coronary vein
     b)   coronary artery
     c)   coronary sinus
     d)   cardiac vein
     e)   azygos vein

b   5.   The second heart sound is most closely associated with the:
     a)   vibration of the ventricular musculature
     b)   closure of the semilunar valves
     c)   closure of the atrioventricular valve
     d)   opening of the semilunar valves
     e)   two of the above

e   6.   Which of these has the thickest wall?
     a)   aorta
     b)   right atrium
     c)   left atrium
     d)   right ventricle
     e)   left ventricle

c   7.   The valve between the left atrium and the left ventricle is the:
   a)   semilunar valve
   b)   pocket valve
   c)   mitral valve
   d)   tricuspid valve
   e)   aortic valve

a   8.   Which of the following would be pierced thirdly by a needle entering the heart from the outside?
   a)   visceral pericardium
   b)   pericardial cavity
   c)   myocardium
   d)   endocardium
   e)   parietal pericardium

a   9.   The pacemaker of the human heart is the:
   a)   sinoatrial node
   b)   atrioventricular node
   c)   left ventricle
   d)   left atrium
   e)   none of the above

d   10.   The maximum rate of decline of the left ventricular volume occurs just after the:
   a)   mitral valve closes
   b)   mitral valve opens
   c)   aortic valve closes
   d)   aortic valve opens

e   11.   Stroke-volume equals:
   a)   end-systolic volume minus end-diastolic volume
   b)   end-diastolic volume minus end-systolic volume
   c)   residual volume
   d)   atrial output minus venous return
   e)   end-diastolic volume minus residual volume

c   12.   The fact that increased diastolic filling increases the force of contraction of the normal heart is known as:
   a)   Marey's law
   b)   Bainbridge reflex
   c)   Starling's law
   d)   cardiac reserve
   e)   cardiac index

c   13.   If the stroke volume is 80 ml and the stroke rate is 72/min, what is the cardiac output?
a)   72 ml/min
b)   80 ml/min
c)   5760 ml/min
d)   1100 ml/min
e)   none of the above is correct

c   14.   The principal advantage of the long cardiac refractory period is that it:
a)   allows the atrium to complete its systole before the ventricle contracts
b)   permits the heart to go into tetanus thus providing a much more vigorous response
c)   prevents the heart from undergoing tetanus
d)   the long refractory period is actually a disadvantage
e)   is directly responsible for the transmission of the action potential from cell to cell

e   15.   The T-wave of an EKG is caused by:
a)   atrial systole
b)   ventricular systole
c)   spread of action potential through the atria
d)   spread of action potentials through the ventricles
e)   repolarization of the ventricular myocardium

e   16.   A lead II EKG records the electrical potential differences between the:
a)   right and left arms
b)   right and left legs
c)   right arm and leg
d)   left arm and leg
e)   right arm and left leg

b   17.   When the heart beats at 60 beats per minute or less, the condition is termed:
a)   paroxysmal atrial tachycardia
b)   bradycardia
c)   atrial ectopic focus
d)   myocardial infarction
e)   flutter

c   18.   It is estimated that the output of the left ventricle is 84 liters in 15 minutes. What is the cardiac minute output?
a)   28 liters
b)   15 liters
c)   5.6 liters
d)   14 liters
e)   5.0 liters

a   19. The term "myogenic" with respect to cardiac muscle contraction refers to the fact that cardiac muscle fibers contract:
   a) spontaneously, without nervous stimulation
   b) rhythmically
   c) in an all-or-none fashion
   d) briefly, but have a long refractory period
   e) in response to autonomic stimulation

a   20. The cardio-inhibitory center slows heartbeat through the:
   a) vagus nerve
   b) phrenic nerve
   c) sympathetic nerves
   d) glossopharyngeal nerves
   e) two of the above are correct

b   21. When ventricular pressure just exceeds atrial pressure the:
   a) atrioventricular valves open
   b) atrioventricular valves close
   c) SA node discharges
   d) semilunar valves close
   e) semilunar valves open

c   22. Which cranial nerve innervates the carotid sinus?
   a) V
   b) VII
   c) IX
   d) X
   e) XII

b   23. The P wave of the EKG correlates with the:
   a) depolarization of the AV node
   b) atrial depolarization
   c) ventricular repolarization
   d) depolarization of the ventricle
   e) none of the above is correct

FINAL EXAMINATION

CHAPTER 19 - BLOOD VESSELS AND LYMPHATICS

b 1. In taking blood pressure, the artery most commonly used is the:
    a)   radial
    b)   brachial
    c)   femoral
    d)   carotid
    e)   none of the above is correct

d 2. Of the arteries named, which is passed through thirdly by a drop of blood?
    a)   right subclavian
    b)   radial
    c)   brachial
    d)   axillary
    e)   brachiocephalic

d 3. Of the five vessels named, which is passed through thirdly by a drop of blood?
    a)   left internal jugular vein
    b)   left brachiocephalic vein
    c)   inferior sagittal sinus
    d)   left transverse sinus
    e)   straight sinus

e 4. Of the five vessels named, which is passed through thirdly by a drop of blood?
    a)   inferior mesenteric vein
    b)   hepatic portal vein
    c)   capillaries of descending colon
    d)   capillaries of liver
    e)   splenic vein

a 5. Which one of the following is passed through thirdly by a drop of blood?
    a)   left gastric artery
    b)   celiac artery
    c)   capillaries of the stomach
    d)   right gastric vein
    e)   aorta

d 6. An embolus resulting from a varicosity of a saphenous vein would lodge in a:
    a)   coronary arteriole
    b)   cerebral arteriole
    c)   saphenous venule
    d)   pulmonary arteriole
    e)   small vessel in liver

a   7.   A large blood vessel with tunica externa thicker than the tunica media and valves would be a:
- a)   vein
- b)   artery
- c)   venule
- d)   arteriole
- e)   capillary

b   8.   The shunting of blood from one vascular bed to another is accomplished primarily by vasoconstriction of which of these vascular components?
- a)   artery
- b)   arteriole
- c)   capillary
- d)   venule
- e)   vein

e   9.   Which one of these factors is <u>least</u> responsible for creating a loss in fluid pressure along a pipe or vessel?
- a)   fluid viscosity
- b)   irregularities on the wall of the vessel
- c)   length of the vessel
- d)   diameter of the vessel
- e)   elasticity of the vessel

d   10.   Afferent fibers from the aortic arch and carotic sinus stretch receptors terminating in the vasomotor center can be classified as _____ fibers.
- a)   vasoconstrictor
- b)   vasodilator
- c)   pressor
- d)   depressor
- e)   none of the above is correct

b   11.   Right atrial pressures are on the order of ____ mm. Hg.
- a)   0
- b)   15
- c)   50
- d)   80
- e)   120

d   12.   The velocity of blood flow is inversely proportional to:
- a)   stroke volume
- b)   cardiac output
- c)   arterial pressure
- d)   total cross-sectional area of all vessels of that class
- e)   cross-sectional area of any one vessel

b   13.   The control of blood pressure by carbon dioxide is <u>primarily</u> due to its effect on the:
- a) vasomotor center
- b) chemoreceptors in the carotid and aortic bodies
- c) arteriole constriction directly
- d) none of the above is correct

b   14.   Which of the following offers the best explanation for the homeostatic regulation of blood pressure which ensures adequate blood to the brain when going from a horizontal to a vertical position?
- a) the minimal blood pressure in the arteries is such that blood flow is adequate regardless of body position
- b) the fall in blood pressure upon arising is detected in the carotid sinus and aortic arch pressoreceptors. This signals increased heart rate and vasoconstrictor response which increases blood pressure
- c) the fall of the blood through the jugular veins fills the right ventricle more, evoking Starling's law of the heart
- d) the otoliths of the macula acoustica have direct neuronal input into both the cardiac control and vasomotor centers, signaling them to increase heart rate and sympathetic tone on arterioles

d   15.   Under normal conditions most of the blood in the vascular tree is found in the:
- a) heart
- b) arteries
- c) capillaries
- d) veins
- e) tissue fluid

a   16.   The flow of blood through the coronary vessels is:
- a) intermittent
- b) determined solely by aortic pressure
- c) increased during ventricular systole
- d) not influenced by intramural pressure

FINAL EXAMINATION

CHAPTER 20 - LYMPHATIC SYSTEM

c 1. How is a lymph capillary like a blood capillary?
       a)   they both contain blood
       b)   they both contain valves
       c)   they both have thin walls
       d)   they are both connected to the vena cava

b 2. The thoracic duct:
       a)   drains the right half of the body
       b)   receives lymph from the digestive tract
       c)   drains into the inferior vena cava
       d)   two of the above
       c)   all of the above

b 3. A large organ, designed to filter blood, which is structurally similar to
       a lymph node is the:
       a)   thymus
       b)   spleen
       c)   Peyer's patch
       d)   pharyngeal tonsil
       e)   adenoid

e 4. The lymphatic system:
       a)   is composed of vessels that are classified as only two types,
            veins and capillaries
       b)   is found in all parts of the body
       c)   moves lymph only toward the heart
       d)   contains lymph nodes
       e)   all of these are correct

d 5. The lymphatic system has the following function(s):
       a)   returns tissue fluid to the blood circulatory system
       b)   absorbs digested fats
       c)   filters out debris, bacteria, etc.
       d)   a, b, and c
       e)   a and c only

c 6. Lymph nodes are involved in all but which of the following?
       a)   production of antibodies
       b)   phagocytosis of bacteria
       c)   production of erythrocytes
       d)   production of lymphocytes
       e)   production of monocytes

d   7.   Lacteals:
- a)   are found in the villi
- b)   are a part of the lymphatic system
- c)   absorb the products of fat digestion
- d)   all of these
- e)   none of these

c   8.   Lymph capillaries are different from blood capillaries in that they:
- a)   have valves
- b)   have walls containing smooth muscle
- c)   are blind-ended
- d)   are permeable
- e)   lymph capillaries are in every way identical to vascular capillaries

e   9.   Lymph enters the bloodstream directly through the _____.
- a)   right lymphatic duct
- b)   thoracic duct
- c)   bronchomediastinal trunks
- d)   all of the above
- e)   both a and b

c   10.   Lymph leaving the cisterna chyli next passes immediately into the:
- a)   right lymphatic duct
- b)   lacteal
- c)   thoracic duct
- d)   left subclavian vein
- e)   subclavian trunks

c   11.   Proteins in the interstitial spaces are normally removed by:
- a)   absorption into vascular capillaries
- b)   breakdown and utilization by body cells
- c)   lymphatic vessels
- d)   none of the above is correct

a   12.   In which organ are many lymphocytes and macrophages located in red pulp?
- a)   spleen
- b)   thymus gland
- c)   tonsils
- d)   all of the above contain red pulp

b   13.   Which organ confers on certain lymphocytes the ability to mature into cells capable of being involved in cell-mediated immunity?
- a)   spleen
- b)   thymus gland
- c)   tonsils
- d)   all of the above

CHAPTER 21 - DEFENSE MECHANISMS

e   1.   Select from the following list all those antibodies which a person with
            type AB, Rh+ blood could have (more than one answer possible).
            a)   anti A
            b)   anti B
            c)   anti Rh+
            d)   anti Rh-
            e)   he could have none of these

b   2.   The plasma of a person with A antigen contains which of these
            antibodies?
            a)   antibody A
            b)   antibody B
            c)   antibodies A and B
            d)   antibody 0
            e)   no antibodies at all for blood antigens

a   3.   The definition of the term "immunity" is:
            a)   the ability of the body to resist many chemicals and organisms
                  that can damage tissues
            b)   the resistance of the skin and mucous membranes to invasion by
                  harmful substances or organisms
            c)   the inflammatory response of injured tissue
            d)   the development of specific substances in response to exposure to
                  foreign material by specialized lymphocytes and cells derived from
                  them
            e)   limited to the phagocytosis of foreign cells or proteins by
                  neutrophils or monocytes

d   4.   The walling-off of the injured area which helps to prevent or delay
            the escape of toxic products or bacteria is created by:
            a)   neutrophils which stick together because of opsonins to form a
                  living barrier
            b)   monocytes which stick together because of opsonins to form a
                  living barrier
            c)   proliferation of newly formed epithelial cells growing down and
                  around the injury
            d)   the movement of fibrinogen out of the plasma and its conversion
                  to fibrin around the injury
            e)   pavementing of leukocytes around the injury

a   5.   The adhering of leukocytes to the endothelial wall of a capillary prior
            to their passage into the tissue spaces is known as:
            a)   pavementing
            b)   opsonization
            c)   margination
            d)   chemotaxis
            e)   degranulation

d   6.   In certain types of allergies, the inflammatory response is characterized by a large number of:
- a) neutrophils
- b) monocytes
- c) lymphocytes
- d) eosinophils
- e) basophils

d   7.   Once a phagocytic cell attaches to a particle and engulfs it, the particle is destroyed by substances produced by the phagocytic vesicle. These substances include:
- a) lysosomal enzymes and HCL
- b) opsonin and lysosomal enzymes
- c) opsonin and hydrogen peroxide
- d) lysosomal enzymes and hydrogen peroxide
- e) opsonin and HCL

c   8.   Chemical mediators of the acute, non-specific inflammatory response are released from or generated by which of the following?
- a) hormones, tissue cells in injured areas and enzyme-catalyzed reactions
- b) neurohumors from injured neurons, hormones, and tissue cells in the injured area
- c) tissue cells in the injured area, enzyme-catalyzed reactions and leukocytes
- d) leukocytes, neurohumors from neurons in the injured area, and enzyme-catalyzed reactions
- e) leukocytes, hormones, and neurohumors from neurons in the injured area.

e   9.   The action of kinins includes which of the following?
- a) dilate arterioles
- b) induce pain
- c) act as powerful chemotactic agents
- d) increase vascular permeability
- e) all of the above

c   10.   In addition to inducing fever, endogenous pyrogen is also responsible for:
- a) stimulating leukopoiesis (production and release of leukocytes)
- b) chelating or detoxifying many microbial toxins
- c) causing iron and other trace metals to move out of the plasma and into the liver. This action may retard the proliferation of bacteria that require much iron to multiply.
- d) "tagging" the surface of the bacterial cell so that leukocytes are chemotactically stimulated to phagocytize them

e 11. The cascade of reactions resulting in the formation of complement proteins is initiated by:
   a) release of histamine from degranulating mast cells
   b) the synthesis of kallikrein which attaches to the microbe causing it to release "factor C" which triggers the entire immune response
   c) pyrogens released by neutrophils increase body temperatures; as a result, properdin is converted to complement C-3 which leads to the formation of other complement components
   d) leukotrienes are released by leukocytes when they ingest microbes. The leukotrienes enzymatically convert prostaglandin to complement component C-3 leading to the formation of other complement components.
   e) none of the above is correct

a 12. One of the most effective inducers of interferon production is:
   a) double-stranded RNA
   b) viral protein shells
   c) messenger RNA
   d) DNA
   e) complement protein

a 13. Antibodies are produced by:
   a) plasma cells derived from B cells
   b) antigens
   c) suppressor T cells
   d) fibroblasts
   e) cytotoxic T cells stimulated by the hormone thymosin upon exposure to antigens

d 14. The binding of an antibody to an antigen specifically involves the antigen linking up to the:
   a) entire antibody molecule
   b) constant region of the light polypeptide chain of the antibody molecule
   c) constant region of the heavy polypeptide chain of the antibody molecule
   d) variable end of the heavy and light chains of the antibody molecule

d 15. The actions of chemicals released by effector delayed hypersensitivity T cells include all but which one of the following?
   a) attracting macrophages
   b) enhancement of phagocytosis
   c) damage or destruction of many cells
   d) all of the above are actions of chemicals released by effector delayed hypersensitivity T cells

b   16. Active immunity can be acquired by either infection with a live
          organism or by:
          a)   being given an antibody from a person or animal that had been
                infected
          b)   vaccination
          c)   both of the above

e   17. The injection of anti-Rh antibodies into a woman who has just given
          birth to an Rh positive child to prevent her from developing anti-Rh
          antibodies is an example of:
          a)   tolerance
          b)   hypersensitivity
          c)   vaccination
          d)   active immunity
          e)   passive immunity

d   18. Which of the following is correct?
          a)   a person who is Rh+ could also have antigen A on his or her
                erythrocytes but not antigen B
          b)   a person who is type O has neither anti-A nor anti-B antibodies
                in his or her plasma
          c)   a person with type AB blood can donate blood to both type A
                people and type B people
          d)   none of the above is correct

d   19. In the clonal suppression theory of tolerance, appropriate exposure to
          an antigen:
          a)   removes or modulates the receptors for the antigen on the
                surfaces of particular cells, especially immature B cells
          b)   destroys those B cells which could inactivate that antigen
          c)   result in the formation of complement which removes the hapten
                from the protein, causing it to no longer be antigenic
          d)   none of the above is correct

b   20. Plasma cells
          a)   are progeny of helper T cells
          b)   produce antibodies
          c)   phagocytize microorganisms at the site of an infection
          d)   secrete active complement components
          e)   none of the above

CHAPTER 22 - THE RESPIRATORY SYSTEM

b   1.   Exhalation is increased in <u>force</u> by contraction of:
   - a)   external intercostal muscles
   - b)   internal intercostal muscles
   - c)   the diaphragm
   - d)   muscles running from the thyroid to the cricoid cartilage
   - e)   muscles of the pharynx wall

a   2.   The amount of air you can still inhale after inhaling normally is termed:
   - a)   inspiratory reserve volume
   - b)   expiratory reserve volume
   - c)   residual volume
   - d)   dead space
   - e)   inspiratory and expiratory reserve volume (combined)

b   3.   If the chest wall is perforated:
   - a)   the lung will tend to be forced out through the wound
   - b)   the lung on that side will collapse
   - c)   breathing will be largely unaffected due to the large reserve capacity of the respiratory muscles
   - d)   pleurisy results immediately
   - e)   the respiratory rate slows drastically

b   4.   Inhalation results when:
   - a)   the lungs expand and make the chest bigger
   - b)   the ribs and diaphragm cause chest cavity to enlarge
   - c)   air pressure within lungs increases
   - d)   the diaphragm relaxes and the muscles between the ribs contract
   - e)   the autonomic nerves fire rhythmically stimulating respiratory muscles

e   5.   The respiratory membrane consists of:
   - a)   a single thickness of epithelial cells
   - b)   choice "a" and a basement membrane
   - c)   two thicknesses of epithelial cells
   - d)   choice "c" plus a basement membrane
   - e)   choice "c" plus two basement membranes

e   6.   Voice is produced by the vibration of the:
   - a)   uvula
   - b)   ventricular folds
   - c)   arytenoid cartilages
   - d)   epiglottis
   - e)   vocal membranes

c   7.   Food is prevented from entering the trachea by the:
   - a)   fauces
   - b)   soft palate
   - c)   epiglottis
   - d)   saliva
   - e)   uvula

d   8.   Of the structures listed, which is passed through thirdly by a molecule of oxygen entering the body:
   - a)   glottis
   - b)   bronchus
   - c)   nasal cavity
   - d)   trachea
   - e)   bronchiole

c   9.   The force responsible for normal expiration at rest is supplied by:
   - a)   muscle contraction of the diaphragm
   - b)   external intercostal muscle contraction
   - c)   elastic recoil of tissues
   - d)   contraction of smooth muscle in air passages

a   10.   The total capacity of the lungs of an average male will be about _____ ml.
   - a)   5900
   - b)   4700
   - c)   4400
   - d)   500

d   11.   The pitch of the voice sounds is determined by:
   - a)   the shape of the glottal opening
   - b)   the length of the vocal cords
   - c)   the tension of the vocal cords
   - d)   all of the above determine the pitch of the voice

d   12.   Surfactant acts to:
   - a)   lower surface tension of the alveoli
   - b)   prevent collapse of the alveoli
   - c)   prevent dehydration of the alveolar surface
   - d)   all of the above are correct

b   13.   Oxygen delivery to tissues is accelerated by:
   - a)   decreased oxygen tension in the blood
   - b)   increased carbon dioxide tension in the blood
   - c)   decreased carbon dioxide tension in the tissues
   - d)   increased oxygen tension in the tissues

d   14.   If the pulmonary baroreceptors were inoperative, breathing would:
          a)   cease
          b)   become deeper and more rapid
          c)   become shallower and more rapid
          d)   become slower and deeper
          e)   none of the above is correct

c   15.   Cutting the nerves from the respiratory center in the medulla to the respiratory muscles will:
          a)   increase the respiratory rate
          b)   decrease the respiratory rate
          c)   completely stop respiration
          d)   none of the above is correct

h   16.   The release of oxygen by hemoglobin for use in body tissues is dependent on all of the following except:
          a)   low partial pressure of oxygen of body tissues
          b)   alkaline character of the interstitial fluid
          c)   heat from metabolic reactions
          d)   high partial pressure of carbon dioxide in the blood

a   17.   In the alveoli of the lungs, the partial pressure of oxygen is about _____ mm. Hg.
          a)   100
          b)   159
          c)   40-46
          d)   dependent on cardiac output

a   18.   Which of the following is a correct statement about the transport of carbon dioxide by the blood? Most of the carbon dioxide is carried as:
          a)   bicarbonate ion
          b)   carbamino compounds
          c)   dissolved gas
          d)   carboxyhemoglobin

c   19.   The Adam's apple is part of the:
          a)   epiglottis
          b)   cricoid cartilage
          c)   thyroid cartilage
          d)   arytenoid cartilage
          e)   hyoid bone

c   20. Since the atmospheric air is 20.9% oxygen, and since the barometric pressure at 10,000 feet is 523 mm. Hg., the partial pressure of oxygen at that elevation in the inspired air is _____ mm. Hg.
- a) 25
- b) 159
- c) 109
- d) 523
- e) cannot be determined from the data given above

a   21. The phase of respiration involving the exchange of oxygen and carbon dioxide between the blood and the cells is:
- a) internal
- b) external
- c) cellular
- d) breathing

c   22. The floor of the nasal fossa is partially formed from the:
- a) vomer
- b) inferior nasal concha
- c) horizontal process of the palatine bone
- d) cribriform plate of the ethmoid
- e) none of the above is correct

a   23. In which way does chloride ion move in deep tissue capillaries?
- a) from the plasma into the red blood cell
- b) from the red cell into the plasma
- c) there is no shift in chloride ion across cell membranes in deep tissue capillaries
- d) none of the above is correct

d   24. Chemoreceptors in the medulla:
- a) react to slight changes in carbon dioxide
- b) respond to hydrogen ion concentration in cerebrospinal fluid
- c) can cause marked increases in breathing rates
- d) all of the above are correct

b   25. The sum of residual volume and the expiratory reserve volume is called:
- a) inspiratory capacity
- b) functional residual capacity
- c) vital capacity
- d) tidal volume

CHAPTER 23 - DIGESTIVE SYSTEM: ANATOMY AND MECHANICAL PROCESSES

b   1.   While food does not pass through this structure, enzymes are secreted here that are capable of digesting carbohydrates, fats and proteins.
- a) liver
- b) pancreas
- c) stomach
- d) small intestine
- e) large intestine

b   2.   The action involved in rhythmic segmentation in the gut serves to:
- a) progress food down the intestine
- b) propel food from mouth in vomiting
- c) aid defecation
- d) churn and mix food with digestive juices
- e) none of the above

e   3.   Which of these regions of the gut displays plicae semilunares?
- a) stomach
- b) duodenum
- c) jejunum
- d) ileum
- e) ascending colon

a   4.   Villi serve to:
- a) absorb nutrients
- b) secrete bile
- c) produce antibodies
- d) synthesize vitamins
- e) move food down gut

d   5.   The common bile duct opens into the:
- a) stomach
- b) ileum
- c) ascending colon
- d) duodenum
- e) jejunum

b   6.   That portion of the stomach in contact with the diaphragm and superior to the entrance of the esophagus is technically known as the:
- a) cervix
- b) fundus
- c) body
- d) cardia
- e) pyloric region

c   7.   The duct of the submaxillary gland drains into the:
        a)   lateral wall of the cheek
        b)   space beside the root of the tongue, posteriorly
        c)   space below the tip of the tongue, to either side of the lingual frenum
        d)   cervices on the surface of the tongue
        e)   the moat around the circumvallate papilla

a   8.   The outer covering of the crown of the tooth is called:
        a)   enamel
        b)   keratin
        c)   dentin
        d)   cement
        e)   gingiva
        f)   peridontal membrane

c   9.   Rhythmic contractions that produce stationary constrictions in the wall of the small intestine are called:
        a)   peristaltic waves
        b)   mass peristalsis
        c)   segmentation
        d)   haustra
        e)   none of the above is correct

d 10.   That portion of the tooth lying below the gum, but above the bone is called the:
        a)   root
        b)   apex
        c)   crown
        d)   neck

e 11.   The temporary folds in the mucosa of the stomach are called:
        a)   plicae circulares
        b)   haustra
        c)   gastric pits
        d)   intestinal glands
        e)   none of the above is correct

a 12.   The act of swallowing is conveniently divided into three stages. Which of these is voluntary?
        a)   only the first
        b)   only the second
        c)   only the third
        d)   the first and second
        e)   all 3 stages are voluntary

d   13.   The mass of food formed by the tongue and swallowed is called a/an:
- a)   chymic swill
- b)   chyle
- c)   chylomicron
- d)   bolus
- e)   eructation

e   14.   The stomach is lined with:
- a)   transitional epithelium
- b)   non-keratinized stratified squamous epithelium
- c)   keratinized stratified squamous epithelium
- d)   ciliated pseudostratified columnar epithelium
- e)   none of the above is correct

c   15.   Blood flow through the liver lobule passes from:
- a)   hepatic artery to sinusoid to portal vein
- b)   hepatic vein to sublobular vein to sinusoid
- c)   portal vein to sinusoid to central vein
- d)   hepatic artery to sublobular vein to sinusoid
- e)   none of the above is correct

d   16.   If a pin were to perforate the wall of the small intestine and enter the lumen of the areas below, which would be passed through third?
- a)   muscularis mucosa
- b)   muscularis externa
- c)   visceral peritoneum
- d)   submucosa
- e)   lamina propria

d   17.   As bile passes from the liver cells, what structure listed below will it pass through thirdly?
- a)   common bile duct
- b)   duodenum
- c)   left or right hepatic duct
- d)   common bile duct
- e)   bile canaliculi

CHAPTER 24 - THE DIGESTIVE SYSTEM: CHEMICAL DIGESTION
AND ABSORPTION

a   1.   In their passage through the digestive tract, starches are first acted
upon by enzymes in the:
    a)   mouth
    b)   stomach
    c)   small intestine
    d)   pancreas
    e)   colon

c   2.   Most water is absorbed by the:
    a)   colon
    b)   esophagus
    c)   small intestine
    d)   anus
    e)   stomach

e   3.   What is one of the roles of HCL in the stomach?
    a)   its presence in the stomach is a symptom of a digestive disorder
resulting in indigestion
    b)   it serves only to cause stomach ulcers
    c)   it is a digestive enzyme that hydrolyzes fats
    d)   it aids in the digestion of starch
    e)   it facilitates protein digestion

d   4.   Duodenal glands produce:
    a)   amylase
    b)   pepsinogen
    c)   HCL
    d)   mucus

b   5.   In their passage through the digestive tract, proteins are first acted
upon by enzymes in the:
    a)   mouth
    b)   stomach
    c)   small intestine
    d)   pancreas
    e)   colon

c   6.   What is the product of the zymogenic cells?
    a)   insulin
    b)   HCL
    c)   pepsinogen
    d)   trypsinogen
    e)   none of the above is correct

e   7. Which one of these is <u>not</u> a component of bile?
        a) cholic acid
        b) bicarbonate ion
        c) bilirubin
        d) lecithin
        e) trypsin

e   8. Which one of the following does not initiate or stimulate gastric secretion?
        a) moderate distension of the stomach by food - the local myenteric reflex
        b) vagovagal reflex
        c) secretogogues
        d) sight or smell of food
        e) duodenal filling when the stomach is full

c   9. Which one of the following is not a function of secretion?
        a) stimulates bicarbonate-rich, water pancreatic secretion
        b) stimulates duodenal glands
        c) stimulates gastric juice secretion when gastrin is actively being produced
        d) stimulates secretion of bile

b   10. Secretion of bile by the liver is stimulated by:
        a) cholecystokinin
        b) choleretics
        c) gastrin
        d) carboxypeptidase

c   11. Which one of the following is necessary for the absorption of vitamin B-12 from the gut?
        a) bile salts
        b) kallikrein
        c) intrinsic factor
        d) enterokinase

b   12. Chyme is:
        a) fluid in a lacteal after a fatty meal has been eaten
        b) the material in the duodenum
        c) the material flowing through the common bile duct
        d) none of the above is correct

e   13. The substances secreted by the stomach include:
        a) gastrin, cholecystokinin, and trypsinogen
        b) pepsinogen, HCL, and chymotrypsinogen
        c) gastrin, HCL, and chymotrypsinogen
        d) pepsinogen, HCL, and gastrin
        e) HCL, trypsinogen, and gastrin

c 14. The exocrine portion of the pancreas secretes:
    a) cholecystokinin
    b) glucagon
    c) pancreatic amylase
    d) pepsinogen
    e) none of the above is correct

e 15. The action of cholecystokinin is to:
    a) stimulate enzyme secretion from the pancreas
    b) neutralize HCL
    c) stimulate contraction of the gallbladder
    d) inhibit digestion of carbohydrates
    e) two of the above are correct

c 16. Enterokinase is important in the digestion of:
    a) fats
    b) carbohydrates
    c) proteins
    d) cholesterol

c,b 17. The parietal cells secrete:
    a) amylase
    b) HCL
    c) pepsinogen
    d) mucus
    e) trypsinogen

c 18. Most of the bile salts are absorbed back into the blood by the:
    a) duodenum
    b) jejunum
    c) ileum
    d) colon
    e) none of the above is correct

d 19. Bile salts:
    a) modify water to make it a fat solvent
    b) hydrolyze fat into water soluble molecules
    c) repel fat, keeping fat droplets separated from one another
    d) none of the above

e 20. Which substances are absorbed in the small intestine?
    a) glucose
    b) amino acids
    c) calcium ions
    d) two of the above are absorbed in the small intestine
    e) all of the above are absorbed in the small intestine

b   21.   Which of the following occur in the large intestine?

        a)   active absorption of water
        b)   secretion of bicarbonate ions
        c)   production of pepsin
        d)   absorption of enterokinase
        e)   none of the above

CHAPTER 25 - METABOLISM, NUTRITION, ETC.

c   1.   Which one of the following processes requires oxygen?
   a)   glycolysis
   b)   lactic acid production
   c)   electron transport system
   d)   none of the above require oxygen

a   2.   When pyruvic acid glucose is converted to acetyl CoA:
   a)   carbon dioxide is produced
   b) NAD^+ is produced
   c) $FADH_2$ is produced
   d)   GDP is produced
   e)   none of the above is correct

c   3.   In the electron transport system, the final electron acceptor is:
   a)   cytochrome
   b)   NADH+H
   c)   oxygen
   d)   pyruvic acid
   e)   acetyl CoA

b   4.   The Krebs' cycle requires the presence of oxygen in order for it to occur, yet no actual oxygen enters the series of reactions constituting that cycle. Which one of the following best explains this apparent dilemma?
   a)   the Krebs' cycle may operate aerobically or anaerobically
   b)   the coenzymes which accept electrons from the Krebs' cycle are in limited amounts in the cell and they must dispose of their electrons in the electron transport system in order to allow continued operation of the Krebs' cycle
   c)   the oxygen needed in the Krebs' cycle is necessary in the glycolysis part of the catabolic breakdown of sugar. Without oxygen, suitable substrates for the Krebs' cycle are not produced.
   d)   oxygen is necessary in transferring acetyl CoA across the mitochondrial membrane. Without oxygen, the acetyls will not be available to link up to oxaloacetic acid.
   e)   none of the above is correct

d   5.   What is gluconeogenesis? It is concerned with the:
   a)   production of glycogen
   b)   deamination of amino acids
   c)   breakdown of glucose
   d)   ability of the liver to form glucose from non-carbohydrate substances
   e)   none of the above is correct

e   6.   When fat is mobilized from adipose tissue, it is transported in the blood in the form of:
a)   chyle
b)   chyme
c)   chylomicrons
d)   micelles
e)   fatty acids

d   7.   Which one of the following is not a monosaccharide?
a)   glucose
b)   fructose
c)   galactose
d)   all of the above are monosaccharides

d   8.   In the postabsorptive period, the body gets most of its energy from the breakdown of:
a)   glucose
b)   amino acids
c)   glycogen
d)   fatty acids
e)   protein

e   9.   Control of body temperature is regulated by centers in the:
a)   medulla
b)   pons
c)   thalamus
d)   cerebellum
e)   none of the above is correct

a   10.   Urea is produced in the:
a)   liver
b)   urinary bladder
c)   kidney
d)   spleen
e)   none of the above is correct

a   11.   Most of the heat lost from the body when it is immersed in cold water is due to heat loss by:
a)   conduction
b)   osmosis
c)   radiation
d)   absorption

e.   12.   The liver:
a)   synthesizes urea
b)   stores glycogen
c)   produces prothrombin
d)   excretes bilirubin
e)   all of the above

d   13. Which one of the following is directly produced by the deamination of
          an amino acid?
          a) fatty acid
          b) an aldehyde
          c) glycerol
          d) keto acid
          e) none of the above is correct

b   14. Which one of the following is not a fat-soluble vitamin?
          a) A
          b) C
          c) D
          d) E
          e) K

c   15. A deficiency of which vitamin is implicated in the occurrence of
          pellagra?
          a) biotin
          b) folic acid
          c) niacin
          d) pantothenic acid
          e) none of the above

a   16. The specific dynamic action of food is to:
          a) increase the metabolic rate
          b) reduce oxygen consumption
          c) pass from the absorptive to the post-absorptive state
          d) lower metabolism to the basal metabolic rate
          e) none of the above is correct

d   17. Amino acids can enter the Krebs' cycle as:
          a) acetyl CoA
          b) urea
          c) ammonia
          d) alpha keto acids
          e) none of the above is correct

d   18. Most organs use glucose as an energy source during the absorptive
          state. Which one of the following organs uses alpha keto acids during
          the absorptive state?
          a) skeletal muscle
          b) brain
          c) pancreas
          d) liver
          e) none of the above is correct

e   19. After all the different kinds of cells have stored all the glucose they can, excess glucose is converted to:
a)   urea
b)   amino acids
c)   keto acids
d)   glycogen
e)   fat

e   20. During the post-absorptive state, nerve cells obtain energy from:
a)   fatty acids
b)   vitamins
c)   glycerol
d)   amino acids
e)   glucose

b   21. During glycolysis:
a)   oxygen is consumed
b)   ATP is formed
c) $NADH+H^+$ is converted to NAD^+
d)   FAD is reduced
e)   none of the above

d   22. The basal metabolic rate is the amount of heat energy that:
a)   will raise the temperature of a liter of water one degree C
b)   will raise the temperature of 1 ml. of water one degree C
c)   the body must produce per minute in order to maintain homeostasis
d)   none of the above is correct

b   23. Malic acid is a component of:
a)   the electron transport system
b)   the Krebs' cycle
c)   the glycolytic pathway
d)   none of the above

CHAPTER 26 - THE URINARY SYSTEM

c   1. The ureter extends downward:
    a) behind the parietal peritoneum and joins the urinary bladder from above
    b) in front of the parietal peritoneum and joins the urinary bladder from above
    c) behind the parietal peritoneum and joins the urinary bladder from below
    d) in front of the parietal peritoneum and joins the urinary bladder from below
    e) none of the above is correct

b   2. Bowman's capsule:
    a) returns material to the blood
    b) filters material from the blood
    c) makes urea
    d) contains urine in its final state
    e) surrounds the kidney (=true capsule)

b   3. Most reabsorption of substances from the filtrate occurs in the:
    a) loop of Henle
    b) proximal tubule
    c) collecting tubule
    d) glomerular capsule
    e) distal tubule

b   4. The liquid that collects in the cavity of Bowman's capsule is:
    a) concentrated urine
    b) blood plasma minus blood proteins
    c) used bile ready for excretion
    d) glycogen and water
    e) the result of pyelonephritis

d   5. If the arteriole that supplies blood to the glomerulus becomes constricted:
    a) hydrostatic pressure in the glomerulus increases
    b) protein concentration in the filtrate increases
    c) blood flow into the efferent arteriole increases
    d) glomerular filtration rate decreases
    e) none of the above

d 6. Urine leaving the distal convoluted tubule passes through which of the following listed structures thirdly:
   a) collecting duct
   b) calyx
   c) ureter
   d) pelvis
   e) urethra
   f) bladder

e 7. The collecting ducts:
   a) are found within both the cortex and medulla
   b) contain dilute urine
   c) receive fluid from several nephrons
   d) empty into the pelvis
   e) all of these

a 8. Of the structures listed, which of these is passed through thirdly by a water molecule?
   a) proximal convoluted tubule
   b) Bowman's capsule
   c) loop of Henle
   d) distal convoluted tubule
   e) glomerulus

a 9. The microscopic anatomical unit of excretion found in the kidney is the:
   a) nephron
   b) glomerular capsule
   c) glomerulus
   d) renal corpuscle
   e) cortex

c 10. Which of these is the correct sequence of blood vessels?
   a) renal artery, capillary bed, afferent arteriole, efferent arteriole, renal vein
   b) efferent arteriole, glomerulus, venule, afferent arteriole, collecting duct
   c) afferent arteriole, glomerulus, efferent arteriole, peritubular capillary, venule, renal vein
   d) renal artery, efferent arteriole, glomerulus, afferent arteriole, peritubular capillaries, venule, renal vein
   e) afferent arteriole, peritubular capillaries, efferent arteriole, glomerulus, renal vein

e   11.   Which of the following are reabsorbed by the proximal convoluted tubule?
- a)   sodium
- b)   potassium
- c)   glucose
- d)   chloride
- e)   all of the above are reabsorbed by the proximal convoluted tubule

c   12.   The process of urination:
- a)   involves contraction of the detrusor muscle
- b)   requires relaxation of the two urethral sphincters
- c)   is a reflex involving sacral segments of the spinal cord
- d)   may either be consciously facilitated or inhibited, but is not under voluntary control in the ordinary sense
- e)   all of the above are correct

a   13.   The counter-current mechanism maintains normal urinary output by:
- a)   maintaining high medullary osmotic pressure in the kidney
- b)   maintaining low medullary osmotic pressure in the kidney
- c)   increasing the rate of glomerular filtration
- d)   increasing the effective glomerular filtration pressure
- e)   two of the above are correct

e   14.   The true capsule of the kidney is formed of:
- a)   transitional epithelium
- b)   adipose tissue
- c)   parietal peritoneum
- d)   visceral peritoneum
- e)   dense fibrous connective tissue

c   15.   Given the following data, calculate the Effective Glomerular Filtration Pressure. Average glomerular hydrostatic (blood) pressure = 67 mm. Hg.; C.O.P of blood entering glomerulus = 28 mm. Hg.; C.O.P. of blood leaving glomerulus = 34 mm. Hg.; hydrostatic pressure of Bowman's capsule = 14 mm. Hg.
- a)   14 mm. Hg.
- b)   19 mm. Hg.
- c)   22 mm. Hg.
- d)   25 mm. Hg.
- e)   none of the above is correct

c   16.   Clearance refers to:
- a)   the amount of fluid passed across all the glomeruli per minute
- b)   the amount of solute passed into the urine per minute
- c)   the volume of plasma from which a substance is entirely removed per minute
- d)   none of the above is correct

b 17. Fluid at the bottom of the loop of Henle is likely to be _____ to that in the blood plasma.
   a) hypotonic
   b) hypertonic
   c) isotonic
   d) none of the above is correct

c 18. ADH production:
   a) is increased following the drinking of large amounts of water
   b) leads to an increased volume of urine output
   c) causes an increased permeability of kidney tubules to water
   d) is decreased by increasing the tonicity of the blood

e 19. Aldosterone production:
   a) decreases tubular reabsorption of sodium
   b) increases the permeability of the distal tubule to water
   c) increases renal reabsorption of potassium
   d) decreases reabsorption of chloride
   e) none of the above is correct

a 20. Which one of the following processes is not used to remove solutes from the fluid in the kidney tubules?
   a) osmosis
   b) dialysis
   c) active transport
   d) all of the above processes remove solutes from the fluid in the kidney tubules

e 21. The epithelium found in the ureters is:
   a) simple columnar
   b) simple squamous
   c) simple cuboidal
   d) stratified squamous
   e) none of the above is correct

CHAPTER 27 - FLUIDS AND ELECTROLYTES - ACID BASE BALANCE

d   1.   Using the sodium bicarbonate-carbonic acid buffer system, what two chemicals are produced as a result of adding HCL?
    a)   water and sodium bicarbonate
    b)   sodium chloride and sodium bicarbonate
    c)   water and sodium chloride
    d)   carbonic acid and sodium chloride
    e)   sodium carbonate and sodium chloride

d   2.   The kidney facilitates the action of the bicarbonate buffer system by:
    a)   conserving bicarbonate if needed
    b)   excreting excessive bicarbonate if needed
    c)   neither a nor b are correct
    d)   both a and b are correct

a   3.   By far, the greatest fluid volume in the body is contained in the:
    a)   cells
    b)   fluids between the cells
    c)   blood plasma and lymph
    d)   cerebrospinal fluid

b   4.   If the blood is too acidic, which of the following would not be likely to occur?
    a)   hydrogen ions would be excreted by the kidneys
    b)   bicarbonate ions would be excreted in increased amounts
    c)   the person's rate of breathing would increase
    d)   ammonia would be secreted by cells of the kidney tubules

c   5.   What is the action of hemoglobin with respect to acid-base balance?
    a)   hemoglobin releases hydroxide ions for neutralization of hydrogen ions
    b)   hemoglobin catalyzes carbon dioxide and water formation from carbonic acid
    c)   hemoglobin picks up hydrogen ions
    d)   hemoglobin has nothing to do with acid-base balance

d   6.   Which of the following would tend to increase the glomerular filtration rate?
    a)   a decrease in plasma volume
    b)   an increase in the activity of sympathetic nerves to the kidneys
    c)   a decreased release of ADH
    d)   an increase in the activity of neurons from circulatory pressure receptors
    e)   none of the above

c   7.   Acid-base buffering systems minimize changes in blood pH by:
         a)   causing an increase in respiratory rate
         b)   synthesizing ammonia
         c)   changing strong acids from any source into weaker acids
         d)   changing weak acids from any source into stronger acids
         e)   stimulating the mechanisms responsible for hydrogen ion secretion

c   8.   The interstitial fluid and the plasma are separated from each other by the:
         a)   intercellular fluid
         b)   cell membranes of tissue cells
         c)   capillary walls
         d)   plasma compartment
         e)   lamina propria

e   9.   Juxtaglomerular cells of the kidneys release:
         a)   angiotensin
         b)   calcitonin
         c)   ADH
         d)   aldosterone
         e)   none of the above

a   10.   Extracellular fluids are:
         a)   high in sodium and low in potassium
         b)   high in both sodium and potassium
         c)   low in both sodium and potassium
         d)   low in sodium and high in potassium

d   11.   Hyperventilation results in:
         a)   metabolic acidosis
         b)   respiratory acidosis
         c)   metabolic alkalosis
         d)   respiratory alkalosis

b   12.   If the blood is alkaline:
         a)   more sodium is reabsorbed
         b)   less bicarbonate ions are reabsorbed
         c)   more bicarbonate ions are reabsorbed
         d)   more water is reabsorbed
         e)   two of the above are correct

d   13.   The major role of the kidney in the maintenance of acid-base balance is to:
         a)   excrete bicarbonate ions
         b)   excrete urea
         c)   excrete sodium in exchange for hydrogen ions
         d)   excrete hydrogen ions in exchange for sodium ions

a 14. Ammonia secretion by kidney tubules increases when:
    a) the pH of the tubular fluid is low
    b) the pH of the tubular fluid is high
    c) the body is breaking amino acids down into urea
    d) there is too much potassium ion
    e) none of the above is correct

e 15. In intracellular fluid, the most abundant cation is:
    a) magnesium
    b) calcium
    c) sodium
    d) hydrogen
    e) none of the above is correct

c 16. A person drinks a large amount of water. Which one of the following changes will not occur?
    a) a short term increase in plasma volume
    b) a decrease in the solute concentration of body fluids
    c) a decrease in aldosterone secretion
    d) a decrease in ADH secretion

d 17. Aldosterone increases blood pressure by:
    a) actively producing vasoconstriction
    b) stimulating renin secretion thus causing increasing amounts of angiotensin to be formed
    c) directly stimulating the heart
    d) promoting sodium retention causing an increase in blood volume
    e) aldosterone actually decreases blood pressure

e 18. A major cause for the secretion of aldosterone is:
    a) increasing sodium ion concentration
    b) increasing calcium ion concentration
    c) increasing hydrogen ion concentration
    d) increasing water volume in the plasma
    e) increasing potassium ion concentration

b 19. The most abundant cation in extracellular fluid is:
    a) potassium
    b) sodium
    c) calcium
    d) hydrogen
    e) none of the above is correct

c 20. Which one of the following is not true of calcium ion balance in the body?
   a) decreasing extracellular calcium concentration stimulates parathormone secretion
   b) extracellular calcium ion levels stimulate the parathyroid gland directly
   c) parathormone increases the amount of calcium ion in the urine
   d) calcitonin lowers calcium levels in the blood

b 21. Which one of the following is not correct?
   a) parathormone decreases the reabsorption of phosphate
   b) aldosterone increases potassium retention and decreases sodium retention
   c) usually all of the filtered potassium is reabsorbed
   d) potassium can be secreted

a 22. Atrial natriuretic factor:
   a) is released by atrial cardiac muscle cells
   b) stimulates the secretion of renin
   c) directly stimulates the adrenal secretion of aldosterone
   d) acts directly at various sites in the kidneys to inhibit sodium excretion
   e) none of the above

CHAPTER 28 - THE REPRODUCTIVE SYSTEM

d 1. The follicle stimulating hormone (FSH):
   a) is secreted by the ovary
   b) has no effect in the male
   c) has no effect in the female
   d) causes the follicle to develop in the female and sperm to develop in the male
   e) is the direct stimulus for ovulation

c 2. Under the influence of LH(ICSH) from the pituitary, the interstitial endocrinocytes of the testes:
   a) produce sperm
   b) secrete estrogen
   c) secrete testosterone
   d) increase in number
   e) release sperm for transport to the epididymis

c 3. Which of the following statements is incorrect?
   a) the follicle produces estrogen
   b) the corpus luteum produces progesterone
   c) FSH produces the egg and breaks down the uterine lining
   d) LH is necessary to the maintenance of the corpus luteum
   e) when estrogen is high, it causes the pituitary to stop producing FSH

c 4. The structures homologous to the scrotum of the male in the female is/are the:
   a) mons veneris
   b) labia minora
   c) labia majora
   d) clitoris
   e) urethral folds

d 5. Of the structures listed, which is passed through thirdly by an ovum?
   a) abdominal cavity
   b) uterus
   c) vagina
   d) oviduct
   e) ostium of oviduct

b 6. The effect of estrogen secretion on the uterine endometrium is specifically to:
   a) cause menstruation
   b) cause elongation of uterine glands
   c) cause enlargement and secretion of the uterine glands
   d) constrict spiral arteries

d   7.   Before puberty, sex differences in humans are maintained by hormones of the:
a)   gonads
b)   anterior lobe of the pituitary
c)   adrenal medulla
d)   adrenal cortex
e)   thyroid gland

d   8.   Which of the following is not a function of progesterone?
a)   it brings about the final preparation of the uterus for implantation
b)   it inhibits ovulation
c)   it stimulates the enlargement of the breasts during pregnancy
d)   it stimulates the contraction of the walls of the uterus during childbirth
e)   all of the above are functions of progesterone

a   9.   The dominant hormone controlling the secretory (proliferative) phase of the uterus is:
a)   progesterone
b)   estrogen
c)   oxytocin
d)   testosterone
e)   prolactin

b   10. Luteinizing hormone promotes:
a)   the secretory activity of the uterine endometrium
b)   secretory activity of the corpus luteum
c)   the onset of follicular maturation
d)   development of the female secondary sex characteristics directly

d   11. A functional advantage of having human testes in the scrotum rather than in the abdomen is:
a)   lack of room in the abdomen
b)   shorter sperm duct
c)   more direct blood supply
d)   lower temperature
e)   at least two of the above

e   12. Of the structures listed, which is passed through thirdly by a sperm cell?
a)   urethra
b)   efferent ducts
c)   seminiferous tubules
d)   ductus deferens
e)   duct of epididymis

e 13. Semen:
   a) contains many sperm in a fluid medium
   b) is ejaculated during copulation
   c) is used for artificial insemination
   d) passes out of an erect penis
   e) all of these

c,d 14. When a sperm enters an egg in mammals: (select two correct answers)
   a) the first polar body is formed
   b) the first meiotic division occurs
   c) the fertilization process is complete
   d) the second meiotic division of the egg takes place
   e) the second meiotic division of the sperm takes place

c 15. Which of these descriptions could be associated with the luteal phase of the menstrual cycle?
   a) low FSH, high estrogen, developing follicle, increase in uterine lining
   b) high LH, high estrogen, developing follicle, uterine lining breakdown
   c) LH is beginning to decrease, progesterone is increasing, adequate corpus luteum, good uterine lining
   d) blood flow to uterine lining is diminishing, progesterone level diminishing, follicular rupture is about to occur

c 16. Of the stages listed, which stage occurs thirdly?
   a) corpus hemorrhagicum
   b) developing follicle
   c) Graafian follicle
   d) corpus luteum
   e) primary follicle

a 17. Between ejaculations the sperm is stored in which one of the following structures?
   a) epididymis
   b) seminiferous tubules
   c) ejaculatory duct
   d) seminal vesicles
   e) ampullary region of the ductus deferens

d 18. LH stimulates the:
   a) sperm ducts of the testis
   b) pituitary to produce ACTH
   c) follicle to produce estrogen
   d) corpus luteum to produce progesterone
   e) growth of mammary gland alveoli

c   19. The function of the sustentacular cells is to:
        a)   produce testosterone
        b)   produce spermatogonia
        c)   sustain spermatids during their maturation
        d)   reject malformed spermatozoa
        e)   produce estrogen

d   20. Female meiosis differs primarily from that in the male in that:
        a)   the cells produced are diploid
        b)   it occurs so much more often
        c)   there are no centrioles or asters in the female cell
        d)   only one functional cell is produced from each female cell
             starting meiosis

e   21. The structure in the female that is homologous to the penis is the:
        a)   labia majora
        b)   mons pubis
        c)   vestibular gland
        d)   ovary
        e)   none of the above is correct

a   22. In the male embryo, which of these structures degenerates shortly
        after the gonads begin to differentiate into testes?
        a)   paramesonephric ducts
        b)   seminiferous tubules
        c)   mesonephric ducts
        d)   genital tubercle

a   23. The structure that surrounds the urethra just below the bladder is the:
        a)   prostate gland
        b)   vas deferens
        c)   seminal vesicle
        d)   epididymis
        e)   bulbourethral gland

d   24. The corpus cavernosum of the male is homologous to the _____ of
        the female.
        a)   bulb of the vestibule
        b)   labia majora
        c)   labia minora
        d)   clitoris

d   25. The greater vestibular glands are homologous to the _____ of the
        male.
        a)   prostate gland
        b)   seminal vesicles
        c)   paraurethral glands
        d)   bulbourethral glands

CHAPTER 29 - PREGNANCY, EMBRYONIC DEVELOPMENT, AND INHERITANCE

a 1. The direct cause of milk secretion is the hormone:
  a) prolactin
  b) estrogen
  c) progesterone
  d) FSH
  e) LH

c 2. During the second month of pregnancy, most of the estrogen and progesterone found in the blood of the mother is derived from the:
  a) pituitary - adenohypophysis
  b) placenta
  c) corpus luteum
  d) uterine myometrium
  e) developing follicle

c 3. The mammalian umbilical cord contains portions of:
  a) maternal arteries
  b) maternal capillaries
  c) the allantois
  d) the chorion
  e) the fetal appendix

c 4. In mammals, the fetal membrane coming into direct contact with maternal blood is the:
  a) allantois
  b) amnion
  c) chorion
  d) yolk sac
  e) lesser omentum

c 5. Before birth, the embryo of a placental mammal lives in an aqueous medium enclosed within the:
  a) chorion
  b) allantois
  c) amnion
  d) yolk sac

c 6. The placenta consists of:
  a) maternal tissues only
  b) embryonic tissues only
  c) maternal tissues and extraembryonic membranes
  d) maternal and embryonic blood as well as the umbilicus including the yolk sac and the allantois

e   7.   Essential changes in fetal circulation at birth or shortly thereafter include obliteration of all but which of these?
    a)   foramen ovale
    b)   ductus arteriosus
    c)   umbilical vein
    d)   umbilical arteries
    e)   all of the above are obliterated

e   8.   Milk letdown is under control of:
    a)   FSH
    b)   prolactin
    c)   glucocorticoids
    d)   estrogen
    e)   oxytocin

d   9.   The corpus luteum of pregnancy is stimulated to continue to secrete past the normal menstrual time by the hormone:
    a)   estrogen
    b)   FSH
    c)   LH
    d)   HCG
    e)   prolactin

d   10.   The developing human egg becomes buried in the lining of the uterus by a process known as _____.
    a)   invagination
    b)   ingestion
    c)   placentation
    d)   implantation
    e)   none of the above

c   11.   The placenta is the region where:
    a)   maternal and fetal blood mix
    b)   fetal blood gives off oxygen
    c)   maternal blood receives carbon dioxide
    d)   fetal blood is produced
    e)   none of the above is correct

d   12.   The extraembryonic membrane that first contacts the uterine wall is the:
    a)   yolk sac
    b)   allantois
    c)   amnion
    d)   chorion
    e)   placenta

e   13. During the fourth month of pregnancy most of the progesterone
          production comes from the:
          a) myometrium
          b) pituitary
          c) developing follicle
          d) corpus luteum
          e) none of the above is correct

a   14. Which of these listed below stimulates the smooth muscle of the
          pregnant uterus?
          a)   oxytocin
          b)   relaxin
          c)   FSH
          d)   prolactin
          e)   none of the above is correct

e   15. Which one of the following is true of fraternal twins?
          a)   they are always of the same sex
          b)   they result when a two-celled embryo separates into two separate
               cells
          c)   their genes are all the same
          d)   they result when one sperm fertilizes two different eggs
          e)   none of the above is correct

b   16. A genotype that has two identical genes for a particular characteristic
          is said to be:
          a)   dominant
          b)   homozygous
          c)   recessive
          d)   heterozygous
          e)   none of the above is correct

a   17. AaBb is crossed with AaBb. In the offspring:
          a)   kinds of genotypes will outnumber kinds of phenotypes
          b)   kinds of phenotypes will outnumber kinds of genotypes
          c)   genotypes will occur in a 9:3:3:1 proportion
          d)   none of the above is correct

d   18. The superficial layer of the thickened endometrium is called the:
          a)   chorion
          b)   syncytiotrophoblast
          c)   cytotrophoblast
          d)   decidua
          e)   none of the above is correct

d   19.   Which one of the following listed occurs thirdly?
           a)   implantation
           b)   late cleavage
           c)   fertilization
           d)   blastocyst
           e)   development of the chorionic villi

e   20.   Which one of the following becomes an umbilical ligament?
           a)   allantois
           b)   ductus arteriosus
           c)   ductus venosus
           d)   foramen ovale
           e)   none of the above is correct

c   21.   At the present rate of population increase, the world population will
           double in _____ years.
           a)   100
           b)   75
           c)   50
           d)   25
           e)   15

e   22.   Embryonic endoderm gives rise to:
           a)   skin
           b)   muscles
           c)   bone
           d)   blood
           e)   none of the above is correct